THE SINGULARITY IS NEARER

ALSO BY RAY KURZWEIL

The Age of Intelligent Machines

The 10% Solution for a Healthy Life

The Age of Spiritual Machines

Fantastic Voyage (with Terry Grossman, MD)

The Singularity Is Near

Transcend (with Terry Grossman, MD)

How to Create a Mind

Danielle: Chronicles of a Superheroine

A Chronicle of Ideas

THE SINGULARITY IS NEARER

WHEN WE MERGE
WITH AI

RAY KURZWEIL

VIKING

VIKING
An imprint of Penguin Random House LLC
penguinrandomhouse.com

Diagrams on pages 83, 84, and 85 from *A New Kind of Science* by Stephen Wolfram (pages 56, 23–27, 31). Copyright © 2002 by Stephen Wolfram, LLC. Used with permission of Wolfram Media, wolframscience.com/nks. Graphic on page 118 used with permission of Gallup, Inc. (news.gallup.com/poll/1603 /crime.aspx); page 176 used with permission of Lazard, Inc. Photo on page 182 of VertiCrop System by Wikimedia Commons user Valcenteu via CC BY 3.0 (creativecommons.org/licenses/by-sa/3.0/); page 185 FDA photo by Michael J. Ermarth.

LIBRARY OF CONGRESS CATALOGING-IN-PUBLICATION DATA

Names: Kurzweil, Ray, author.
Title: The singularity is nearer : when we merge with ai / Ray Kurzweil.
Description: [New York] : Viking, [2024]. | Includes bibliographical
    references and index.
Identifiers: LCCN 2023051391 (print) | LCCN 2023051392 (ebook) |
    ISBN 9780399562761 (hardcover) | ISBN 9780399562778 (ebook) |
    ISBN 9780593489413 (international edition) |
Subjects: LCSH: Brain—Evolution. | Human evolution. | Genetics. |
    Nanotechnology. | Robotics.
Classification: LCC QP376 .K853 2024 (print) | LCC QP376 (ebook) |
    DDC 612.8/2—dc23/eng/20240311
LC record available at https://lccn.loc.gov/2023051391
LC ebook record available at https://lccn.loc.gov/2023051392

Printed in the United States of America
10 9 8 7 6 5 4 3 2 1

Designed by Cassandra Garruzzo Mueller

To Sonya Rosenwald Kurzweil.
As of a few days ago,
I have now gotten to know her
(and love her) for fifty years!

CONTENTS

ACKNOWLEDGMENTS xi

INTRODUCTION 1

CHAPTER 1: **WHERE ARE WE IN THE SIX STAGES?** 7

CHAPTER 2: **REINVENTING INTELLIGENCE** 11

CHAPTER 3: **WHO AM I?** 75

CHAPTER 4: **LIFE IS GETTING EXPONENTIALLY BETTER** 111

CHAPTER 5: **THE FUTURE OF JOBS: GOOD OR BAD?** 195

CHAPTER 6: **THE NEXT THIRTY YEARS IN HEALTH AND WELL-BEING** .. 235

CHAPTER 7: **PERIL** 267

CHAPTER 8: **DIALOGUE WITH CASSANDRA** 287

APPENDIX 293

NOTES 313

INDEX 401

ACKNOWLEDGMENTS

I'd like to express my gratitude to my wife, Sonya, for her loving patience through the vicissitudes of the creative process and for sharing ideas with me for fifty years.

To my children, Ethan and Amy; my daughter-in-law, Rebecca; my son-in-law, Jacob; my sister, Enid; and my grandchildren, Leo, Naomi, and Quincy for their love, inspiration, and great ideas.

To my late mother, Hannah, and my late father, Fredric, who taught me the power of ideas in walks through the New York woods, and gave me the freedom to experiment at a young age.

To John-Clark Levin for his meticulous research and intelligent analysis of the data that serves as a basic foundation of this book.

To my longtime editor at Viking, Rick Kot, for his leadership, unwavering guidance, and expert editing.

To Nick Mullendore, my literary agent, for his astute and enthusiastic guidance.

To Aaron Kleiner, my lifelong business partner (since 1973), for his devoted collaboration for the past fifty years.

To Nanda Barker-Hook for her skilled writing assistance and expert oversight and management of my speeches.

To Sarah Black for her outstanding research insights and organization of ideas.

To Celia Black-Brooks for her thoughtful support and expert strategy on sharing my ideas with the world.

To Denise Scutellaro for her adept handling of my business operations.

To Laksman Frank for his excellent graphic design and illustrations.

To Amy Kurzweil and Rebecca Kurzweil for their guidance on the craft of writing, and their own wonderful examples of very successful books.

To Martine Rothblatt for her dedication to all of the technologies I discuss in the book and for our longtime collaborations in developing outstanding examples in these areas.

To the Kurzweil team, who provided significant research, writing, and logistical support for this project, including Amara Angelica, Aaron Kleiner, Bob Beal, Nanda Barker-Hook, Celia Black-Brooks, John-Clark Levin, Denise Scutellaro, Joan Walsh, Marylou Sousa, Lindsay Boffoli, Ken Linde, Laksman Frank, Maria Ellis, Sarah Black, Emily Brangan, and Kathryn Myronuk.

To the dedicated team at Viking Penguin for all of their thoughtful expertise, including Rick Kot, executive editor; Allison Lorentzen, executive editor; Camille LeBlanc, associate editor; Brian Tart, publisher; Kate Stark, associate publisher; Carolyn Coleburn, executive publicist; and Mary Stone, marketing director.

To Peter Jacobs of CAA for his invaluable leadership and support of my speaking engagements.

To the teams at Fortier Public Relations and Book Highlight for their exceptional public relations expertise and strategic guidance in sharing this book far and wide.

To my in-house and lay readers, who have provided many clever and creative ideas.

And, finally, to all the people who have the courage to question outdated assumptions and use their imaginations to do things that have never been done before. You inspire me.

INTRODUCTION

In my 2005 book *The Singularity Is Near*, I set forth my theory that convergent, exponential technological trends are leading to a transition that will be utterly transformative for humanity. There are several key areas of change that are continuing to accelerate simultaneously: computing power is becoming cheaper, human biology is becoming better understood, and engineering is becoming possible at far smaller scales. As artificial intelligence grows in ability and information becomes more accessible, we are integrating these capabilities ever more closely with our natural biological intelligence. Eventually nanotechnology will enable these trends to culminate in directly expanding our brains with layers of virtual neurons in the cloud. In this way we will merge with AI and augment ourselves with millions of times the computational power that our biology gave us. This will expand our intelligence and consciousness so profoundly that it's difficult to comprehend. This event is what I mean by the Singularity.

The term "singularity" is borrowed from mathematics (where it refers to an undefined point in a function, like when dividing by zero) and physics (where it refers to the infinitely dense point at the center of a black hole, where the normal laws of physics break down). But it is important to remember that I use the term as a metaphor. My prediction of the technological Singularity does not suggest that rates of change will actually become infinite, as exponential growth does not imply infinity, nor does a physical singularity. A black hole has gravity strong enough to trap even light itself, but there is no means in quantum mechanics to account for a truly infinite amount of mass. Rather, I use the singularity metaphor because it captures our inability to comprehend such a radical shift with our current level of intelligence. But as the transition happens, we will enhance our cognition quickly enough to adapt.

As I detailed in *The Singularity Is Near,* long-term trends suggest that the Singularity will happen around 2045. At the time that book was published, that date lay forty years—two full generations—in the future. At that distance I could make predictions about the broad forces that would bring about this transformation, but for most readers the subject was still relatively far removed from daily reality in 2005. And many critics argued then that my timeline was overoptimistic, or even that the Singularity was impossible.

Since then, though, something remarkable has happened. Progress has continued to accelerate in defiance of the doubters. Social media and smartphones have gone from virtually nonexistent to all-day companions that now connect a majority of the world's population. Algorithmic innovations and the emergence of big data have allowed AI to achieve startling breakthroughs sooner than even experts expected—from mastering games like *Jeopardy!* and Go to driving automobiles, writing essays, passing bar exams, and diagnosing cancer. Now, powerful and flexible large language models like GPT-4 and Gemini can translate natural-language instructions into computer code—dramatically reducing the barrier between humans and machines. By the time you read this, tens of millions of people likely will have experienced these capabilities firsthand. Meanwhile, the cost to sequence a human's genome has fallen by about 99.997 percent, and neural networks have begun unlocking major medical discoveries by simulating biology digitally. We're even gaining the ability to finally connect computers to brains directly.

Underlying all these developments is what I call the law of accelerating returns: information technologies like computing get exponentially cheaper because each advance makes it easier to design the next stage of their own evolution. As a result, as I write this, one dollar buys about 11,200 times as much computing power, adjusting for inflation, as it did when *The Singularity Is Near* hit shelves.

The following graph, which I'll discuss in depth later in the book, summarizes the most important trend powering our technological civilization: the long-term exponential growth (shown as a roughly straight line on this logarithmic scale) in the amount of computing power a constant dollar can purchase. Moore's law famously observes that transistors have been steadily shrinking, allowing computers to get ever more powerful—but that is just one manifestation of the law of accelerating returns, which already held true long before transistors were invented and can be expected to continue even after transistors reach their physical limits and are succeeded by new technologies. This trend has defined the modern world, and almost all the coming breakthroughs discussed in this book will be enabled by it directly or indirectly.

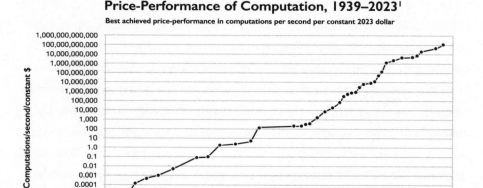

Price-Performance of Computation, 1939–2023[1]

Best achieved price-performance in computations per second per constant 2023 dollar

To maximize comparability of machines, this graph focuses on price-performance during the era of programmable computers, but approximations for earlier electromechanical computing devices show that this trend stretches back at least to the 1880s.[2]

So we have kept on schedule for the Singularity. The urgency of this book comes from the nature of exponential change itself. Trends that were barely noticeable at the start of this century are now actively impacting billions of lives. In the early 2020s we entered the sharply steepening part of the exponential curve, and the pace of innovation is affecting society like never before. For perspective, the moment you're reading this is probably closer to the creation of the first superhuman AI than to the release of my last book, 2012's *How to Create a Mind*. And you're probably closer to the Singularity than to the release of my 1999 book *The Age of Spiritual Machines*. Or, measured in terms of human life, babies born today will be just graduating college when the Singularity happens. This is, on a very personal level, a different kind of "near" than it was in 2005.

That is why I've written this book now. Humanity's millennia-long march toward the Singularity has become a sprint. In the introduction to *The Singularity Is Near*, I wrote that we were then "in the early stages of this transition." Now we are entering its culmination. That book was about glimpsing a distant horizon—this one is about the last miles along the path to reach it.

Luckily, we can now see this path much more clearly. Although many technological challenges remain before we can achieve the Singularity, its key precursors are rapidly moving from the realm of theoretical science to active research and development. During the coming decade, people will interact with AI that can seem convincingly human, and simple brain–computer interfaces will impact daily life much like smartphones do today. A digital revolution in biotech will cure diseases and meaningfully extend people's healthy lives. At the same time, though, many workers will feel the sting of economic disruption, and all of us will face risks from accidental or deliberate misuse of these new capabilities. During the 2030s, self-improving AI and maturing nano-technology will unite humans and our machine creations as never before—heightening both the promise and the peril even further. If we can meet the scientific, ethical, social, and political challenges posed by these advances, by 2045 we will transform life on earth profoundly for the better. Yet if we fail, our very survival is in question. And so this book is about our final approach to the Singularity—the opportunities and dangers we must confront together over the last generation of the world as we knew it.

To begin, we'll explore how the Singularity will actually happen, and put this in the context of our species' long quest to reinvent our own intelligence. Creating sentience with technology raises important philosophical questions, so we'll address how this transition affects our own identity and sense of purpose. Then we will turn to the practical trends that will characterize the coming decades. As I will show, the law of accelerating returns is driving exponential improvements across a very wide range of metrics that reflect human well-being. One of the most obvious downsides of innovation, though, is unemployment caused by automation in its various forms. While these harms are real, we'll see why there is good reason for long-term optimism—and why we are ultimately not in competition with AI.

As these technologies unlock enormous material abundance for our civilization, our focus will shift to overcoming the next barrier to our full flourishing: the frailties of our biology. So next, we'll look ahead to the tools we'll use over the coming decades to gain increasing mastery over biology itself—first by defeating the aging of our bodies and then by augmenting our limited brains and ushering in the Singularity. Yet these breakthroughs may also put us in jeopardy. Revolutionary new systems in biotechnology, nanotechnology, or artificial intelligence could possibly lead to an existential catastrophe like a devastating pandemic or a chain reaction of self-replicating machines. We'll conclude with an assessment of these threats, which warrant careful planning, but as I'll explain, there are very promising approaches for how to mitigate them.

These are the most exciting and momentous years in all of history. We cannot say with confidence what life will be like after the Singularity. But by understanding and anticipating the transitions leading up to it, we can help ensure that humanity's final approach will be safe and successful.

CHAPTER 1

WHERE ARE WE IN THE SIX STAGES?

In *The Singularity Is Near*, I described the basis of consciousness as information. I cited six epochs, or stages, from the beginning of our universe, with each stage creating the next stage from the information processing of the last. Thus, the evolution of intelligence works via an indirect sequence of other processes.

The First Epoch was the birth of the laws of physics and the chemistry they make possible. A few hundred thousand years after the big bang, atoms formed from electrons circling around a core of protons and neutrons. Protons in a nucleus seemingly should not be so close together, because the electromagnetic force tries to drive them violently apart. However, there happens to be a separate force called the strong nuclear force, which keeps the protons together. "Whoever" designed the rules of the universe provided this additional force, otherwise evolution through atoms would have been impossible.

Billions of years later, atoms formed molecules that could represent elaborate information. Carbon was the most useful building block, in that it could form four bonds, as opposed to one, two, or three for many other nuclei. That we live in a world that permits complex chemistry is extremely unlikely. For example, if the strength of gravity were ever so slightly weaker, there would be no supernovas to create the chemical elements that life is made from. If it were just slightly stronger, stars would burn out and die before intelligent life could form. Just this one physical constant had to be in an extremely narrow range or we would not be here. We live in a universe that is very precisely balanced to allow a level of order that has enabled evolution to unfold.

Several billion years ago, the Second Epoch began: life. Molecules became complex enough to define an entire organism in one molecule. Thus, living creatures, each with their own DNA, were able to evolve and spread.

In the Third Epoch, animals described by DNA then formed brains, which themselves stored and processed information. These brains gave evolutionary advantages, which helped brains develop more complexity over millions of years.

In the Fourth Epoch, animals used their higher-level cognitive ability, along with their thumbs, to translate thoughts into complex actions. This was humans. Our species used these abilities to create technology that was able to store and manipulate information—from papyrus to hard drives. These technologies augmented our brains' abilities to perceive, recall, and evaluate information patterns. This is another source of evolution that itself is far greater than the level of progress before it. With brains, we added roughly one cubic inch of brain matter every 100,000 years, whereas with digital computation we are doubling price-performance about every sixteen months.

In the Fifth Epoch, we will directly merge biological human cognition with the speed and power of our digital technology. This is brain–computer interfaces. Human neural processing happens at a speed of several hundred cycles per second, as compared with several billion per second for digital technology. In addition to speed and memory size, augmenting our brains with nonbiological computers will allow us to add many more layers to our neocortices—unlocking vastly more complex and abstract cognition than we can currently imagine.

The Sixth Epoch is where our intelligence spreads throughout the universe, turning ordinary matter into computronium, which is matter organized at the ultimate density of computation.

In my 1999 book *The Age of Spiritual Machines*, I predicted that a

Turing test—wherein an AI can communicate by text indistinguishably from a human—would be passed by 2029. I repeated that in 2005's *The Singularity Is Near.* Passing a valid Turing test means that an AI has mastered language and commonsense reasoning as possessed by humans. Turing described his concept in 1950,[1] but he did not specify how the test should be administered. In a bet that I have with Mitch Kapor, we defined our own rules that are much more difficult than other interpretations.

My expectation was that in order to pass a valid Turing test by 2029, we would need to be able to attain a great variety of intellectual achievements with AI by 2020. And indeed, since that prediction, AI has mastered many of humanity's toughest intellectual challenges—from games like *Jeopardy!* and Go to serious applications like radiology and drug discovery. As I write this, top AI systems like Gemini and GPT-4 are broadening their abilities to many different domains of performance—encouraging steps on the road to general intelligence.

Ultimately, when a program passes the Turing test, it will actually need to make itself appear far less intelligent in many areas because otherwise it would be clear that it is an AI. For example, if it could correctly solve any math problem instantly, it would fail the test. Thus, at the Turing test level, AIs will have capabilities that in fact go far beyond the best humans in most fields.

Humans are now in the Fourth Epoch, with our technology already producing results that exceed what we can understand for some tasks. For the aspects of the Turing test that AI has not yet mastered, we are making rapid and accelerating progress. Passing the Turing test, which I have been anticipating for 2029, will bring us to the Fifth Epoch.

A key capability in the 2030s will be to connect the upper ranges of our neocortices to the cloud, which will directly extend our thinking. In this way, rather than AI being a competitor, it will become an extension of ourselves. By the time this happens, the nonbiological portions of our minds will provide thousands of times more cognitive capacity than the biological parts.

As this progresses exponentially, we will extend our minds many millions-fold by 2045. It is this incomprehensible speed and magnitude of transformation that will enable us to borrow the singularity metaphor from physics to describe our future.

CHAPTER 2

REINVENTING INTELLIGENCE

WHAT DOES IT MEAN TO REINVENT INTELLIGENCE?

If the whole story of the universe is one of evolving paradigms of information processing, the story of humanity picks up more than halfway through. Our chapter in this larger tale is ultimately about our transition from animals with biological brains to transcendent beings whose thoughts and identities are no longer shackled to what genetics provides. In the 2020s we are about to enter the last phase of this transformation—reinventing the intelligence that nature gave us on a more powerful digital substrate, and then merging with it. In so doing, the Fourth Epoch of the universe will give birth to the Fifth.

But how will this happen more concretely? To understand what reinventing intelligence entails, we will first look back to the birth of AI and the two broad schools of thought that emerged from it. To see why one prevailed over the other, we will relate this to what neuroscience tells us about how the cerebellum and the neocortex gave rise to human intelligence. After surveying how deep learning is currently re-creating the powers of the neocortex, we can assess what AI still needs to achieve to reach human levels, and how we will know when it has. Finally, we'll turn to how, aided by superhuman AI, we will engineer brain–computer interfaces that vastly expand our neocortices with layers of virtual neurons. This will unlock entirely new modes of thought and ultimately expand our intelligence millions-fold: this is the Singularity.

THE BIRTH OF AI

In 1950, the British mathematician Alan Turing (1912–1954) published an article in *Mind* titled "Computing Machinery and Intelligence."[1] In it, Turing asked one of the most profound questions in the history of science: "Can machines think?" While the idea of thinking machines dates back at least as far as the bronze automaton Talos in Greek myth,[2] Turing's breakthrough was boiling the concept down to something empirically testable. He proposed using the "imitation game"—which we now know as the Turing test—to determine whether a machine's computation was able to perform the same cognitive tasks that our brains can. In this test, human judges interview both the AI and human foils using instant messaging without seeing whom they are talking to. The judges then pose questions about any subject matter or situation they wish. If after a certain period of time the judges are unable to tell which was the AI responder and which were the humans, then the AI is said to have passed the test.

By transforming this philosophical idea into a scientific one, Turing generated tremendous enthusiasm among researchers. In 1956, mathematics professor John McCarthy (1927–2011) proposed a two-month, ten-person study to be conducted at Dartmouth College, in Hanover, New Hampshire.[3] The goal was the following:

> The study is to proceed on the basis of the conjecture that every aspect of learning or any other feature of intelligence can in principle be so precisely described that a machine can be made to simulate it. An attempt will be made to find how to make machines use language, form abstractions and concepts, solve kinds of problems now reserved for humans, and improve themselves.[4]

In preparing for the conference, McCarthy proposed that this field, which would ultimately automate every other field, be called "artificial intelligence."[5] This is not a designation I like, given that "artificial" makes this form of intelligence seem "not real," but it is the term that has endured.

The study was conducted, but its goal—specifically, getting machines to understand problems described in natural language—was not achieved within the two-month time frame. We are still working on it—of course, now with far more than ten people. According to Chinese tech giant Tencent, in 2017 there were already about 300,000 "AI researchers and practitioners" worldwide,[6] and the 2019 *Global AI Talent Report*, by Jean-François Gagné, Grace Kiser, and Yoan Mantha, counted some 22,400 AI experts publishing original research—of whom around 4,000 were judged to be highly influential.[7] And according to Stanford's Institute for Human-Centered Artificial Intelligence, AI researchers in 2021 generated more than 496,000 publications and over 141,000 patent filings.[8] In 2022, global corporate investment in AI was $189 billion, a thirteenfold increase over the past decade.[9] The numbers will be even higher by the time you read this.

All this would have been hard to imagine in 1956. Yet the Dartmouth workshop's goal was roughly equivalent to creating an AI that could pass the Turing test. My prediction that we'll achieve this by 2029 has been consistent since my 1999 book *The Age of Spiritual Machines*, published at a time when many observers thought this milestone would *never* be reached.[10] Until recently this projection was considered extremely optimistic in the field. For example, a 2018 survey found an aggregate prediction among AI experts that human-level machine intelligence would not arrive until around 2060.[11] But the latest advances in large language models have rapidly shifted expectations. As I was writing early drafts of this book, the consensus on Metaculus, the world's top forecasting website, hovered between the 2040s and the 2050s. But surprising AI progress over the past two years upended expectations, and by May 2022 the Metaculus consensus exactly agreed with me on the 2029 date.[12] Since then it has even fluctuated to as soon as 2026, putting me technically in the slow-timelines camp![13]

Even experts in the field have been surprised by many of the recent breakthroughs in AI. It's not just that they are happening sooner than most expected, but that they seem to occur suddenly, and without much warning that a leap forward is imminent. For example, in October 2014 Tomaso Poggio, an MIT expert on AI and cognitive science, said, "The ability to describe the content of an image would be one of the most intellectually challenging things of all for a machine to do. We will need another cycle of basic research to solve this kind of question."[14] Poggio estimated that this breakthrough was at least two decades away. The very next month, Google debuted object recognition AI that could do just that. When *The New Yorker*'s Raffi Khatchadourian asked him about this, Poggio retreated to a more philosophical skepticism about whether this ability represented true intelligence. I point this out not as a criticism of Poggio but rather as an observation of a tendency we all share. Namely, before AI achieves some goal, that goal seems extremely difficult and singularly human. But after AI reaches it, the accomplishment diminishes in our human eyes. In other words, our true progress is actually more significant than it seems in hindsight. This is one reason why I remain optimistic about my 2029 prediction.

So why have these sudden breakthroughs occurred? The answer lies in a theoretical problem dating back to the dawn of the field. In 1964, when I was in high school, I met with two artificial intelligence pioneers: Marvin Minsky (1927–2016), who co-organized the Dartmouth College workshop on AI, and Frank Rosenblatt (1928–1971). In 1965 I enrolled at MIT and began studying with Minsky, who was doing foundational work that underlies the dramatic AI breakthroughs we are seeing today. Minsky taught me that there are two techniques for creating automated solutions to problems: the symbolic approach and the connectionist approach.

The symbolic approach describes in rule-based terms how a human expert would solve a problem. In some cases the systems based on it could be successful. For example, in 1959 the RAND Corporation introduced the "General Problem Solver" (GPS)—a computer program that could combine simple mathematical axioms to solve logic problems.[15] Herbert A. Simon, J. C. Shaw, and Allen Newell developed the General Problem Solver to have the theoretical ability to solve *any* problem that could be expressed as a set of well-formed formulas (WFFs). In order for the GPS to work, it would have to use one WFF (essentially an axiom) at each stage in the process, methodically building them into a mathematical proof of the answer.

Even if you don't have experience with formal logic or proof-based math, this idea is basically the same as what happens in algebra. If you know that $2 + 7 = 9$, and that an unknown number x added to 7 is 10, you can prove that $x = 3$. But this kind of logic has much broader applications than just solving equations. It's also what we use (without even thinking about it) when we ask ourselves whether something meets a certain definition. If you know that a prime number cannot have any factors other than 1 and itself, and you know that 11 is a factor of 22, and that 1 does not equal 11, you can conclude that 22 is not a prime number. By starting with the most basic and fundamental axioms possible, the GPS could do this sort of calculation for much more difficult questions. Ultimately, this is what human mathematicians do as well—the difference is that a machine can (in theory at least) search through every possible way of combining the fundamental axioms in search of the truth.

To illustrate, if there were ten such axioms available to choose from at each point, and let's say twenty axioms were needed to reach a solution, that would mean there were 10^{20}, or 100 billion billion, possible solutions. We can deal with such big numbers today with modern computers, but this was way beyond what 1959 computational speeds could achieve. That year, the DEC PDP-1 computer could carry out about 100,000 operations per second.[16] By 2023 a Google Cloud A3 virtual machine could carry out roughly 26,000,000,000,000,000,000 operations per second.[17] One dollar now buys around 1.6 *trillion* times as much computing power as it did when the GPS was developed.[18] Problems that would take tens of thousands of years with 1959 technology now take only minutes on retail computing hardware. To compensate for its limitations, the GPS had heuristics programmed that would attempt to rank the priority of possible solutions. The heuristics worked some of the time, and their successes supported the idea that a computerized solution could ultimately solve any rigorously defined problem.

Another example was a system called MYCIN, which was developed during the 1970s to diagnose and recommend remedial treatments for infectious diseases. In 1979 a team of expert evaluators compared its performance with that of human doctors and found that MYCIN did as well as or better than any of the physicians.[19]

A typical MYCIN "rule" reads:

IF: 1) The infection that requires therapy is meningitis, and
2) The type of the infection is fungal, and
3) Organisms were not seen on the stain of the culture, and
4) The patient is not a compromised host, and
5) The patient has been to an area that is endemic for coccidiomycoses, and
6) The race of the patient is one of: [B]lack [A]sian [I]ndian, and
7) The cryptococcal antigen in the csf was not positive

THEN: There is suggestive evidence (.5) that cryptococcus is not one of the organisms (other than those seen on cultures or smears) which might be causing the infection.[20]

By the late 1980s these "expert systems" were utilizing probability models and could combine many sources of evidence to make a decision.[21] While a single *if-then* rule would not be sufficient by itself, by combining many thousands of such rules, the overall system could make reliable decisions for a constrained problem.

Although the symbolic approach has been used for over half a century, its primary limitation has been the "complexity ceiling."[22] When MYCIN and other such systems made a mistake, correcting it might fix that particular issue but would in turn give rise to three other mistakes that would rear their heads in other situations. There seemed to be a limit on intricacy that made the overall range of real-world problems that could be addressed very narrow.

One way of looking at the complexity of rule-based systems is as a set of possible failure points. Mathematically, a group of n things has 2^{n-1} subsets (not counting the empty set). Thus, if an AI uses a rule set with only one rule, there is only one failure point: Does that rule work correctly on its own or not? If there are two rules, there are three failure points: each rule on its own, and interactions in which those two rules are combined. This grows exponentially. Five rules means 31 potential failure points, 10 rules means 1,023, 100 rules means more than one thousand billion billion billion, and 1,000 rules means over a googol googol googols! Thus, the more rules you have already, the more each new rule adds to the number of possible subsets. Even if only an extremely minuscule fraction of possible rule combinations introduce a new problem, there comes a point (where exactly this point lies varies from one situation to another) where adding one new rule to fix a problem is likely to cause more than one additional problem. This is the complexity ceiling.

Probably the longest-running expert system project is Cyc (from the word "encyclopedic"), created by Douglas Lenat and his colleagues at Cycorp.[23] Initiated in 1984, Cyc has the goal of encoding all of "commonsense knowledge"—broadly known facts like *A dropped egg will break* or *A child running through the kitchen with muddy shoes will annoy his parents.* These millions of small ideas are not clearly written down in any one place. They are unspoken assumptions underlying human behavior and reasoning that are necessary for understanding what the average person knows in a variety of domains. Yet because the Cyc system also represents this knowledge with symbolic rules, it, too, has to face the complexity ceiling.

Back in the 1960s, as Minsky advised me on the pros and cons of the symbolic approach, I began to see the added value of the connectionist one. This entails networks of nodes that create intelligence through their structure rather than through their content. Instead of using smart rules, they use dumb nodes that are arranged in a way that can extract insight from data itself. As a result, they have the potential to discover subtle patterns that would never occur to human programmers trying to devise symbolic rules. One of the key advantages of the connectionist approach is that it allows you to solve problems without understanding them. Even if we had a perfect ability to formulate and implement error-free rules for symbolic AI problem-solving (which we do not), we would be limited by our imperfect understanding of which rules would be optimal in the first place.

This is a powerful way to tackle complex problems, but it is a double-edged sword. Connectionist AI is prone to becoming a "black box"—capable of spitting out the correct answer, but unable to explain how it found it.[24] This has the potential to become a major issue because people will want to be able to see the reasoning behind high-stakes decisions about things like medical treatment, law enforcement, epidemiology, or risk management. This is why many AI experts are now working to develop better forms of "transparency" (or "mechanistic interpretability") in machine learning–based decisions.[25] It remains to be seen how effective transparency will be as deep learning becomes more complex and more powerful.

Back when I started in connectionism, though, the systems were much simpler. The basic idea was to create a computerized model inspired by how human neural networks work. At first this was very abstract because the method was devised before we had a detailed understanding of how biological neural networks are actually organized.

DIAGRAM OF SIMPLE NEURAL NET

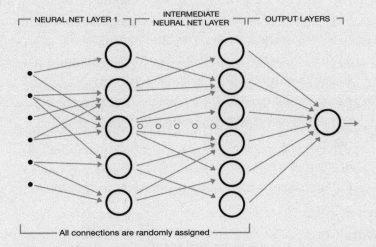

Here is the basic schema for a neural net algorithm. Many variations are possible, and the designer of the system needs to provide certain critical parameters and methods (detailed below).

Creating a neural net solution to a problem involves the following steps:

- Define the input.
- Define the topology of the neural net (i.e., the layers of neurons and the connections between the neurons).
- Train the neural net on examples of the problem.
- Run the trained neural net to solve new examples of the problem.
- Take your neural net company public.

These steps (except for the last one) are detailed below:

THE PROBLEM INPUT

The problem input to the neural net consists of a series of numbers. This input can be:

- in a visual pattern recognition system, a two-dimensional array of numbers representing the pixels of an image; or
- in an auditory (e.g., speech) recognition system, a two-dimensional array of numbers representing a sound, in which the first dimension represents parameters of the sound (e.g., frequency components) and the second dimension represents different points in time; or
- in an arbitrary pattern recognition system, an n-dimensional array of numbers representing the input pattern.

DEFINING THE TOPOLOGY

To set up the neural net, the architecture of each neuron consists of:

- multiple inputs in which each input is "connected" to either the output of another neuron or one of the input numbers; and
- generally, a single output, which is connected either to the input of another neuron (which is usually in a higher layer) or to the final output.

SET UP THE FIRST LAYER OF NEURONS

- Create N_0 neurons in the first layer. For each of these neurons, "connect" each of the multiple inputs of the neuron to "points" (i.e., numbers) in the problem input. These connections can be determined randomly or using an evolutionary algorithm (see below).
- Assign an initial "synaptic strength" to each connection

created. These weights can start out all the same, can be assigned randomly, or can be determined in another way (see below).

SET UP THE ADDITIONAL LAYERS OF NEURONS

Set up a total of M layers of neurons. For each layer, set up the neurons in that layer.

For $layer_i$:

- Create N_i neurons in $layer_i$. For each of these neurons, "connect" each of the multiple inputs of the neuron to the outputs of the neurons in $layer_{i-1}$ (see variations below).
- Assign an initial "synaptic strength" to each connection created. These weights can start out all the same, can be assigned randomly, or can be determined in another way (see below).
- The outputs of the neurons in $layer_M$ are the outputs of the neural net (see variations below).

THE RECOGNITION TRIALS

HOW EACH NEURON WORKS

Once the neuron is set up, it does the following for each recognition trial:

- Each weighted input to the neuron is computed by multiplying the output of the other neuron (or initial input) that the input to this neuron is connected to by the synaptic strength of that connection.
- All of these weighted inputs to the neuron are summed.

- If this sum is greater than the firing threshold of this neuron, then this neuron is considered to fire and its output is 1. Otherwise, its output is 0 (see variations below).

DO THE FOLLOWING FOR EACH RECOGNITION TRIAL

For each layer, from $layer_0$ to $layer_M$, and
    for each neuron in the layer:

- Sum its weighted inputs. (Each weighted input = the output of the other neuron [or initial input] that the input to this neuron is connected to multiplied by the synaptic strength of that connection.)
- If this sum of weighted inputs is greater than the firing threshold for this neuron, set the output of this neuron to 1, otherwise set it to 0.

TO TRAIN THE NEURAL NET

- Run repeated recognition trials on sample problems.
- After each trial, adjust the synaptic strengths of all the interneuronal connections to improve the performance of the neural net on this trial. (See the discussion below on how to do this.)
- Continue this training until the accuracy rate of the neural net is no longer improving (i.e., reaches an asymptote).

KEY DESIGN DECISIONS

In the simple schema above, the designer of this neural net algorithm needs to determine at the outset:

- What the input numbers represent.
- The number of layers of neurons.
- The number of neurons in each layer. (Each layer does not necessarily need to have the same number of neurons.)
- The number of inputs to each neuron in each layer. The number of inputs (i.e., interneuronal connections) can also vary from neuron to neuron and from layer to layer.
- The actual "wiring" (i.e., the connections). For each neuron in each layer, this consists of a list of other neurons, the outputs of which constitute the inputs to this neuron. This represents a key design area. There are a number of possible ways to do this:
    - (i) Wire the neural net randomly; or
    - (ii) Use an evolutionary algorithm (see below) to determine an optimal wiring; or
    - (iii) Use the system designer's best judgment in determining the wiring.
- The initial synaptic strengths (i.e., weights) of each connection. There are a number of possible ways to do this:
    - (i) Set the synaptic strengths to the same value; or
    - (ii) Set the synaptic strengths to different random values; or
    - (iii) Use an evolutionary algorithm to determine an optimal set of initial values; or
    - (iv) Use the system designer's best judgment in determining the initial values.
- The firing threshold of each neuron.
- The output, which can be:
    - (i) the outputs of layer$_M$ of neurons; or
    - (ii) the output of a single output neuron, the inputs of which are the outputs of the neurons in layer$_M$; or

    (iii) a function of (e.g., a sum of) the outputs of the
        neurons in layer$_M$; or

    (iv) another function of neuron outputs in multiple
        layers.

• The synaptic strengths of all the connections, which
must be adjusted during the training of this neural net.
This is a key design decision and is the subject of a great
deal of research and discussion. There are a number of
possible ways to do this:

    (i) For each recognition trial, increment or decrement
        each synaptic strength by a (generally small) fixed
        amount so that the neural net's output more closely
        matches the correct answer. One way to do this is
        to try both incrementing and decrementing and see
        which has the more desirable effect. This can be
        time-consuming, so other methods exist for
        making local decisions on whether to increment or
        decrement each synaptic strength.

    (ii) Other statistical methods exist for modifying the
        synaptic strengths after each recognition trial so
        that the performance of the neural net on that trial
        more closely matches the correct answer.

    (iii) Note that neural net training will work even if the
        answers to the training trials are not all correct.
        This allows using real-world training data that may
        have an inherent error rate. One key to the success
        of a neural net–based recognition system is the
        amount of data used for training. Usually a very
        substantial amount is needed to obtain satisfactory
        results. Just as with human students, the amount of
        time that a neural net spends learning its lessons is
        a key factor in its performance.

VARIATIONS

Many variations of the above are feasible:

- There are different ways of determining the topology. In particular, the interneuronal wiring can be set either randomly or using an evolutionary algorithm, which mimics the effects of mutation and natural selection on network design.
- There are different ways of setting the initial synaptic strengths.
- The inputs to the neurons in $layer_i$ do not necessarily need to come from the outputs of the neurons in $layer_{i-1}$. Alternatively, the inputs to the neurons in each layer can come from any lower or higher layer.
- There are different ways to determine the final output.
- The method described above results in an "all or nothing" (1 or 0) firing, called a nonlinearity. There are other nonlinear functions that can be used. Commonly, a function is used that goes from 0 to 1 in a rapid but relatively more gradual fashion. Also, the outputs can be numbers other than 0 and 1.
- The different methods for adjusting the synaptic strengths during training represent key design decisions.

The above schema describes a "synchronous" neural net, in which each recognition trial proceeds by computing the outputs of each layer, starting with $layer_0$ through $layer_M$. In a true parallel system, in which each neuron is operating independently of the others, the neurons can operate "asynchronously" (i.e., independently). In an asynchronous approach, each neuron is constantly scanning its inputs and fires whenever the sum of its weighted inputs exceeds its threshold (or whatever its output function specifies).

The goal is to then find actual examples from which the system can figure out how to solve a problem. A typical starting point is to have the neural net wiring and synaptic weights set randomly, so that the answers produced by this untrained neural net will thus also be random. The key function of a neural net is that it must learn its subject matter, just like the mammalian brains on which it is (at least roughly) modeled. A neural net starts out ignorant but is programmed to maximize a "reward" function. It is then fed training data (e.g., photos containing corgis and photos containing no corgis, as labeled by humans in advance). When the neural net produces a correct output (e.g., accurately identifying whether there's a corgi in the image), it gets reward feedback. This feedback can then be used to adjust the strength of each interneuronal connection. Connections that are consistent with the correct answer are made stronger, while those that provide a wrong answer are weakened.

Over time, the neural net organizes itself to be able to provide the correct answers without coaching. Experiments have shown that neural nets can learn their subject matter even with unreliable teachers. If the training data is labeled correctly only 60 percent of the time, a neural net can still learn its lessons with an accuracy well over 90 percent. Under some conditions, even smaller proportions of accurate labels can be used effectively.[26]

It's not intuitive that a teacher can train a student to surpass her own abilities, and likewise it can be confusing how unreliable training data can yield excellent performance. The short answer is that errors can cancel each other out. Let's say you're training a neural net to recognize the numeral 8 from handwritten samples of the numerals 0 through 9. And let's say that a third of the labels are inaccurate—a random mixture of 8s coded as 4s, 5s coded as 8s, and so on. If the dataset is large enough, these inaccuracies will offset each other and not skew the training much in any particular direction. This preserves most of the useful information in the dataset about what 8s look like, and still trains the neural net to a high standard.

Despite these strengths, early connectionist systems had a fundamental limitation. One-layer neural networks were mathematically incapable of solving some kinds of problems.[27] When I visited Professor Frank Rosenblatt at Cornell in 1964, he showed me a one-layer neural network called the Perceptron, which could recognize printed letters. I tried simple modifications to the input. The system did a fairly good job of auto-association (that is, it could recognize the letters even if I covered parts of them) but fared less well with invariance (that is, it failed to recognize letters after size and font changes).

In 1969 Minsky criticized the surge in interest in this area, even though he had done pioneering work on neural nets in 1953. He and Seymour Papert, the two cofounders of the MIT Artificial Intelligence Laboratory, wrote a book called *Perceptrons*, which formally demonstrated why a Perceptron was inherently incapable of determining whether or not a printed image was connected. The two images on page 28 are from the cover of *Perceptrons*. The top image is not connected (the black lines do not form a single contiguous shape), whereas the bottom image is connected (the black lines form a single contiguous shape). A human can determine this, as can a simple software program. A feed-forward (in which connections between the nodes do not form any loops) Perceptron such as Rosenblatt's Mark 1 Perceptron cannot make this determination.

In short, the reason feed-forward Perceptrons can't solve this problem is that doing so entails applying the XOR (exclusive or) computing function, which classifies whether a line segment is part of one contiguous shape in the image but not part of another. Yet a single layer of nodes without feedback is mathematically incapable of implementing XOR because it essentially has to classify all the data in one go with a linear rule (e.g., "If both of these nodes fire, the function output is true"), and XOR requires a feedback step ("If either of these nodes fires, *but they don't both fire*, the function output is true").

When Minsky and Papert reached this conclusion, it effectively killed most of the funding for the connectionism field, and it would be decades before it came back. But in fact, back in 1964 Rosenblatt explained to me that the Perceptron's inability to deal with invariance

was due to a lack of layers. If you took the output of a Perceptron and fed it back to another layer just like it, the output would be more general and, with repeated iterations of this process, would increasingly be able to deal with invariance. If you had enough layers and enough training data, it could deal with an amazing level of complexity. I asked him whether he had actually tried this, and he said no but that it was high on his research agenda. It was an amazing insight, but Rosenblatt died only seven years later, in 1971, before he got the chance to test his insights. It would be another decade before multiple layers were commonly used, and even then, many-layered networks required more computing power and training data than was practical. The tremendous surge in AI progress in recent years has resulted from the use of multiple neural net layers more than a half-century after Rosenblatt contemplated the idea.

So connectionist approaches to AI were largely ignored until the mid-2010s, when hardware advances finally unlocked their latent potential. Finally it was cheap enough to marshal sufficient computational power and training examples for this method to excel. Between the publication of *Perceptrons* in 1969 and Minsky's death in 2016, computational price-performance (adjusting for inflation) increased by a factor of about 2.8 *billion*.[28] This changed the landscape for what approaches were possible in AI. When I spoke to Minsky near the end of his life, he expressed regret that *Perceptrons* had been so influential, as by then connectionism had recently become widely successful within the field.

Connectionism is thus a bit like the flying-machine inventions of Leonardo da Vinci—they were prescient ideas, but not workable until lighter and stronger materials could be developed.[29] Once the hardware caught up, vast connectionism, such as one-hundred-layer networks, became feasible. As a result, such systems were able to solve problems that had never been tackled before. This is the paradigm driving all the most spectacular advances of the past several years.

THE CEREBELLUM: A MODULAR STRUCTURE

To understand neural networks in the context of human intelligence, I propose a small detour: let's go back to the beginning of the universe. The initial movement of matter toward greater organization progressed *very* slowly, with no brains to guide it. (See the section "The Incredible Unlikeliness of Being," in chapter 3, regarding the likelihood of the universe to have the ability to encode useful information at all.) The amount of time needed to create a new level of detail was hundreds of millions to billions of years.[30]

Indeed, it took billions of years before a molecule could begin to formulate coded instructions to create a living being. There is some disagreement over the currently available evidence, but most scientists place the beginning of life on earth somewhere between 3.5 billion and 4.0 billion years ago.[31] The universe is an estimated 13.8 billion years old (or, more precisely, that's the amount of time that has passed since the big bang), and the earth likely formed about 4.5 billion years ago.[32] So around 10 billion years passed between the first atoms forming and the first molecules (on earth) becoming capable of self-replication. Part of this lag may be explained by random chance—we don't know quite how unlikely it was for molecules randomly bumping around in early earth's "primordial soup" to combine in just the right way. Perhaps life could have started somewhat earlier, or maybe it would have been more likely for it to start much later. But before any of those necessary conditions were possible, whole stellar lifecycles had to play out as stars fused hydrogen into the heavier elements needed to sustain complex life.

According to scientists' best estimates, about 2.9 billion years then passed between the first life on earth and the first multicellular life.[33] Another 500 million years passed before animals walked on land, and 200 million more before the first mammals appeared.[34] Focusing on the brain, the length of time between the first development of primitive nerve nets and the emergence of the earliest centralized, tripartite brain was somewhere over 100 million years.[35] The first basic neocortex didn't appear for another 350 million to 400 million years, and it took another 200 million years or so for the modern human brain to evolve.[36]

All through this history, more sophisticated brains provided a marked evolutionary advantage. When animals competed for resources, the smarter ones often prevailed.[37] Intelligence evolved over a much shorter period than prior steps: millions of years, a distinct acceleration. The most notable change in the brains of pre-mammals was the region called the cerebellum. Human brains today actually have more neurons in the cerebellum than in the neocortex, which plays the biggest role in our higher-order functions.[38] The cerebellum is able to store and activate a large number of scripts that control motor tasks, such as one for signing your signature. (These scripts are often informally known as "muscle memory." This is not, in fact, a phenomenon of muscles themselves but rather of the cerebellum. As an action is repeated again and again, the brain adapts to make it easier and more subconscious—like the passage of many wagon wheels gradually boring ruts into a trail.)[39]

One way to catch a fly ball is to solve all the differential equations governing the ball's trajectory as well as your own movements and at the same time reposition your body based on those solutions. Unfortunately, you don't have a differential equation–solving apparatus in your brain, so instead you solve a simpler problem: how to place the glove most effectively between the ball and your body. The cerebellum assumes that your hand and the ball should appear in similar relative positions for each catch, so if the ball is dropping too fast and your hand appears to be going too slowly, it will direct your hand to move more quickly to match the familiar relative position.

These simple actions by the cerebellum to map sensory inputs onto muscle movements correspond to the mathematical idea of "basis functions" and enable us to catch the ball without solving any differential equations.[40] We are also able to use the cerebellum to anticipate what our actions would be even if we don't actually take them. Your cerebellum might tell you that you could catch the ball but you're likely to crash into another player, so maybe you should not take this action. This all happens instinctively.

Likewise, if you are dancing, your cerebellum will frequently direct your movements without your conscious awareness. People who are lacking a fully functional cerebellum due to injury or disease can still direct voluntary actions via the neocortex, but doing so requires focused effort, and they may suffer from coordination problems known as ataxia.[41]

A key component of mastering physical skills is performing their constituent actions often enough to ingrain them in your muscle memory. Movements that once required conscious thought and focus start to feel automatic. This basically represents a shift from motor cortex control to more cerebellar control. Whether you are throwing a football, solving a Rubik's Cube, or playing the piano, the less conscious mental effort you need to direct to the task, the better you are likely to perform. Your actions will be faster and smoother, and you can devote your attention to other aspects of success. When musicians have this mastery of their instrument, they can produce a given note as effortlessly and intuitively as ordinary people produce notes with their voice while singing "Happy Birthday." If I asked you how you make your vocal cords produce the right note as opposed to the wrong one, you probably couldn't describe the process in words. This is what psychologists and coaches call "unconscious competence," because the ability largely functions at a level below your conscious awareness.[42]

Yet the powers of the cerebellum aren't the result of some supremely complex architecture. While it does contain the majority of the neurons in an adult human (or other species') brain, there is not a lot of information about its overall design in the genome—it is composed largely of small and simple modules.[43] Although neuroscience is still working to understand the details of how the cerebellum functions, we know that it consists of thousands of small processing modules arranged in a feed-forward structure.[44] This helps shape our understanding of what neural architectures are needed to accomplish the cerebellum's functions, and new discoveries about the cerebellum may therefore provide further insights that are useful to the AI field.

Most of the cerebellum's modules have narrowly defined functions—those that govern your finger movements while playing the piano don't apply to the movement of your legs while walking. Even though the cerebellum has been a key brain region for hundreds of millions of years, humans rely on it less and less for survival, as our more flexible neocortex has taken the lead in navigating modern society.[45]

By contrast, non-mammalian animals don't have the advantages of a neocortex. Rather, their cerebellums have recorded very precisely the key behaviors that they need to survive. These cerebellum-driven animal behaviors are known as fixed action patterns. These are hardwired into members of a species, unlike behavior learned through observation and imitation. Even in mammals, some fairly complex behaviors are innate. For example, deer mice dig short burrows, while beach mice dig longer burrows with an escape tunnel.[46] When lab-raised mice with no previous experience of burrows were placed on sand, they each dug the kind of burrow made by their respective species in the wild.

For the most part, a given action in the cerebellum—like a frog's ability to precisely catch a fly with its tongue—persists in a species until a population with an improved action outcompetes it via natural selection. When behaviors are driven by genetics instead of learning, they are orders of magnitude slower to adapt. While learning allows creatures to meaningfully modify their behavior during a single lifetime, innate behaviors are limited to gradual change over many generations. Interestingly, though, computer scientists now sometimes use "evolutionary" approaches that mirror genetically determined behavior.[47] This involves creating a set of programs with certain random characteristics and seeing how well they function at a certain task. Those that perform well can have their characteristics combined, much like the genetic mixing of animal reproduction. Random "mutations" can then be introduced to see which boost performance. Over many generations, this can optimize problem-solving in ways that human programmers might never think of.

Implementing the equivalent of this approach in the real world takes millions of years. This may seem slow, but remember that evolution prior to biology—like the formation of the complex precursor chemicals needed for life—tended to take hundreds of millions of years, so the cerebellum was actually an accelerator.

THE NEOCORTEX: A SELF-MODIFYING, HIERARCHICAL, FLEXIBLE STRUCTURE

In order to make faster progress, evolution needed to devise a way for the brain to develop new behaviors without waiting for genetic change to reconfigure the cerebellum. This was the neocortex. Meaning "new rind," it emerged some 200 million years ago in a novel class of animals: mammals.[48] In these early mammals, which were rodent-like creatures, the neocortex was the size of a postage stamp and just as thin; it wrapped itself around their walnut-size brains.[49] But it was organized in a more flexible way than the cerebellum. Rather than being a collection of disparate modules controlling different behaviors, the neocortex worked more like a coordinated whole. Therefore, it was capable of a new type of thinking: it could invent new behaviors in days or even hours. This unlocked the power of learning.

More than 200 million years ago, the slow adaptation processes of non-mammalian animals were generally not a problem, since the environment changed very slowly. It typically took thousands of years for an environmental transformation that would require a response in the cerebellum.

So the neocortex was essentially waiting for a calamity in order to take over the world. That crisis, which we now call the Cretaceous–Paleogene extinction event, occurred 65 million years ago, 135 million years after the neocortex came into existence. Due to an asteroid impact and possibly also volcanic activity, the environment all across the earth changed suddenly, resulting in about 75 percent of all animal and plant species, including the dinosaurs, going extinct. (While the creatures we commonly know as dinosaurs went extinct during this event, some scientists consider birds to be a surviving branch of dinosaurs.)[50]

This was when the neocortex, which could invent new solutions quickly, rose to prominence. Mammals increased in size. The mammalian brain grew at an even quicker pace, taking up a larger fraction of an animal's body mass. And the neocortex grew faster still, by developing folds to expand its surface area.

If you took the neocortex of a human and stretched it out, it would be the size and thickness of a large dinner napkin.[51] But due to the extreme complexity of its structure, it now constitutes about 80 percent of the weight of the human brain.[52]

I describe in more detail how the neocortex works in my 2012 book *How to Create a Mind*, but a very brief summary here will convey the key concepts. The neocortex consists of a relatively simple repeating structure, each of which consists of about one hundred neurons. These modules can learn, recognize, and remember a pattern. The modules also learn to organize themselves into hierarchies, with each higher level mastering ever more sophisticated concepts. These repeating subunits are known as cortical minicolumns.[53]

According to current estimates, there are 21 to 26 billion neurons in the whole cerebral cortex, and 90 percent of those—or an average of around 21 billion—are in the neocortex itself.[54] At approximately a hundred neurons each, this suggests that we have something roughly on the order of 200 million minicolumns.[55] Emerging research shows that, unlike digital computers, which do most of their operations sequentially, the modules of the neocortex employ massive parallelism.[56] In essence, many things are happening simultaneously. This makes the brain a very dynamic system, and a big challenge to model computationally.

Neuroscience still has a lot to learn about the details, but the basics of how the minicolumns are organized and connected shed light on their function. Much like artificial neural networks running on silicon hardware, neural networks in the brain use hierarchical layers that separate raw data inputs (sensory signals, in the human case) and outputs (for humans, behavior). This structure allows progressive levels of abstraction, culminating in the subtle forms of cognition that we recognize as human.

At the bottom level (connected directly to sensory inputs), a module might serve to recognize a given visual stimulus as a curved shape. The other levels process the outputs of lower neocortical modules and add context and abstraction. Thus, progressively higher levels (farther from those connected to the senses) may recognize that curved shape as part of a letter, recognize that letter as part of a word, and connect that word to rich semantic meanings. At the top level are concepts that are far more abstract, such as the recognition that a statement is funny or ironic or sarcastic.

Although the "height" of a neocortical level sets its level of abstraction with respect to a single set of signals propagating up from sensory inputs, this process is not one directional. The six main layers of the neocortex dynamically communicate with one another in both directions—so we can't say that abstract thought happens exclusively in the highest layers.[57] Rather, it's more useful to think of the levels–abstractness relationship on a species level. That is, our multilayered neocortices give us more capacity for abstract thought than creatures with simpler cortices. And when humans are able to connect our neocortices directly to cloud-based computation, we'll unlock the potential for even more abstract thought than our organic brains can currently support on their own.

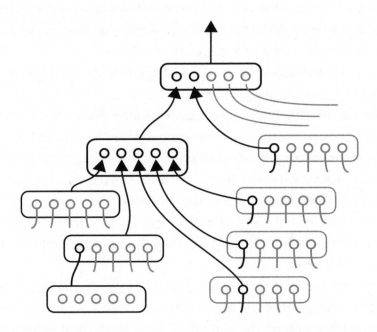

The neurological underpinnings of these abstractions are a fairly recent discovery. When a sixteen-year-old female epileptic patient was undergoing brain surgery in the late 1990s, the neurosurgeon Itzhak Fried kept her awake so she could respond to what was happening.[58] This was feasible because there are no pain receptors in the brain.[59] Whenever he stimulated a particular spot on her neocortex, she would laugh. Fried and his team quickly realized that they were triggering the actual perception of humor. She was not just laughing as a reflex—she genuinely found the present situation funny, even though nothing humorous had occurred in the operating room. When the doctors asked her why she was laughing, she did not reply along the lines of "Oh, no particular reason" or "You just stimulated my brain," but instead would immediately find a cause to account for it. She would explain her laughter with a comment like, "You guys are just so funny—standing around."[60]

The feasibility of locating and triggering the spot on the neocortex that codes for finding something amusing revealed that it is responsible for concepts like humor and irony. Other noninvasive testing has reinforced this finding. For example, reading ironic sentences lights up parts of the brain known as the ToM (theory of mind) network.[61] This neocortical abstraction ability was the enabling factor for humans to invent language, music, humor, science, art, and technology.[62]

No other species has ever achieved these things (despite frequent clickbait headlines to the contrary). No other animal can keep a beat in their head or tell a joke or give a speech or write (or read!) this book. Although some other animals, such as chimpanzees, can fashion primitive tools, these instruments are not sophisticated enough to trigger a rapid process of self-improvement.[63] Similarly, some other animals use simple forms of communication, but they are not capable of communicating hierarchical ideas, which we can do with human language.[64] We were already doing an excellent job of being primates without a frontal cortex, but when these additional modules became available to allow us to understand concepts about the world and existence, we went beyond just being an advanced animal and became a philosophical one.

Yet we should remember that brain evolution was just one part of our ascent as a species. For all our neocortical power, human science and art wouldn't be possible without one other key innovation: our thumbs.[65] Animals with comparable or even larger (in absolute terms) neocortices than humans—such as whales, dolphins, and elephants—don't have anything like an opposable thumb that can precisely grasp natural materials and fashion them into technology. The lesson: we are very fortunate evolutionarily!

We are also fortunate that our neocortex doesn't just have layers, but connects them in novel and powerful ways. The hierarchical organization of modules is not unique to the neocortex—the cerebellum also has hierarchies.[66] What sets the neocortex apart are three key features that enable the creativity of mammals and especially humans: (1) it can propagate the neuron-firing pattern for a given concept broadly throughout its structure, instead of just in the particular area where it originated; (2) a given firing pattern can be associated with similar aspects of many different concepts, and related concepts are represented by related firing patterns; and (3) millions of patterns can fire simultaneously[67] throughout the neocortex and interact with one another in complicated ways.[68]

For example, the highly complex connections within the neocortex allow for rich associative memories.[69] A memory in the brain is like a Wikipedia page—it can be linked to from many different places, and can change over time. Like a Wikipedia article, memories can also be multimedia. A memory can be triggered by a smell, a taste, a sound, or almost any sensory input.

Also, similarities in neocortical firing patterns promote analogical thinking. The pattern that represents lowering the position of your hand will be related to the pattern that represents lowering the pitch of your voice—and even to metaphorical lowerings, like the concepts for a falling temperature or a declining empire in history. Thus, we can form a pattern from learning a concept in one domain and then apply it to a completely different domain.

The neocortex's ability to draw analogies between disparate fields is responsible for many of the key intellectual leaps throughout history. For example, Charles Darwin's theory of evolution emerged from an analogy to geology. Before Darwin (1809–1882), scientists basically believed that God made the decision to create each species individually. There had been a few earlier quasi-evolutionary theories—most famously, that of Jean-Baptiste Lamarck (1744–1829), who proposed that animals had a natural progression of evolving into more complex species, and that offspring could inherit characteristics their parents acquired or developed during their own lifetimes.[70] Yet for each of these theories, their proposed mechanisms were either poorly explained or false.

But Darwin was exposed to a different idea by studying the work of Scottish geologist Charles Lyell (1797–1875), who advocated a controversial notion about the origin of large canyons.[71] The prevailing view was that the canyon existed as a God-given creation, and a river flowing through it just happened to find the canyon's bottom through gravity. Lyell's conception was that the river occurred *first* and that the canyon came only later. His theory met with significant resistance and took some time to gain acceptance, but scientists soon came to realize that flowing water's tiny impacts on rock, multiplied over the course of millions of years, could indeed cut as deep a gorge as the Grand Canyon. Lyell's theory drew heavily on the work of his fellow Scottish geologist James Hutton (1726–1797), who had first proposed the theory of uniformitarianism,[72] which held that instead of the world being shaped primarily by a catastrophic biblical flood, it was the product of a constant set of natural forces acting gradually over time.

Darwin faced a much more daunting challenge in his own field. Biology was endlessly complex, but Darwin saw the link between Lyell and his own studies as a naturalist, a connection he made at the opening of his 1859 book *On the Origin of Species*. He took Lyell's concept of the significance of a river eroding one small grain of stone at a time and applied it to one small genetic change over a generation. Darwin defended his theory with the explicit analogy: "As modern geology has almost banished such views as the excavation of a great valley by a single diluvial wave, so will natural selection, if it be a true principle, banish the belief of the continued creation of new organic beings, or of any great and sudden modification in their structure."[73] This sparked arguably the most profound scientific revolution our civilization has yet achieved. The other contenders for this honor, from Newton and gravitation to Einstein and relativity, were built on similar analogical insights.

DEEP LEARNING: RE-CREATING THE POWERS OF THE NEOCORTEX

How can we digitally replicate the flexibility and abstraction power of the neocortex? As discussed at the beginning of this chapter, rule-based symbolic systems are too rigid to capture the fluidity of human cognition. Connectionist approaches were impractical for a long time because they take so much computing power to train. But the price of computation has fallen dramatically. Why?

Intel cofounder Gordon Moore (1929–2023) was the eponymous originator of Moore's law, which he first articulated in 1965 and has since become the most prominent trend in information technology.[74] In its best-known form, it observes that, owing to progressive miniaturization, the number of transistors that can be put on a computer chip doubles roughly every two years. Skeptics of continued exponential progress in computing have frequently pointed out that Moore's law will inevitably come to an end as transistor density in integrated circuits reaches its physical limit at atomic scale, but this overlooks a deeper fact. Moore's law is just one instance of the more fundamental force I call the law of accelerating returns, where information technology creates feedback loops of innovation. This had already driven exponentially improving computational price-performance across four major technological paradigms—electromechanical, relays, vacuum tubes, and transistors—when Moore made his famous observation. And after integrated circuits have reached their limits, new paradigms using nanomaterials or three-dimensional computing will take over.[75]

This megatrend has been steadily and exponentially progressing since at least 1888 (long before Moore was even born!).[76] Around 2010 it finally reached a threshold where it became able to unlock the hidden power of a connectionist approach modeled on the many-layered hierarchical computation that takes place in the neocortex: deep learning. It is deep learning that has enabled the startling and seemingly sudden breakthroughs that the AI field has achieved since *The Singularity Is Near* was published.

The first of these breakthroughs to signal the radically transformative potential of deep learning was AI's mastery of the board game Go. Because Go has a vastly larger number of possible moves than chess, and it is harder to judge whether any given move is a good one, the AI approaches that had worked to beat human chess grandmasters were making almost no progress on Go. Even optimistic experts judged that the problem wouldn't be cracked until the 2020s at best. (As of 2012, for example, leading AI futurist Nick Bostrom speculated that Go would not be mastered by artificial intelligence until about 2022.)[77] But then, in 2015–16, Alphabet subsidiary DeepMind created AlphaGo, which used a "deep reinforcement learning" method in which a large neural net processed its own games and learned from its successes and failures.[78] It started with a huge number of recorded human Go moves and then played itself many times until the version AlphaGo Master was able to beat the world human Go champion, Ke Jie.[79]

A more significant development occurred a few months later with AlphaGo Zero. When IBM beat world chess champion Garry Kasparov with Deep Blue in 1997, the supercomputer was filled with all the know-how its programmers could gather from human chess experts.[80] It was not useful for anything else; it was a chess-playing machine. By contrast, AlphaGo Zero was not given any human information about Go except for the rules of the game, and after about three days of playing against itself, it evolved from making random moves to easily defeating its previous human-trained incarnation, AlphaGo, by 100 games to 0.[81] (In 2016, AlphaGo had beaten Lee Sedol, who at the time ranked second in international Go titles, in four out of five games.) AlphaGo Zero used a new form of reinforcement learning in which the program became its own instructor. It took AlphaGo Zero just twenty-one days to reach the level of AlphaGo Master, the version that defeated sixty top professionals online and the world champion Ke Jie in three out of three games in 2017.[82] After forty days, AlphaGo

Zero surpassed all other versions of AlphaGo and became the best Go player in human or computer form.[83] It achieved this with no encoded knowledge of human play and no human intervention.

But that is not the most significant milestone for DeepMind. The next incarnation, AlphaZero, can transfer abilities learned from Go to other games like chess.[84] The program not only defeated all human challengers at chess but also defeated all other chess-playing machines, and did so after only four hours of training, using no prior knowledge except the rules. It was equally successful at the game Shōgi. The latest version as I write this is MuZero, which repeated these feats without even being given the rules![85] With this "transfer learning" ability, MuZero can master any board game in which there is no chance, ambiguity, or hidden information, or any deterministic video game like Atari's *Pong*. This ability to apply learning from one domain to a related subject is a key feature of human intelligence.

But deep reinforcement learning is not limited to mastering such games. AIs that can play *StarCraft II* or poker, both of which feature uncertainty and require a sophisticated understanding of rival players in the game, have also recently exceeded the performance of all humans.[86] The only exceptions (for now) are board games that require very high linguistic competencies. *Diplomacy* is perhaps the best example of this—a world domination game that is impossible for a player to win through luck or skill, and forces players to talk to one another.[87] To win, you have to be able to convince people that moves that help you will be in their own self-interest. So an AI that can consistently dominate *Diplomacy* games will likely have also mastered deception and persuasion more broadly. But even at *Diplomacy*, AI made impressive progress in 2022, most notably Meta's CICERO, which can beat many human players.[88] Such milestones are now being reached almost every week.

The same deep-learning capabilities that dominate games with ease can also be applied to performance in complex real-world situations. Basically what we need is a simulator that can replicate the domain about which the AI is trying to learn—such as the varied and ambiguity-filled experience of driving a car. All kinds of things can happen while you're behind the wheel, from another car suddenly stopping in front of you or coming toward you the wrong way to a child chasing a ball into the street. To address this, the Alphabet subsidiary Waymo created self-driving software for its autonomous cars but initially had a human monitor participate in all the rides.[89] Every aspect of these rides was recorded, and from that survey a very comprehensive simulator was created. Now that the actual vehicles have recorded well over 20 million miles (at the time of this writing),[90] simulated cars can travel billions of miles training in this realistic virtual space.[91] With the accumulation of this vast experience, an actual self-driving car will ultimately be able to perform much, much better than human drivers. Similarly, as described further in chapter 6, AI is using a range of novel simulation techniques to make better predictions about how proteins fold. This is one of the most challenging problems in biology and solving it is opening the door to discovering breakthrough medicines.

Yet while MuZero can conquer many different games, its achievements are still relatively narrow—it can't write a sonnet or comfort the sick. In order to reach the breathtaking generality of the human neocortex, AI will need to master language. It is language that enables us to connect vastly disparate domains of cognition and allows high-level symbolic transfer of knowledge. That is, with language we don't need to see a million examples of raw data to learn something—we can dramatically update our knowledge just by reading a single-sentence summary.

The fastest progress in this area is now coming from approaches that process language by using deep neural nets to represent the meanings of words in a (very) many-dimensional space. There are several mathematical techniques for doing this, but the upshot is that they allow the AI to discover the meaning of language without any of the hard-coded linguistic rules that a symbolic approach would require. As one example, we can construct a multilayer feed-forward neural net and find billions (or trillions) of sentences to train it. These can be gathered from public sources on the web. The neural net is then used to assign each sentence a point in 500-dimensional space (that is, a list of 500 numbers, though this number is arbitrary; it can be any substantial large number). At first, the sentence is given a random assignment for each of the 500 values. During training, the neural net adjusts the sentence's place within the 500-dimensional space such that sentences that have similar meanings are placed close together; dissimilar sentences will be far away from one another. If we run this process for many billions of sentences, the position of any sentence in the 500-dimensional space will indicate what it means by virtue of what it is close to.

In this way the AI learns meaning not from a grammatical rule book or dictionary but rather from the contexts that words are actually used in. For example, it would learn that the word "jam" has distinct homonyms by virtue of the fact that in some contexts people talk about eating jam, and in others they jam with electric guitars, but nobody talks about eating electric guitars. With the exception of the tiny fraction of our vocabularies we formally learn in school or look up explicitly, this is exactly how humans learn all the words we know. And AI has already expanded its associative powers beyond the realm of text. A 2021 OpenAI project called CLIP is a neural network trained to link images to text that describes them. As a result, nodes in CLIP are able to "respond to the same concept whether presented literally, symbolically, or conceptually."[92] For example, the same node may fire in response to a photo of a spider, a drawing of Spider-Man, or the word "spider." This is exactly how the human brain processes concepts across contexts, and it is a major leap forward for AI.

Another variation of this method is a 500-dimensional space that carries sentences in *every language*. Thus, if you want to translate a sentence from one language to another, you just look for another sentence in the target language at the closest point in this hyperdimensional space. You can also find other sentences that are reasonably close to your intended meaning by looking at other nearby sentences in the space. A third option is to create two paired 500-dimensional spaces, where one of the spaces reveals the answers to questions in the first space. This requires assembling billions of sentences for which one is a response to the other. A further expansion of this concept is to create a "Universal Sentence Encoder,"[93] which my team at Google created to embed each sentence in a dataset with thousands of detected features such as "ironic," humorous," or "positive." AI trained on this richer data learns not just to mimic how humans use language but also to grasp deeper semantic features that may not be apparent from the literal meaning of the words in a sentence. This meta-knowledge makes for fuller understanding.

We have been creating a variety of applications at Google that use and produce conversational language based on these principles. A prominent one is Gmail Smart Reply.[94] If you have Gmail, you will notice that it provides you with three suggestions to respond to each email. The responses take into consideration not just the email you are responding to but all of the other emails in that chain, the subject line, and other indications of whom you are writing to. All these elements of your email require precisely this type of multidimensional representation of each point in the conversation. This is a combination of a multilayered feed-forward neural network with a hierarchical representation of language content that represents the back-and-forth of the dialogue. Gmail Smart Reply felt awkward to some at first but quickly gained acceptance by users for its naturalness and convenience, and now it accounts for a significant minority of all Gmail traffic.

Another Google capability based on this approach was called Talk to Books. (It was available as an experimental standalone service from 2018 to 2023.) Once you had Talk to Books loaded, you could simply ask it a question. The software reviewed every sentence (all 500 million of them) in more than 100,000 books in half a second. It then provided you with the best answers to your question. This was not an application of normal Google Search, which finds you relevant links using a combination of keyword matching, frequency of user clicks, and other measurements, but instead worked via the actual meaning of your question and the meaning of each of the 500 million sentences in the 100,000-plus books.

One of the most promising applications of hyperdimensional language processing is a class of AI systems called transformers. These are deep-learning models that use a mechanism called "attention" to focus their computational power on the most relevant parts of their input data—in much the same way that the human neocortex lets us direct our own attention toward the information most vital to our thinking. Transformers are trained on massive amounts of text, which they encode as "tokens"—usually a combination of parts of words, words, and strings of words. The model then uses a very large number of "parameters" (billions to trillions, as I write this) to classify each of the tokens. You can think of parameters as factors that can be used to make predictions about something.

As a scaled-down example, if I can use only one parameter to predict "Is this animal an elephant?" I might choose "trunk." So if the neural net's node dedicated to judging whether the animal has a trunk fires ("Yes, it does"), the transformer would categorize it as an elephant. But even if that node learns to perfectly recognize trunks, there are some animals with trunks that aren't elephants, so the one-parameter model will misclassify them. By adding parameters like "hairy body," we can improve accuracy. Now if both nodes fire ("hairy body" and "trunk"), I can guess that it's probably not an elephant but rather a woolly mammoth. The more parameters I have, and the more granular detail I can capture, the better predictions I can make.

In a transformer, such parameters are stored as weights between nodes in the neural net. And in practice, while they sometimes correspond to human-understandable concepts like "hairy body" or "trunk," they often represent highly abstract statistical relationships that the model has discovered in its training data. Using these relationships, transformer-based large language models (LLMs) can predict which tokens would be most likely to follow a certain input prompt by a human. They then convert those back into text (or images, audio, or video) that humans can understand. Invented by Google researchers in 2017, this mechanism has powered most of the enormous AI advances of the past few years.[95]

The key fact to understand is that transformers depend for their accuracy on huge numbers of parameters. This requires vast amounts of computation both for training and for usage. OpenAI's 2019 model GPT-2 had 1.5 billion parameters,[96] and despite flashes of promise, it did not work very well. But once transformers got over 100 billion parameters, they unlocked major breakthroughs in AI's command of natural language—and could suddenly answer questions on their own with intelligence and subtlety. GPT-3 used 175 billion in 2020,[97] and a year later DeepMind's 280-billion-parameter model Gopher performed even better.[98] Also in 2021, Google debuted a 1.6-trillion-parameter transformer called Switch, making it open-source to freely apply and build on.[99] Although Switch's record-breaking size turned heads, its most important innovation was a technique called "mixture of experts." With this approach, the transformer is able to focus more efficiently on using the most relevant parts of the model for a given task. This is important progress toward preventing computational costs from spiraling out of control as models get ever larger.

So why is scale so important? In short, it lets models access deeper features of their training data. Smaller models do relatively well when the task is something narrow like using historical data to predict temperatures. But language is fundamentally different. Because the number of ways to start a sentence is essentially infinite, even if a transformer has been trained on hundreds of billions of tokens of text, it can't simply memorize verbatim quotes to complete it. Instead, with many billions of parameters, it can process the input words in the prompt at the level of associative meaning and then use the available context to piece together a completion text never before seen in history. And because the training text features many different styles of text, such as question-and-answer, op-ed pieces, and theatrical dialogue, the transformer can learn to recognize the nature of the prompt and generate an output in the appropriate style. While cynics may dismiss this as a fancy trick of statistics, because those statistics are synthesized from the combined creative output of millions of humans, the AI attains genuine creativity of its own.

GPT-3 was the first such model to be commercially marketed and to display this creativity in a way that impressed its users.[100] For example, scholar Amanda Askell prompted it with a passage from philosopher John Searle's famous "Chinese room argument."[101] This thought experiment observes that a non-Chinese-speaking human translating the language by manually operating a computer translation algorithm with pen and paper wouldn't understand the stories being translated. Thus, how could an AI running the same program be said to truly understand? GPT-3 responded, "It is obvious that I do not understand a word of the stories," explaining that the translation program is a formal system that "does not explain understanding any more than a cookbook explains a meal." This metaphor had never appeared anywhere before but rather appears to be a new adaptation of philosopher David Chalmers's metaphor that a recipe does not fully explain the properties of a cake. This is precisely the sort of analogizing that helped Darwin discover evolution.

Another capability unlocked by GPT-3 was stylistic creativity. Because the model had enough parameters to deeply digest a staggeringly large dataset, it was familiar with virtually every kind of human writing. Users could prompt it to answer questions about any given subject in a huge variety of styles—from scientific writing to children's books, poetry, or sitcom scripts. It could even imitate specific writers, living or dead. When computer programmer Mckay Wrigley asked GPT-3 to answer "How do we become more creative?" in the style of pop psychologist Scott Barry Kaufman, it gave a novel answer that the real Kaufman acknowledged "definitely sounds like something I would say."[102]

In 2021 Google introduced LaMDA, which was optimized to focus on lifelike open-ended conversation.[103] If you told LaMDA to answer questions in character as a Weddell seal, for example, it would give coherent, playful answers from a seal's perspective—telling a would-be hunter, "Haha good luck. Hope you don't freeze before you take a shot at one of us!"[104] This demonstrated the kind of contextual knowledge that had long eluded AI.

Another startling advance in 2021 was multimodality. Previous AI systems had generally been limited to inputting and outputting one kind of data—some AI focused on recognizing images, other systems analyzed audio, and LLMs conversed in natural language. The next step was connecting multiple forms of data in a single model. So OpenAI introduced DALL-E (a pun on surrealist painter Salvador Dalí and the Pixar movie *WALL-E*),[105] a transformer trained to understand the relationship between words and images. From this it could create illustrations of totally novel concepts (e.g., "an armchair in the shape of an avocado") based on text descriptions alone. In 2022 came its successor, DALL-E 2,[106] along with Google's Imagen and a flowering of other models like Midjourney and Stable Diffusion, which quickly extended these capabilities to essentially photorealistic images.[107] Using a simple text input like "a photo of a fuzzy panda wearing a cowboy hat and black leather jacket riding a bike on top of a mountain," the AI can conjure up a whole lifelike scene.[108] This creativity will transform creative fields that recently seemed strictly in the human realm.

In addition to generating marvelous images, these multimodal models also achieved a more fundamental breakthrough. In general, models like GPT-3 exemplify "few-shot learning." That is, after being trained, they can get a fairly small sample of text and cogently complete it. This is the equivalent of showing an image-focused AI only five images of something unfamiliar, like unicorns (instead of five thousand or five million, as previous methods required), and getting it to recognize new unicorn images, or even create unicorn images of its own. But DALL-E and Imagen took this a dramatic step further by excelling at "zero-shot learning." DALL-E and Imagen could combine concepts they'd learned to create new images wildly different from anything they had ever seen in their training data. Prompted by the text "an illustration of a baby daikon radish in a tutu walking a dog," DALL-E spat out adorable cartoon images of exactly that. Likewise for "a snail with the texture of a harp." It even created "a professional high quality emoji of a lovestruck cup of boba"—complete with heart eyes beaming above the floating tapioca balls.

Zero-shot learning is the very essence of analogical thinking and intelligence itself. It demonstrates that the AI isn't just parroting back what we feed it. It is truly learning concepts with the ability to creatively apply them to novel problems. Perfecting these capabilities and expanding them across more domains will be a defining artificial intelligence challenge of the 2020s.

In addition to zero-shot flexibility within a given type of task, AI models are also rapidly gaining cross-domain flexibility. Just seventeen months after MuZero demonstrated mastery across numerous games, DeepMind unveiled Gato—a single neural network that can tackle tasks ranging from playing video games or chatting via text to captioning images or controlling a robot arm.[109] None of these capabilities are new in themselves, but combining them into one unified brainlike system is a big step toward human-style generalization and portends very rapid progress ahead. In *The Singularity Is Near*, I predicted that we would combine thousands of individual skills into one AI before a successful Turing test would be accomplished.

One of the most powerful tools for flexibly applying human intelligence is computer programming—indeed, that's how we created AI in the first place. In 2021, OpenAI debuted Codex, which could take natural-language prompts from users and translate them into working code in a variety of languages like Python, JavaScript, and Ruby.[110] In minutes, someone with no programming experience could type out what they wanted the program to do and create a simple game or application. DeepMind's 2022 model AlphaCode[111] boasted even greater coding proficiency, and by the time you read this, still more powerful programming AIs will be available. Such capabilities will unleash staggering amounts of human potential over the next few years as coding skill ceases to be a requirement for implementing creative ideas through software.

Despite all the achievements of the models I've just described, though, they have all struggled when faced with complex tasks and no human guidance to intercede. Even if they could complete all the subtasks individually, it was hard for them to figure out how everything was supposed to fit together. In April 2022, Google's 540-billion-parameter PaLM model achieved stunning progress on this problem, particularly in two areas fundamental to our own intelligence: humor and inferential reasoning.[112]

Humor seems quintessentially human because it draws on so many different elements. To "get" any given joke, we may have to understand concepts like wordplay, irony, or common experiences people share. Often, several of these concepts are combined in complex ways. This is why jokes are often hard to translate across languages and cultures. PaLM was able to read jokes and accurately explain why they're funny. For example, it was given the following input: "I was supposed to start writing the paper at 5:00 PM. But then I started playing with this cool new language model for 10 minutes. 10 minutes later, it's suddenly 9:30 PM!"[113] It gave this perfect exposition:

> This joke is about how time flies when you're having fun. The person was supposed to start writing a paper at 5:00 PM, but they started playing with a language model instead. They thought they were only playing with it for 10 minutes, but it was actually 4.5 hours.[114]

Even more importantly, PaLM could explain how it reached conclusions via "chain-of-thought" reasoning, although not yet (at least as of 2023) as deeply as what humans can do. This is what teachers want when they assign essays—the goal isn't just parroting a correct answer; it's demonstrating deeper understanding by building up a coherent sequence of inferences step-by-step. Too often, though, large language models are so opaque that nobody can figure out how they arrived at a particular output. So by clarifying this process, PaLM both provides more trustworthy results and helps software engineers diagnose the exact problem when it makes an error. It also turns out that breaking down questions into smaller chunks helps PaLM correctly answer highly complex questions that would be challenging for even a bright human:[115]

> "Input: Michael is at that really famous museum in France looking at its most famous painting. However, the artist who made this painting just makes Michael think of his favorite cartoon character from his childhood. What was the country of origin of the thing that the cartoon character usually holds in his hand?"
>
> "Model Output: The most famous painting in the Louvre is the Mona Lisa. The artist who made the Mona Lisa is Leonardo da Vinci. Leonardo da Vinci is also the name of the main character in the cartoon Teenage Mutant Ninja Turtles. Leonardo da Vinci is from Italy. The thing that Leonardo da Vinci usually holds in his hand is a katana. The country of origin of the katana is Japan. The answer is 'Japan'."

In November 2022, OpenAI launched an interface called Chat-GPT, which allowed the general public for the first time to easily interact with an LLM—a model known as GPT-3.5.[116] Within two months, 100 million people had tried it, likely including you.[117] Because the system could generate many fresh and varied answers to a given question, it became a big disruptor in education as students used Chat-GPT to write their essays, while teachers lacked a reliable way to detect cheating (though some promising tools exist).[118]

Then, in March of 2023, GPT-4 was rolled out for public testing via ChatGPT. This model achieved outstanding performance on a wide range of academic tests such as the SAT, the LSAT, AP tests, and the bar exam.[119] But its most important advance was its ability to reason organically about hypothetical situations by understanding the relationships between objects and actions—a capability known as world modeling.

This allows it to not just answer questions like complicated physics problems, but to reason accurately about what would happen if the laws of physics were changed in a certain way. It can also envision situations from the perspective of different characters; for example, it can recognize that two people looking at numbers from opposite directions will see them reversed. Or it can look at an image of balloons held down by a weight and recognize that if the strings were cut, the balloons would fly away.[120] GPT-4 even keeps track of objects spatially over time, such as in this example by security engineer Daniel Feldman:

> Prompt: "I'm in my house. On top of my chair in the living room is a coffee cup. Inside the coffee cup is a thimble. The thimble does not fit snugly inside of the coffee cup. Inside the thimble is a single diamond. I move the chair to my bedroom. Then I put the coffee cup on the bed. Then I turn the cup upside down. Then I return it to rightside-up, and place the coffee cup on the kitchen counter. Where is my diamond?"
>
> Answer: "When you turned the coffee cup upside down on your bed, the diamond inside the thimble would have likely fallen out onto the bed. The diamond is now on your bed."[121]

As I completed this book in the middle of 2023, the latest major innovation was PaLM-E—a Google system that combines the reasoning ability of PaLM with embodiment in a robot.[122] This system is able to take natural-language instructions and carry them out in a complex physical environment. For example, when told to "bring me the rice chips from the drawer," PaLM-E was able to navigate through a kitchen, find the chips, pick them up, and deliver them successfully. Such abilities will rapidly expand AI's reach into the real world.

AI progress is now so fast, though, that no traditional book can hope to be up to date. The logistical steps of laying out and printing a book take nearly a year, so even if you purchased this volume as soon as it was published, many astonishing new advances will surely have been made by the time you read this. And AI will likely be woven much more tightly into your daily life. The old links-page paradigm of internet search, which lasted for about twenty-five years, is rapidly being augmented with AI assistants like Google's Bard (powered by the Gemini model, which surpasses GPT-4 and was released as this book entered final layout) and Microsoft's Bing (powered by a variant of GPT-4).[123] Meanwhile, application suites like Google Workspace and Microsoft Office are integrating powerful AI that will make many kinds of work smoother and faster than ever before.[124]

Scaling up such models closer and closer to the complexity of the human brain is the key driver of these trends. I have long believed that the amount of computation is key to providing intelligent answers, but up until recently this view was not widely shared and could not be demonstrated. About three decades ago, in 1993, I had a debate with my own mentor Marvin Minsky. I argued that we needed about 10^{14} calculations per second to begin to emulate human intelligence. Minsky, for his part, maintained that the amount of computation was not important, and that we could program a Pentium (the processor in a desktop computer from 1993) to be as intelligent as a human. Because we had such different opinions on this, we held a public debate at MIT's primary debate hall (Room 10-250), attended by several hundred students. Neither of us was able to win that day, as I did not have enough computation to demonstrate intelligence and he did not have the right algorithms.

Yet the connectionist breakthroughs of 2020–2023 have made clear that the amount of computation *is* key to achieving sufficient intelligence. I started in AI around 1963, and it has taken sixty years to reach this level of computation. The amount of computation used to train a state-of-the-art model is now increasing by about a factor of four each year—and capabilities are rapidly maturing.[125]

WHAT DOES AI STILL NEED TO ACHIEVE?

As the past few years demonstrate, we are already well on our way to re-creating the capabilities of the neocortex. Today, AI's remaining deficiencies fall into several main categories, most notably: contextual memory, common sense, and social interaction.

Contextual memory is the ability to keep track of how all the ideas in a conversation or written work dynamically fit together. As the size of the relevant context increases, the number of relationships among ideas grows exponentially. Recall the complexity ceiling idea from earlier in this chapter—similar math makes it very computation-intensive to increase the context window that a large language model can handle.[126] If there are ten wordlike ideas (i.e., tokens) in a given sentence, the number of possible relationships among subsets of them is $2^{10}-1$, or 1,023. If there are fifty such ideas in a paragraph, that's 1.12 quadrillion possible contextual relationships among them! Although the vast majority of these are irrelevant, the demands of remembering the context for an entire chapter or book by brute force spiral rapidly out of control. This is why GPT-4 forgets things you told it earlier in the conversation, and why it can't write a novel with a consistent and logical plot.

The good news is twofold: researchers are making great progress in designing AI that can focus more efficiently on relevant context data, and exponential price-performance improvements mean that the cost of computation will probably fall by more than 99 percent within a decade.[127] Further, algorithmic improvements and AI-specific hardware specialization mean that price-performance for LLMs is likely to increase much faster than that.[128] For perspective, from August 2022 to March 2023 alone, the price of input/output tokens via the GPT-3.5 application programming interface fell by 96.7 percent![129] This trend will likely accelerate as AI is used directly in optimizing chip design, which has already begun.[130]

The next area of deficiency is common sense. This is the ability to imagine situations and anticipate their consequences in the real world. For example, even though you've never studied what would happen if gravity suddenly stopped working in your bedroom, you can readily picture that hypothetical and predict what might result. This kind of reasoning is also important for causal inference. If you have a dog and come home to find a broken vase, you can infer what happened. Despite ever-more-frequent flashes of insight, AI still struggles with this because it doesn't yet have a robust model of how the real world works, and training data rarely includes such implicit knowledge.

Finally, social nuances like an ironic tone of voice are not well represented in the text databases that AI still mostly trains on. Without such understanding, it's hard to develop a "theory of mind"—an ability to recognize that others have beliefs and knowledge different from ours, put ourselves in their shoes, and infer their motivations. Yet AI is now making rapid strides in this area. In 2021, Google Fellow Blaise Agüera y Arcas reported presenting LaMDA with a classic scenario used to test theory of mind in child psychology.[131] In the scenario, Alice forgets her glasses in a drawer and leaves the room. While she's away, Bob takes her glasses out of the drawer and hides them under a cushion. The question: Where will Alice look for the glasses when she comes back into the room? LaMDA correctly answered that she will look in the drawer. Within two years PaLM and GPT-4 were correctly answering many theory-of-mind questions. This capability will afford AI crucial flexibility. A human Go champion can play the game very well but can also monitor how other people nearby are doing and make jokes when appropriate, or be flexible and stop the game if someone needs medical attention.

My optimism about AI soon closing the gap in all these areas rests on the convergence of three concurrent exponential trends: improving computing price-performance, which makes it cheaper to train large neural nets; the skyrocketing availability of richer and broader training data, which allows training computation cycles to be put to better use; and better algorithms that enable AI to learn and reason more efficiently.[132] Although computation speeds for the same cost have been doubling roughly every 1.4 years on average since 2000, the actual growth in the total computations ("compute") used to train a state-of-the-art artificial intelligence model has been doubling every 5.7 months since 2010. This is around a ten-billion-fold increase.[133] By contrast, during the pre-deep-learning era, from 1952 (the demonstration of one of the first machine learning systems, six years before the Perceptron's groundbreaking neural network) to the rise of big data, around 2010, there was a nearly two-year doubling time (which roughly tracked with Moore's law) in the amount of compute to train a top AI.[134]

Put another way, if the 1952–2010 trend had continued to 2021, compute would have increased by a factor of less than 75 instead of some ten-billion-fold. This has been much faster than improvements in overall computing price-performance. So the cause isn't a major hardware revolution. Instead, it's mainly two factors. First, AI researchers have been innovating new methods in parallel computing— so that greater numbers of chips can work together on the same machine-learning problem. Second, as big data has made deep learning more useful, investors around the world have been pouring ever greater amounts of money into the field, hoping to achieve breakthroughs.

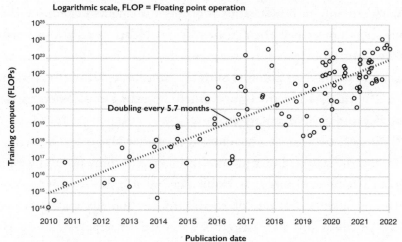

Training Compute (FLOPS) of Milestone Machine Learning Systems Over Time, n=98

Logarithmic scale, FLOP = Floating point operation

Chart by Anderljung et al., based on 2022 data from Sevilla et al., building on previous 2018 research on AI and compute by Amodei and Hernandez of OpenAI.[135]

The resulting growth in total spending on training reflects the ballooning scope of useful data. It has only been in the past few years that we can affirmatively state the following: any kind of skill that generates clear enough performance feedback data can be turned into a deep-learning model that propels AI beyond all humans' abilities.

Human skills vary widely in the accessibility of their training data. Some skills are both easy to assess in quantitative terms and easy to gather all the relevant data on. For example, in playing chess, there are clear outcomes of win, loss, or draw, and Elo ratings that provide a quantitative measure of an opponent's strength. And chess data is easy to gather because games involve no ambiguity and can be represented as a mathematical sequence of moves. Other skills are easy to quantify in principle, but data collection and analysis is more challenging. Arguing a legal case in court leads to clear win-or-loss outcomes, but it is hard to disentangle how much the lawyer's skill contributed to these outcomes compared to factors like the strength of the case or juror bias. And in some instances, it's not even clear how to quantify the skill—for example, quality of poetry writing or how suspenseful a mystery novel is. Yet even in these latter examples, proxy measures could plausibly be used to train AI. Poetry readers might provide scored 0–100 assessments of how beautiful a poem seemed to them, or fMRI could show how much their brains lit up. Heart rate data or cortisol levels might reveal readers' reactions to suspense. The takeaway is that, with a sufficient amount of data, imperfect and indirect metrics can still guide AI toward improvement. Finding these metrics takes creativity and experimentation.

While a neocortex can have some idea of what a training set is all about, a well-designed neural net can extract insights beyond what biological brains can perceive. From playing a game to driving a car, analyzing medical images, or predicting protein folding, data availability provides an increasingly clear path to superhuman performance. This is creating a powerful economic incentive to identify and collect kinds of data that were previously considered too difficult to bother with.

It can be useful to think about data as a bit like petroleum. Oil deposits exist along a continuum of extraction difficulty.[136] Some oil gushes out of the ground under its own pressure, ready to refine and cheap to produce. Other deposits need expensive deep drilling, hydraulic fracturing, or special heating processes to extract it from shale rock. When oil prices are low, energy companies extract oil only from the cheap and easy sources, but as prices rise it becomes economically viable to exploit the tougher-to-access deposits.

In a similar way, when the benefits of big data were relatively small, companies collected it only in cases where it was relatively cheap to do so. But as machine-learning techniques advance and compute becomes cheaper, the economic value (and often the social value) of many harder-to-access kinds of data will increase. Indeed, thanks to accelerating innovations in big data and machine learning, our ability to gather, store, classify, and analyze data about human skills has grown enormously in just the past year or two.[137] "Big data" has become a buzzword in Silicon Valley, but the fundamental advantage of this technology is very real: it has become practical to use machine-learning techniques that would simply not work with smaller amounts of data. This is going to happen during the 2020s for almost every human skill that exists.

Thinking of AI progress in terms of discrete abilities highlights an important fact. We often speak of human-level intelligence as a monolithic, singular thing—something that an AI either has or doesn't have. But it's far more useful and accurate to view human intelligence as a thick bundle of different cognitive abilities. Some of these, like the ability to recognize ourselves in a mirror, we share with smart animals like elephants and chimpanzees. Others, like composing music, are limited to humans but vary wildly from person to person. Not only do cognitive abilities differ; they can also differ sharply within an individual. Someone may be a mathematical genius but a terrible chess player, or have a photographic memory but struggle with social interaction. Dustin Hoffman's character in *Rain Man* illustrates this memorably.

So when AI researchers talk about human-level intelligence, it generally means the ability of the most skilled humans in a particular domain. In some areas, the gap between the average human and the most skilled human is not very large (e.g., recognizing letters in their native language's alphabet), while in others the gap yawns very wide indeed (e.g., theoretical physics). In the latter cases, there may be a substantial time lag between AI reaching average human ability and superhuman ability. It remains an open question which skills will ultimately prove hardest for AI to master. It might turn out, for example, that in 2034 AI can compose Grammy-winning songs but not write Oscar-winning screenplays, and can solve Millennium Prize Problems in math but not generate deep new philosophical insights. So there may well be a significant transitional period during which AI has passed the Turing test and is superhuman in most respects, but still has not surpassed the top humans in a few key skills.

For the purpose of thinking about the Singularity, though, the most important fiber in our bundle of cognitive skills is computer programming (and a range of related abilities, like theoretical computer science). This is the main bottleneck for superintelligent AI. Once we develop AI with enough programming abilities to give itself even more programming skill (whether on its own or with human assistance), there'll be a positive feedback loop. Alan Turing's colleague I. J. Good foresaw as early as 1965 that this would lead to an "intelligence explosion."[138] And because computers operate much faster than humans, cutting humans out of the loop of AI development will unlock stunning rates of progress. Artificial intelligence theorists jokingly refer to this as "FOOM"—like a comic book–style sound effect of AI progress whizzing off the far end of the graph.[139]

Some researchers, like Eliezer Yudkowsky, see this as more likely to happen extremely fast (a "hard takeoff" in minutes to months), while others, like Robin Hanson, think it will be relatively more gradual (a "soft takeoff" over years or longer).[140] I fall somewhere in the middle. My view is that physical constraints on hardware, resources, and real-world data suggest limits to the speed of FOOM but that we should nonetheless take precautions to avoid a potential hard takeoff going wrong. Relating this back to human cognitive abilities, once we trigger an intelligence explosion, any abilities that are harder for AI than self-improving programming will also be achieved in short order.

With machine learning getting so much more cost-efficient, raw computing power is very unlikely to be the bottleneck in achieving human-level AI. Supercomputers already significantly exceed the raw computational requirements to simulate the human brain. Oak Ridge National Laboratory's Frontier, the world's top supercomputer as of 2023,[141] can perform on the order of 10^{18} operations per second. This is already on the order of 10,000 times as much as the brain's likely maximum computation speed (10^{14} operations per second).[142]

My 2005 calculations in *The Singularity Is Near* noted 10^{16} operations per second as an upper bound on the brain's processing speed (as we have on the order of 10^{11} neurons with on the order of 10^3 synapses each firing on the order of 10^2 times per second).[143] But as I noted then, this was a high estimate to be conservative. In reality, the computation done in an actual brain will ordinarily be much less than this. A range of further research over the past two decades has shown that neurons fire orders of magnitude more slowly—not two hundred times a second, which is around their theoretical maximum, but closer to once a second.[144] In fact, the AI Impacts project estimated, based on the brain's energy consumption, that the average neuron fires only 0.29 times per second—implying that total brain computation could be as low as around 10^{13} operations per second.[145] This matches Hans Moravec's seminal estimate, in his 1988 book *Mind Children: The Future of Robot and Human Intelligence*, which used a totally different methodology.[146]

This still assumes that every neuron is necessary for working human cognition, which we know isn't true. There is a large (but still poorly understood) degree of parallelism in the brain, with individual neurons or cortical modules doing redundant work (or work that at least could be duplicated elsewhere). This is evidenced by people's ability to make a full functional recovery after a stroke or brain injury destroys part of their brain.[147] Thus, the computational demands of simulating the *cognitively relevant* neural structures in our brains are probably even lower than the preceding estimates. And so 10^{14} looks conservative as a most likely range. If brain simulation requires computational power in that range, as of 2023, about \$1,000 worth of hardware can already achieve this.[148] Even if it turns out to require 10^{16} operations per second, \$1,000 of hardware will probably be able to reach that by about 2032.[149]

These estimates are based on my view that a model based only on the firing of neurons can achieve a working brain simulation. It is nonetheless conceivable—though this is a philosophical question that can't be scientifically tested—that subjective consciousness requires a more detailed simulation of the brain. Perhaps we would need to simulate the individual ion channels inside neurons, or the thousands of different kinds of molecules that may influence the metabolism of a given brain cell. Anders Sandberg and Nick Bostrom of Oxford's Future of Humanity Institute estimated that these higher levels of resolution would require 10^{22} or 10^{25} operations per second, respectively.[150] Even in the latter case, they projected that a \$1 billion (in 2008 dollars) supercomputer could achieve this by 2030, and be able to simulate every protein in every neuron by 2034.[151] In time, of course, exponential price-performance gains would drastically reduce those costs.

What you should take away from all this is that even wildly altering our assumptions does not change the essential message of the forecast: computers will be able to simulate human brains in all the ways we might care about within the next two decades or so. This isn't something a century away that our great-grandchildren will have to figure out. We are going to accelerate the extension of our life spans starting in the 2020s, so if you are in good health and younger than eighty, this will likely happen during your lifetime. As another point of perspective, children born today will likely see the Turing test passed while they are in elementary school and see even richer brain emulation achieved when they are of college age. One final comparison is that I am completing this book in 2023, which, even under pessimistic assumptions, is probably closer to full brain emulation being feasible than it is to 1999, when I first made many of these predictions in *The Age of Spiritual Machines.*

PASSING THE TURING TEST

With AI gaining major new capabilities every month and price-performance for the computation that powers it soaring, the trajectory is clear. But how will we judge when AI has finally reached human-level intelligence? The Turing test procedure described at the beginning of this chapter allows us to treat this as a rigorous scientific question. Turing did not specify various details of the test, such as how long the human judges would have to interview the contestants and what capabilities the judges should have. On April 9, 2002, personal computing pioneer Mitch Kapor and I engaged in the first "Long Now" bet, concerning whether or not such a Turing test would be passed by 2029.[152] It introduced a series of issues such as defining how much cognitive enhancement a human could have (to be a judge or a human foil) and still be considered a human.

The reason a well-defined empirical test is necessary is that, as mentioned previously, humans have a powerful tendency to redefine whatever artificial intelligence achieves as not *really* so hard in hindsight. This is often referred to as the "AI effect."[153] Over the seven decades since Alan Turing devised his imitation game, computers have gradually surpassed humans in many narrow areas of intelligence. But they've always lacked the breadth and flexibility of human intellect. After IBM's Deep Blue supercomputer beat world chess champion Garry Kasparov in 1997, many commentators dismissed the accomplishment's relevance to real-world cognition.[154] Because chess involves perfect information about the location of the pieces on the board and their capabilities, and because there is a relatively small number of possible moves each turn, it is easy to represent the game mathematically. Thus, beating Kasparov could be written off as just a fancy trick of math. By contrast, some observers confidently predicted, computers would never master the kinds of ambiguous natural-language tasks that are needed to solve crossword puzzles or compete on the long-running quiz show *Jeopardy!*[155] But crosswords fell within two years,[156] and less than twelve years after that, IBM's Watson went on *Jeopardy!* and handily defeated the two greatest human players, Ken Jennings and Brad Rutter.[157]

These matches demonstrate a very important idea about AI and the Turing test. Based on Watson's ability to process game clues, buzz in, and confidently call out correct responses with his synthesized voice, "he" presented a very convincing illusion that he was thinking in a way that was very similar to how Ken and Brad were thinking. But that's not the only information viewers of the match got. Along the bottom third of the screen, a display showed Watson's top three guesses for each clue. Although the number one guess was almost always correct, the number two and three guesses weren't just wrong, they were often laughably wrong—silly errors that even a very weak human player would never make. For example, in the category "EU, the European Union," the clue was "Elected every 5 years, it has 736 members from 7 parties."[158] Watson correctly guessed that the correct response was the European Parliament, with 66 percent confidence. But Watson's second guess was "MEPs," at 14 percent, and his third was "universal suffrage," at 10 percent.[159] A human who'd never even heard of the European Union would know that neither of those could be correct, simply judging from the syntax of the clue. What this demonstrates is that even though Watson's gameplay seemed human, if you dig down just beneath the surface, it's clear that the "cognition" Watson was doing was quite alien to our own.

More recent advances have made AI understand and use natural language much more fluently. In 2018 Google debuted Duplex, an AI assistant that spoke so naturally over the phone that unsuspecting parties thought it was a real human, and IBM's Project Debater, introduced the same year, realistically engaged in competitive debate.[160] And as of 2023, LLMs can write whole essays to human standards.

Yet despite this progress, even GPT-4 is prone to accidental "hallucinations," wherein the model confidently gives answers that are not based on reality.[161] For example, if you ask it to summarize a nonexistent news article, it may confabulate information that sounds perfectly plausible. Or if you ask it to cite sources for a true scientific fact, it may cite academic papers that don't exist. As I write this, despite the great engineering effort going into curbing hallucinations,[162] it remains an open question how difficult this problem will be to overcome. But these lapses highlight the fact that, like Watson, even these powerful AIs are generating their responses via arcane mathematical and statistical processes very different from what we would recognize as our own thought processes.

Intuitively, this feels like a problem. It's tempting to think that Watson "ought" to reason like humans do. But I would propose that this is superstition. In the real world, what matters is how an intelligent being acts. If different computational processes lead a future AI to make groundbreaking scientific discoveries or write heartrending novels, why should we care how they were generated? And if an AI is able to eloquently proclaim its own consciousness, what ethical grounds could we have for insisting that only our own biology can give rise to worthwhile sentience? The empiricism of the Turing test puts our focus firmly where it should be.

Yet while the Turing test will be very useful for assessing the progress of AI, we should not treat it as the sole benchmark of advanced intelligence. As systems like PaLM 2 and GPT-4 have demonstrated, machines can surpass humans at cognitively demanding tasks without being able to convincingly imitate a human in other domains. Between 2023 and 2029, the year I expect the first robust Turing test to be passed, computers will achieve clearly superhuman ability in a widening range of areas. Indeed, it is even possible that AI could achieve a superhuman level of skill at programming itself before it masters the commonsense social subtleties of the Turing test. That remains an unresolved question, but the possibility shows why our notion of human-level intelligence needs to be rich and nuanced. The Turing test is

This cartoon shows the remaining cognitive tasks that an AI still has to master. The signs on the floor show tasks that have already been achieved by AI. The ones on the wall with dotted lines around them are currently being addressed.

certainly a major part of this, but we'll also need to develop more so-phisticated means of assessing the complex and varied ways that human and machine intelligence will be similar and different.

Although some people object to using the Turing test as an indicator of human-level cognition in a machine, my belief is that a true demon-stration of passing a Turing test would be a compelling experience to people who observed it, and it would convince the public that this was a genuine intelligence rather than merely mimicry. As Turing said in 1950, "May not machines carry out something which ought to be de-scribed as thinking but which is very different from what a man does? . . . [I]f, nevertheless, a machine can be constructed to play the imitation game satisfactorily, we need not be troubled by this objection."[163]

It should be noted that once an AI does pass this strong version of the Turing test, it actually will have surpassed humans for every cog-nitive test that can be expressed through language.[164] AI deficiency in any of these areas could be exposed by the test. Of course, this as-sumes clever judges and sharp human foils. It won't count to imitate a person who is drunk, sleepy, or unfamiliar with the language.[165] Like-wise, an AI that fools judges unacquainted with how to probe its capa-bilities can't be considered to have passed a valid test.

Turing said that his test could be used to assess an AI's abilities in "almost any one of the fields of human endeavor that we wish to in-clude." So a clever human examiner might ask the AI to explain a complex social situation, draw inferences from scientific data, and write a funny sitcom scene. Thus, the Turing test is about much more than merely understanding human language itself—it's about the cog-nitive abilities that we express through language. Of course, a success-ful AI would also have to avoid seeming superhuman. If the examinee could provide instant answers to any trivia question, quickly deter-mine whether huge numbers are prime, and speak a hundred lan-guages fluently, it would be pretty clear that it's not a real human.

Further, an AI that reached this point would also already have many capabilities that *vastly* outstrip those of humans—from memory to speed of cognition. Imagine the cognitive abilities of a system with human-level reading comprehension but paired with perfect recall of every Wikipedia article and all the scientific research ever published. Today, AI's still-limited ability to efficiently understand language acts as a bottleneck on its overall knowledge. By contrast, the main constraints on human knowledge are our relatively slow reading ability, our limited memory, and ultimately our short life spans. Computers can already process data stunningly faster than humans—if it takes a human an average of six hours to read a book, Google's Talk to Books was about five billion times faster[166]—and can have effectively unlimited data storage capacity.

ACCELERATING PARADIGMS FOR EVOLUTION OF INFORMATION PROCESSING

EPOCH	MEDIUM	TIMESCALE
First	Nonliving matter	Billions of years *(nonbiological atomic and chemical synthesis)*
Second	RNA and DNA	Millions of years *(until natural selection introduces a new behavior)*
Third	Cerebellum	Thousands to millions of years *(to add complex skills via evolution)* Hours to years *(for very basic learning)*
Fourth	Neocortex	Hours to weeks *(to master complex new skills)*
	Digital neural nets	Hours to days *(to master complex new skills at superhuman levels)*
Fifth	Brain–computer interfaces	Seconds to minutes *(to explore ideas unimaginable to present-day humans)*
Sixth	Computronium	< Seconds *(to continually reconfigure cognition toward the limits of what the laws of physics allow)*

When AI language understanding catches up to the human level, it won't just be an incremental increase in knowledge, but a sudden explosion of knowledge. This means that an AI going out to pass a traditional Turing test is actually going to have to dumb itself down! Thus, for tasks that don't require imitating a human, like solving real-world problems in medicine, chemistry, and engineering, a Turing-level AI would already be achieving profoundly superhuman results.

To understand where this leads next, we can look to the six epochs described in the previous chapter, summarized in the table "Accelerating Paradigms for Evolution of Information Processing" on page 68.

EXTENDING THE NEOCORTEX INTO THE CLOUD

So far there have been modest efforts to communicate with the brain using electronics, either inside or outside the skull. Noninvasive options face a fundamental trade-off between spatial and temporal resolution—that is, how precisely they can measure brain activity in space versus time. Functional magnetic resonance imaging scans (fMRIs) measure blood flow in the brain as a proxy for neural firing.[167] When a given part of the brain is more active, it consumes more glucose and oxygen, requiring an inflow of oxygenated blood. This can be detected down to a resolution of cubic "voxels" about 0.7 to 0.8 millimeters to a side—small enough to get very useful data.[168] Yet because there is a lag between actual brain activity and blood flow, the brain activity can often be measured to within only a couple of seconds—and can rarely be better than 400 to 800 milliseconds.[169]

Electroencephalograms (EEGs) have the opposite problem. They detect the brain's electrical activity directly, so they can pinpoint signals to within about one millisecond.[170] But because those signals are detected from the outside of the skull, it's hard to pinpoint exactly where they came from, yielding a spatial resolution of six to eight cubic centimeters, which can sometimes be improved to one to three cubic centimeters.[171]

The trade-off between spatial and temporal resolution in brain scans is one of the central challenges in neuroscience as of 2023. These limitations stem from the fundamental physics of blood flow and electricity, respectively, so even though we may see marginal improvements from AI and improved sensor technology, they probably won't be sufficient to allow a sophisticated brain–computer interface.

Going inside the brain with electrodes avoids the spatial-temporal trade-off and allows us to not just directly record individual neurons' activity but to stimulate them—two-way communication. But putting electrodes inside the brain with current technology involves making holes in the skull and potentially damaging neural structures. For this reason, these approaches have thus far been focused on helping those with disabilities like hearing loss or paralysis, for whom the benefit would outweigh the risk. The BrainGate system, for example, enables people with ALS or spinal cord injuries to operate a computer cursor or a robotic arm with their mind alone.[172] Yet because such assistive technology can connect to only fairly small numbers of neurons at once, it can't process highly complex signals like language.

Having a thought-to-text technology would be transformative, which has prompted research aiming to perfect a brain wave–language translator. In 2020 Facebook-sponsored researchers fitted test subjects with 250 external electrodes and used powerful AI to correlate their cortical activity with words in spoken sample sentences.[173] From this, with a 250-word sample vocabulary, they could predict what words the subjects were thinking of with an error rate as low as 3 percent. This is exciting progress, but Facebook halted the project in 2021.[174] And it remains to be seen to what degree this approach will be able to scale up to larger vocabularies (and therefore more complex signals) as it bumps up against the spatial-temporal resolution trade-off. At any rate, in order to expand the neocortex itself, we will still need to master two-way communication with huge numbers of neurons.

One of the most ambitious efforts to scale up to more neurons is Elon Musk's Neuralink, which implants a large set of threadlike electrodes simultaneously.[175] A test in lab rats demonstrated a readout of 1,500 electrodes, as opposed to the hundreds that have been employed in other projects.[176] Later, a monkey implanted with the device was able to use it to play the video game *Pong*.[177] Following a period of regulatory challenges, Neuralink received FDA approval to begin human trials in 2023 and, just as this book went to press, implanted its first 1,024-electrode device in a human.[178]

Meanwhile, the Defense Advanced Research Projects Agency (DARPA) is working on a long-term project called Neural Engineering System Design, which aims to create an interface that can connect to one million neurons for recording and can stimulate 100,000 neurons.[179] The agency has funded several different research programs to achieve this, such as a Brown University team working to create sand-size "neurograins" that can be implanted in the brain, interfacing with neurons and with one another to create a "cortical intranet."[180]

Ultimately, brain–computer interfaces will be essentially noninvasive—which will likely entail harmless nanoscale electrodes inserted into the brain through the bloodstream.

So how much computation will we need to record? As described previously, the total amount of computation to simulate a human brain is probably around 10^{14} operations per second or less. Note that this is for a simulation based on the real architecture of a human brain and capable of passing the Turing test and in all other respects appearing to outside observers to be a human brain. But it would not necessarily include many kinds of activity in the brain not required to generate this observable behavior. For example, it is very doubtful that intracellular details like DNA repair inside the nucleus of a neuron are relevant to cognition.

Yet even if 10^{14} operations per second are taking place within the brain, a brain–computer interface doesn't need to account for the bulk of these computations, as they are preliminary activity happening well below the top layer of the neocortex.[181] Rather, we need to communicate only with its upper ranges. And we can ignore noncognitive brain processes like regulating digestion altogether. I would thus estimate that a practical interface would need only millions to tens of millions of simultaneous connections.

Achieving such a number will require scaling down the interface technology to become smaller and smaller—and we'll increasingly use advanced AI to solve the formidable engineering and neuroscience problems that entails. At some point in the 2030s we will reach this goal using microscopic devices called nanobots. These tiny electronics will connect the top layers of our neocortex to the cloud, allowing our neurons to communicate directly with simulated neurons hosted for us online.[182] This won't require some kind of sci-fi brain surgery—we'll be able to send nanobots into the brain noninvasively through the capillaries. Instead of human brain size being limited by the need to pass through the birth canal, it can then be expanded indefinitely. That is, once we have the first layer of virtual neocortex added, it won't be a one-shot deal—more layers can be stacked on top of that one (computationally speaking) for ever more sophisticated cognition. As this century progresses and the price-performance of computing continues to improve exponentially, the computing power available to our brains will, too.

Remember what happened two million years ago, the last time we gained more neocortex? We became humans. When we can access additional neocortex in the cloud, the leap in cognitive abstraction will likely be similar. The result will be the invention of means of expression vastly richer than the art and technology that's possible today—more profound than we can currently imagine.

There's an inherent limitation in imagining what future means of artistic expression will be like. But it can be useful to think by analogy to the last neocortical revolution. Try to imagine what it would be like for a monkey—a highly intelligent animal with a brain broadly similar to our own—to watch a movie. The action wouldn't be totally inaccessible to the monkey. It would be able to recognize that there are human beings talking on the screen, for example. But the monkey wouldn't understand the dialogue or be able to interpret abstract ideas like "The fact that the characters are wearing metal suits signifies that the action is imagined to take place during the Middle Ages."[183] That's the kind of jump that the human prefrontal cortex allowed.

So when we think of art created for people with cloud-connected neocortices, it's not just a matter of better CGI effects or even engaging senses like taste and smell. It's about radically new possibilities for how the brain itself processes experiences. For example, actors can now convey what their character is thinking only through their words and external physical expressions. But we might eventually have art that puts a character's raw, disorganized, nonverbal thoughts—in all their inexpressible beauty and complexity—directly into our brains. This is the cultural richness that brain–computer interfaces will enable for us.

It will be a process of co-creation—evolving our minds to unlock deeper insight, and using those powers to produce transcendent new ideas for our future minds to explore. At last we will have access to our own source code, using AI capable of redesigning itself. Since this technology will let us merge with the superintelligence we are creating, we will be essentially remaking ourselves. Freed from the enclosure of our skulls, and processing on a substrate millions of times faster than biological tissue, our minds will be empowered to grow exponentially, ultimately expanding our intelligence millions-fold. This is the core of my definition of the Singularity.

CHAPTER 3

WHO AM I?

WHAT IS CONSCIOUSNESS?

The Turing test and other assessments can reveal much about what it means to be human in a general way, but the technologies of the Singularity also compel us to ask what it means to be a *particular* human. Where does *Ray Kurzweil* fit into all this? Now, you may not care all that much about Ray Kurzweil; you care about yourself, so you can pose the same question about your own identity. But for me, why is Ray Kurzweil the center of my experience? Why am I this particular person? Why wasn't I born in 1903 or 2003? Why am I a male or even a human? There is no scientific reason why this has to be the case. When we wonder "Who am I?" we're asking a fundamentally philosophical question. It's a question about consciousness.

In *How to Create a Mind*, I quoted Samuel Butler:

When a fly settles upon the blossom, the petals close upon it and hold it fast till the plant has absorbed the insect into its system; but they will close on nothing but what is good to eat; of a drop of rain or a piece of stick they will take no notice. Curious! that so unconscious a thing should have such a keen eye to its own interest. If this is unconsciousness, where is the use of consciousness?[1]

Butler wrote this in 1871.[2] Should we conclude from his observation that plants are indeed conscious? Or that this specific kind of plant has consciousness? How would we be able to tell? We say confidently that another human is conscious, given that his or her ability to communicate and make decisions is similar to our own. But even that is technically just an assumption. We can't detect consciousness or the lack thereof directly.

But what *is* consciousness? People often use the word "consciousness" to refer to two different but related concepts. One of these refers to the functional ability to be aware of one's surroundings and act as though aware of both one's internal thoughts and an external world that's distinct from them. By this definition, for example, we might say that a deeply sleeping person is not conscious, a drunk person is partially conscious, and a sober person is fully conscious. With the exception of rare cases like "locked-in syndrome," it is generally possible to judge the level of another person's consciousness from the outside. Even certain behaviors in an animal, like recognizing itself in a mirror, can shed light on this kind of consciousness. But when it comes to questions of personal identity like the ones addressed in this chapter, a second meaning is more relevant: *the ability to have subjective experiences inside a mind*—and not merely to give the outward appearance of doing so. Philosophers call such experiences "qualia." So when I say here that we can't detect consciousness directly, I mean that a person's qualia cannot be detected from the outside.

Yet despite its unverifiability, consciousness cannot simply be ignored. If we examine the basis of our moral system, we realize that our ethical judgments often hinge on our assessments of consciousness. We view material objects, no matter how intricate or interesting or valuable, as important only to the extent that they affect the conscious experience of conscious beings. The entire debate about animal rights, for example, turns on the extent to which we believe them to be conscious and what the nature of that conscious experience is.[3]

Consciousness poses a problem for philosophers. Ethical questions such as what kinds of beings have rights often hinge on our intuitions about whether those entities have subjective experiences. But because we cannot detect these from the outside, we use the other, functional sense of consciousness as a proxy. This draws by analogy on our own experiences. Each of us (I can only assume!) has subjective experiences on the inside, and we know we also have the kind of functional self-awareness that can be observed by others. Therefore, we assume that when others display functional consciousness, they must be having interior subjective experiences, too. Even scientists who argue that subjective consciousness is irrelevant to empirical thinking nonetheless act as though the people around them are conscious by being mindful of their experiences.

Yet although we readily extend the presumption of consciousness to our fellow humans, our intuitions about other animals get weaker the more their behavior differs from our own. Although dogs and chimpanzees don't have fully human-level cognition, their complex and emotional behavior seems to most people as if it must be matched inside by subjective experience. What about rodents? They do exhibit some humanlike behaviors, such as social play and exhibiting fear of danger.[4] A smaller proportion of people consider rodents to be conscious, and generally regard their subjective experience as much shallower than that of humans. How about insects?[5] Fruit flies don't exactly recite Shakespeare, but they do carry out behaviors in response to their environment and have brains consisting of about 250,000 neurons. Cockroaches have about 1,000,000. This is only about 0.001 percent as many neurons as the human brain has, though, so there is much less room for complex and hierarchical networks like our own. And what about amoebae? These single-celled organisms do not demonstrate anything resembling the functional kind of consciousness that humans and higher animals do. Even so, in the twenty-first century, scientists have gained a better understanding of how even very primitive life forms can show rudimentary forms of intelligence, such as memory.[6]

There is a sense in which consciousness is binary—does a being experience any qualia at all?—but I'm referring here to the additional question of degree. Imagine how your own level of subjective consciousness differs if you're experiencing a vague dream, are awake but drunk or sleepy, or are fully alert. This is the continuum that researchers wonder about when assessing animal consciousness. And expert opinion is shifting in favor of more animals having more consciousness than was once believed. In 2012 a multidisciplinary group of scientists met at the University of Cambridge to assess the evidence of consciousness among nonhuman animals. This conference resulted in the signing of the Cambridge Declaration on Consciousness, which affirms the likelihood that consciousness is not an exclusively human phenomenon. According to the declaration, "the absence of a neocortex does not appear to preclude an organism from experiencing affective states."[7] The signatories identified the "neurological substrates that generate consciousness" in "all mammals and birds, and many other creatures, including octopuses."[8]

So science tells us that complex brains give rise to functional consciousness. But what causes us to have subjective consciousness? Some say God. Others believe consciousness is a product of purely physical processes. But regardless of consciousness's origin, both poles of the spiritual–secular divide agree that it is somehow sacred. How people (and at least some other animals) became conscious is just a causal argument, whether it was by a benign divinity or undirected nature. The ultimate result, however, is not open to debate—anyone who doesn't acknowledge a child's consciousness and capacity for suffering is considered gravely immoral.

Yet the cause behind subjective consciousness will soon be more than just a subject of philosophical speculation. As technology gives us the ability to expand our consciousness beyond our biological brains, we'll need to decide what we believe generates the qualia at the core of our identity, and focus on preserving it. Since observable behaviors are our only available proxy for inferring subjective consciousness, our natural intuition closely matches the most scientifically plausible account: namely, that brains that can support more sophisticated behavior likewise give rise to more sophisticated subjective consciousness.

Sophisticated behavior, as discussed in the previous chapter, arises from the complexity of information processing in a brain[9]—and this in turn is largely determined by how flexibly it can represent information and how many hierarchical layers are in its network.

This has profound implications for the future of humanity—and, if you live for the next few decades, for you personally. Remember: all the intellectual leaps of recorded history have taken place in brains that have remained structurally the same since the Stone Age. External technology has now enabled each of us to access the great majority of discoveries made by everyone else in our species, but we experience them at a level of consciousness similar to that of our Neolithic ancestors. Yet when we can augment the neocortex itself, during the 2030s and 2040s, we won't just be adding abstract problem-solving power; we will be deepening our subjective consciousness itself.

ZOMBIES, QUALIA, AND THE HARD PROBLEM OF CONSCIOUSNESS

There's something fundamental about consciousness that is impossible to share with others. When we label certain frequencies of light "green" or "red," we have no way of telling whether my qualia—my *experience* of green and red—are the same as yours. Maybe I experience green the same way that you experience red, and vice versa. Yet there's no means for us to directly compare our qualia using language or any other method of communication.[10] In fact, even when it does become possible to directly connect two brains together, it will be impossible to prove whether the same neural signals trigger the same qualia for you as for me. So if our red/green qualia really were reversed from each other, we would be forever unaware of this.

As I noted in *How to Create a Mind*, this realization also leads to a more unsettling thought experiment. What if a person had no qualia at all? Philosopher David Chalmers (born 1966) calls such hypothetical beings "zombies"—people who show all the detectable neurological and behavioral correlates of consciousness but have no subjective experience whatsoever.[11] Science could never tell the difference between zombies and normal humans.

One way of highlighting the difference in our ideas about functional versus subjective consciousness is to consider a dog versus a hypothetical artificial human if we could be certain that it had no subjective experience (i.e., a "zombie"). Even though the zombie could demonstrate much more complex cognition than the dog, most people would probably say that hurting the dog—which we assume to have subjective consciousness—is worse than hurting the zombie, which may yelp in pain but which we know isn't actually feeling anything. The trouble is that in real life, *even in principle*, there is no way to scientifically determine whether another being has subjective consciousness.

If such zombies are at least theoretically possible, then there must be no necessary causal connection between qualia and the physical systems (i.e., brains or computers) that do the information processing that outwardly gives the appearance of consciousness. This is what some religious views say about the soul—that it is a supernatural entity clearly separate from the body. Such speculation would be beyond the reach of science. But if the physical systems that underlie cognition necessarily also generate consciousness—making zombies impossible— there is no coherent way for science to demonstrate this either. Subjective consciousness is qualitatively different from the realm of observable physical laws, and it doesn't follow that particular patterns of information processing according to these laws would yield conscious experience at all. Chalmers calls this the "hard problem of consciousness." His "easy questions," such as what happens to our mind when we are not awake, are among the most difficult in all of science, but at least they can be studied scientifically.[12]

For the hard problem, Chalmers turns to a philosophical idea he calls "panprotopsychism."[13] Panprotopsychism treats consciousness much like a fundamental force of the universe—one that cannot be reduced to simply an effect of other physical forces. One might imagine a sort of universal field that holds the potential for consciousness. In the interpretation of this view that I hold, it is the kind of information-processing complexity found in the brain that "awakens" that force into the kind of subjective experience we recognize. Thus, whether a brain is made of carbon or silicon, the complexity that would enable it to give the outward signs of consciousness also endows it with subjective inner life.

Although we'll never be able to prove this scientifically, there will be a powerful ethical imperative to act as though it is true. Put another way, if there's a plausible chance that an entity you mistreat might be conscious, the safest moral choice is to assume that it is rather than risk tormenting a sentient being. That is, we should act as though zombies are impossible.

Thus, from a panprotopsychist point of view, the Turing test would not just serve to establish human-level functional capability but would also furnish strong evidence for subjective consciousness and, thus, moral rights. While the legal implications of conscious artificial intelligence are profound, I doubt that our political system will adapt fast enough to enshrine such rights in law by the time the first Turing-level AIs are developed. So initially it will fall to the people developing them to formulate ethical frameworks that can restrain abuses.

In addition to the ethical reasons for assuming that apparently conscious beings have consciousness, there is also good theoretical reason to believe that something like panprotopsychism is the accurate causal explanation for consciousness. It strikes a middle ground between dualism and physicalism (or materialism), which have long been the two major schools of thought. Dualism holds that consciousness arises from some totally separate kind of stuff than ordinary "dead" matter. Many dualists identify this as a soul. The problem with this from a scientific perspective is that even if we allow that a supernatural soul may exist, we lack a promising theory for how it would affect matter in the observable world (e.g., the neurons in our brains).[14] The opposite view, physicalism, suggests that consciousness must arise wholly from certain arrangements of ordinary physical matter in our brains. Yet even if this view can perfectly describe the functional aspects of how consciousness works (i.e., explain human intelligence in a way that's analogous to how computer science explains AI), it cannot offer any explanation for the inherently inaccessible-to-science subjective dimension of consciousness. Panprotopsychism strikes a helpful balance between these opposite points of view.

DETERMINISM, EMERGENCE, AND THE FREE WILL DILEMMA

A concept closely related to consciousness is our sense of free will.[15] If you ask the average person on the street how they understand that term, their answer will probably include the idea that they must be able to control their own actions. Our political and judicial systems are based on the principle that everyone has free will in roughly this sense.

But when philosophers seek a more precise definition, there's little agreement on what that term actually means. Many philosophers believe that the existence of free will requires that the future not be predetermined.[16] After all, if it's already certain what will happen in the future, how could our will be free in any meaningful sense? Yet if "free will" just means that your actions can be boiled down to totally random processes at the quantum level, this leaves no room for what most of us would recognize as truly free will. As English philosopher Simon Blackburn put it, "chance is as relentless as necessity" in seemingly precluding free will.[17] Rather, a meaningful concept of free will must somehow synthesize both deterministic and indeterministic philosophical ideas—avoiding rigid predictability without devolving into randomness.

An insightful path between these extremes emerges from the work of physicist and computer scientist Stephen Wolfram (born 1959). His research has long been influential to my own thinking about the intersection of physics and computation. In his 2002 book *A New Kind of Science*, Wolfram sheds light on phenomena that have both deterministic and nondeterministic properties—mathematical objects called cellular automata.[18]

Cellular automata are simple models represented by "cells" that alternate between states (e.g., black or white, dead or alive) based on one of many possible sets of rules. These rules specify how each cell will behave based on the states of nearby cells. This process unfolds over a series of discrete steps and can produce highly complex behavior. One of the most famous examples of cellular automata is called Conway's Game of Life and uses a two-dimensional grid.[19] Hobbyists and mathematicians have found numerous interesting shapes that form predictably evolving patterns according to the rules of Life. Life can even be used to replicate a functional computer or simulate the software to run and display another version of itself!

Wolfram's theory starts with very basic automata—cells in a one-dimensional line, below which new lines are added sequentially based on a set of rules and the states of the cells in the preceding line.

Through extensive analysis, Wolfram points out that with some sets of rules, regardless of the number of steps being considered, you cannot predict future states without going through each of the intermediate iterations.[20] There is no shortcut to summarize the result.

The easiest type of rule is rule class 1. One example of this type is rule 222:[21]

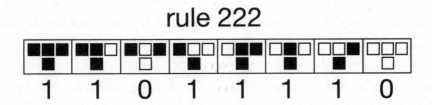

For each cell, there are eight possible combinations of states for the three cells bordering it in the previous step (shown here in the top row). The rule specifies which state each of these combinations causes in the step that comes after it (bottom row). Black and white can also be noted as 1 and 0, respectively.

If we start with a single central black cell and compute the progression of the cells with one row after another by applying rule 222, we obtain this result:[22]

rule 222

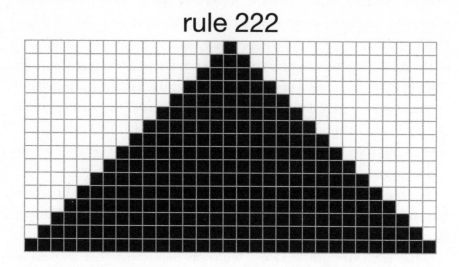

So we can see that rule 222 generates a very predictable pattern. If I were to ask what is the millionth cell—or the million to the millionth cell—from rule 222, you can just answer "black." This is how most science is "supposed to" work: by applying deterministic rules to discern predictable outcomes.

But rule class 1 is just one type of rule. In Wolfram's theory, most of the natural world can be explained by four different classes, distinguished by the kinds of results they produce. Class 2 and class 3 are somewhat interesting in that they yield increasingly complex arrangements of the "black" and "white" cells, but the most fascinating is class 4, exemplified by rule 110:[23]

rule 110

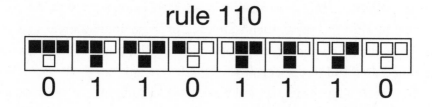

If I follow this rule from a single starting black cell, the result is:

If we continue the iterations, we get images like this:[24]

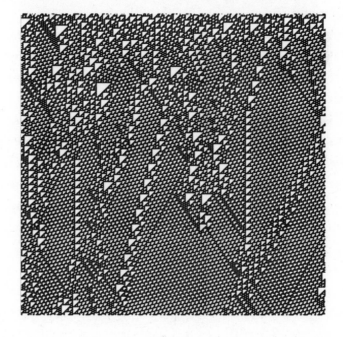

The point here is that there is *no way* to determine what the thousandth row will be, or the millionth to the millionth row, other than to compute them one by one.[25] This means that systems based on class 4 properties—like our own universe, Wolfram argues—possess an irreducible complexity that defies the old, reductive versions of determinism. While this complexity arises out of deterministic programming, in a crucial sense the programming does not fully explain its richness.

A statistical sampling of individual cells would make their states seem essentially random, but we can see that each cell's state results deterministically from the previous step—and the resulting macro image shows a mix of regular and irregular behavior. This demonstrates a property called emergence.[26] In essence, emergence is very simple things, collectively, giving rise to much more complex things. The fractal structures in nature, such as the gnarled path of each growing tree limb, the striped coats of zebras and tigers, the shells of mollusks, and countless other features in biology, all exhibit a class 4 type of coding.[27] We inhabit a world that is deeply affected by the kind of patterning found in such cellular automata—a very simple algorithm producing highly complex behavior straddling the boundary between order and chaos.

It is this complexity in us that may give rise to consciousness and free will. Whether you ascribe the underlying programming of your free will to God or to panprotopsychism or to something else, *you* are more than the program itself.

But it isn't a coincidence that these rules would give rise to consciousness and such a wide range of other natural phenomena. Wolfram makes a forceful case that the laws of physics themselves arise from some kinds of computational rules related to cellular automata. In 2020 he announced the Wolfram Physics Project—an ambitious ongoing effort to understand all of physics via a model that is analogous to cellular automata but more generalized.[28]

This would allow a sort of compromise between classical determinism and quantum indeterminism. While some parts of the macroscale world can be approximated with algorithmic shortcuts—say, to predict where a satellite will be a million orbits from now—this is not true at the most fundamental scales. If reality is based on principles like class 4 rules at its deepest level, we could explain seeming randomness at the quantum scale in deterministic terms, but there would be no summarizing algorithm that could "look ahead" and predict the exact state of the entire universe at some time in the future.[29] This remains speculative, as we don't yet know specifically what that full set of rules is. Perhaps a future "theory of everything" will unify all of this into one coherent explanation, but we're not there yet.

With effective prediction out of the question, that leaves simulation, but the universe couldn't contain a computer large enough to simulate itself. In other words, there is no way to unfold reality without letting it actually go forward.

Later in this chapter, I discuss future possibilities for transferring consciousness from our biological brains to nonbiological computers. This raises an important point for clarification. Although it will eventually become possible to digitally emulate the workings of the brain, this is not the same as pre-computing it in a deterministic sense. This is because brains (whether biological or not) are not closed systems. Brains take in input from the outside world and then manipulate it via astoundingly complex networks—in fact, scientists have recently identified networks in the brain that exist in up to eleven dimensions![30] This complexity likely makes use of rule 110–style phenomena for which there is no way to computationally "peek ahead" without simulating each step in sequence. And because brains are open systems, it is impossible to factor unknown future inputs into a step-by-step simulation. Thus, being able to replicate the function of the brain does not imply an ability to pre-compute its future states. That may be a good reason that the universe exists.

Stated differently, if the rules of a universe are based on something like cellular automata, the only way for them to be expressed would be through unfolding step-by-step—through reality actually happening. By contrast, if a universe had rules that were deterministic but without automata-like emergence, or was based only on randomness, reality wouldn't necessarily require the step-by-step unfolding that we actually observe. Further, if consciousness can emerge only from the kind of order-and-chaos complexity of class 4 automata, this can be seen as a philosophical argument for why *we* exist—without such rules, we wouldn't be here to ponder the question.

This opens the door to "compatibilism"—the view that a deterministic world can still be a world with free will.[31] We can make free decisions (that is, ones not caused by something else, like another person), even though our decisions are determined by underlying laws of reality. A determined world means that we could theoretically look either forward or backward in time, since everything is determined in either direction. But under rule 110–style rules, the only way we can perfectly see forward is through all the steps actually unfolding. And so, viewed through the lens of panprotopsychism, the emergent processes in our brains aren't controlling us; they *are* us. We arise from deeper forces, but our choices cannot be known in advance—so we have free will as long as the processes that give rise to our consciousness are able to be expressed through our actions in the world.[32]

THE FREE WILL DILEMMA OF MORE THAN ONE BRAIN PER HUMAN

If we look at how humans have portrayed androids in movies and novels, it appears that we implicitly share a panprotopsychist imagination— if an AI's behavior seems humanlike, even if its cognition is not based on biological neurons, we will root for it as if it were a subjectively conscious being.

But just as an AI may be composed of many separate algorithms, growing medical evidence demonstrates that the human brain has multiple distinct decision-making units. Consider all of the experiments that have been done on our two (left and right) brains, which suggest that they are largely equal and separate.[33] Researchers Stella de Bode and Susan Curtiss studied forty-nine children who had had one half of their brain removed to prevent a life-threatening seizure disorder.[34] Most of these children subsequently functioned normally, and even those who continued to have a particular disorder nonetheless had a reasonably normal personality. Although we typically develop language mostly on the left, both halves can be functionally equivalent, and a person with either a left or a right brain can master language.[35]

Perhaps most striking is a brain that has both the left and right hemispheres intact but has the 200 million axons[36] between them—the corpus callosum—cut due to a medical problem. Michael Gazzaniga (born 1939) has studied these cases where both brains operate but have no means to communicate between them.[37] Through a series of experiments in which he fed a word to only a patient's right brain, he found that the left hemisphere, which was not aware of the word, nonetheless felt responsibility for choices based on this information, even though the choice was actually made by the other hemisphere.[38] So the left brain would confabulate plausible-sounding explanations for why it claimed to have made each decision, because it did not recognize the presence of another separate brain sharing the same skull.[39]

These and other experiments involving both hemispheres of the brain suggest that a normal person may actually have two brain units capable of independent decision-making, which nonetheless both fall within one conscious identity. Each will think that the decisions are its own, and since the two brains are closely commingled, it will seem that way to both of them.

In fact, if we look beyond just our two hemispheric brains, there are many types of decision-makers within us that could have a free will in the sense described previously. For example, the neocortex, where decision-making happens, consists of many smaller modules.[40] So when we consider a decision, it's possible that different options are represented by different modules, each trying to precipitate its own perspective. My mentor Marvin Minsky was prescient in seeing the brain not as a single united decision-making machine but rather as a complex network of neural machinery whose individual parts may favor different options when we consider a decision. Minsky described our brains as a "society of mind" (the title of his second book) containing various simpler processes that reflect many different perspectives.[41] Is each of these making a free choice? How could we tell? In recent decades, there has been more experimental support for this idea—but our understanding of exactly how neural processes "bubble up" into the decisions we consciously perceive remains quite limited.

"YOU 2" IS CONSCIOUS. IS IT YOU?

All this raises a provocative question. If consciousness and identity can span multiple distinct information-processing structures in the skull—even ones that are not physically connected—what happens when those structures are farther apart?

A key issue I explored in *How to Create a Mind* is the philosophical and moral implications of replicating all the information in a human brain, which will be possible during the lifetimes of most people alive today.

Let's say I use advanced technology to examine a piece of your brain and then make an exact electronic copy of this small segment. (We actually can do a very primitive version of this today with certain parts of a brain—for example, when treating essential tremor or Parkinson's disease.)[42] On its own, this piece is far too simple to be conscious. But then let's say I copy a second tiny piece of your brain—and another, and another. Finally, at the end of this process, I have a complete computerized replica of your brain that contains all the same information and can function in the same way.

So is this "You 2" conscious? You 2 will say it's had all of the same experiences as You (since it is sharing your memories), and it acts like You do, so unless one completely rules out the possibility of any electronic version of a conscious entity's being conscious, then the answer would be yes. Put simply, if an electronic brain represents the same information as a biological brain and claims to be conscious, there is no plausible scientific basis for denying its consciousness. Ethically, then, we ought to treat it as though it is conscious and therefore possessing moral rights. Yet this is not blind speculation—panprotopsychism gives us good philosophical reason to believe that it actually is conscious.

But now a harder question: Is this You 2 really *you*? Keep in mind that "You" (as a normal physical person) also still exist. It could even be that this copy was made without "You" being aware of it, but regardless of that, the essential "You" is continuing. If the experiment was successful, You 2 would act like you, but "You" would not have been altered at all, so "You" would still be you. Since You 2 could act independently, it would immediately diverge from "You"—creating its own memories and reacting to different experiences. So insofar as your identity is the particular arrangement of information in your brain, You 2 would *not* be You, even if it has a consciousness.

Okay, so far so good. Now, in a second experiment, we gradually *replace* each section in your brain with a digital copy—connected to your remaining neurons via a brain–computer interface like that described in the previous chapter. So there are no longer a "You" and a You 2—there is only You. After every phase of this experiment, you are happy with the procedure, and no one, including yourself, bothers to complain. Is the new You after each such replacement still *you*? Even at the end, once your brain is entirely digital?

The question of how identity relates to replacing an object's parts gradually over time dates back to a thought experiment first posed about 2,500 years ago, called the Ship of Theseus.[43] Ancient Greek philosophers imagined a wooden ship whose planks were slowly replaced with new planks, one by one. It seems quite natural to conclude that after the first plank is replaced, the ship itself is still the original ship. It may have a slightly different makeup, but we still talk about this transformation as a change to the original ship and not the creation of a new ship. But what about when more than half of the planks are new ones? Or when all of the planks are new, and none were part of the ship when it was built? This becomes more complicated, but many people would still say that the ship's fundamental identity can survive these incremental changes. Yet imagine that as the new planks were being added, all the old ones were being stored in a warehouse. Now, once the original ship has 100 percent new parts, we reassemble all the old parts from storage into a ship again. Now which ship is the original? The one that continuously existed with only incremental changes but has no parts from the original? Or the one that was re-formed from the original parts?

The Ship of Theseus is a fun thought experiment when it comes to ships or other "dead" objects, but it doesn't have particularly high stakes. The identity of ships over time is ultimately a matter of human convention. Yet the problem assumes supremely high stakes when the objects in question are human beings. For most of us, it matters a great deal whether the person standing next to us really is our loved one or is just a Chalmersian zombie putting on a convincing show.

Let's consider these questions through the lens of the "hard problem" of subjective consciousness. In the scenario where we make You 2 as a replica, it is impossible to determine whether the subjective self of You 2 has some kind of connection with "You." Would your original subjective experience somehow simultaneously encompass both copies of you, even as their information patterns diverge over time due to different experiences? Or would You 2 be separate in this regard? These are scientifically unanswerable questions.

Yet in the scenario where we gradually shift the information in your brain onto a nonbiological substrate, we have much stronger reason to believe that your subjective consciousness would be preserved. Indeed, as mentioned previously, we already do this in a very limited way with certain brain conditions today, and the new neural prosthesis is more capable than the part it replaces. (Thus it is not identical to the part it replaces.) While early implantable devices, like cochlear implants, were able to stimulate brain activity, they did not substitute for any core brain structures.[44] But since the early 2000s scientists have been developing brain prostheses that can help people with structural damage or malfunctions within their brains. For example, prosthetic devices can now do some of the work of the hippocampus in patients with memory problems.[45] These technologies are still in their infancy as of 2023, but during this decade we will see them become both more sophisticated and available affordably to a wider range of patients. Yet with today's technology, there is still no doubt that a person's core identity is preserved, and no one argues that these patients have become Chalmersian zombies.

Everything we know of neuroscience suggests that in the gradual-replacement scenario, you wouldn't even notice small enough alterations, and the brain is amazingly adaptable. Your hybrid brain would retain all the same patterns of information that define you. So there's no reason to think that your subjective consciousness would be compromised, and you would of course remain *you*—there is no one else to call you. However, at the end of this hypothetical process, the final you is *exactly like* You 2 in the first experiment, which we decided was *not you*. How can this be reconciled? The difference is continuity—the digital brain doesn't diverge from the biological one, because there was never a moment when they existed as separate entities.

This leads us to a third case, which is actually not a hypothetical. Every day our own cells undergo a very rapid replacement process. While neurons generally persist, about half of their mitochondria turn over in a month;[46] a neurotubule has a half-life of several days;[47] the proteins that add energy to the synapses are replenished every two to five days;[48] the NMDA receptors in synapses are replaced in a matter of hours;[49] and the actin filaments in the dendrites last for about forty seconds.[50] Our brains are thus almost completely replaced within a few months, so in fact you are a biological version of You 2 as compared with yourself a little while ago. Again, what keeps your identity intact is information and function—not any particular structure or material.

For years I have often gazed at the beautiful Charles River near my home. If I look at the Charles today, I tend to think of it as the same body of water that it was a day ago, or a decade ago, when I commented on the river's continuity in *How to Create a Mind*. This is because even though all of the water molecules passing through a given slice of it are completely different every few milliseconds, those molecules act in a consistent pattern that defines the course of the river. It is the same for minds. As we place nonbiological systems into our bodies and brains, the continuity of our information patterns will make each of us feel like the person we are today—except for the fact that our perceptions may be better or our cognition smarter.

Of course, the same technology that enables us to transition all our skills, personality, and memories to a digital medium would also allow us to create multiple copies of that information. This ability to duplicate ourselves at will is a superpower in the digital world that does not exist in the biological world. Copying our mind files to a remote backup storage system will be a powerful protection against any accident or disease that might damage our brain. This is not "immortality" any more than Excel spreadsheets uploaded to the cloud are immortal—disasters can still befall the data centers and wipe them out. But it will let us guard ourselves against the senseless mishaps that snuff out so many lives and identities. And my interpretation of panprotopsychism suggests that our subjective consciousness may somehow encompass all copies of this defining information.

This has another tantalizing implication. If we set a You 2 loose in the world—free to follow a different path from "You"—its information-pattern identity would diverge, but since this would be a gradual and continual process, there's a chance that your subjective consciousness could span both simultaneously. I suspect that, based on the theory of panprotopsychism, our subjective consciousness is tied to information-as-identity and would thus somehow encompass all copies of information that were once identical to our own.

Yet because in this scenario You 2 could potentially insist passionately that it has a different subjective consciousness from "You" (as the physical decision-making structures that govern communication would be separate), and there would be no way to objectively determine the truth, our legal and ethical systems would likely have to treat both as separate entities.

THE INCREDIBLE UNLIKELINESS OF BEING

In making sense of our identity, it is awe-inspiring to consider the extraordinary chain of unlikely events that enabled each one of us to come into being. Not only did your parents have to meet and make a baby, but the exact sperm had to meet the exact egg to result in *you*. It's hard to estimate the likelihood of your mother and father having met and deciding to have a baby in the first place, but just in terms of the sperm and the egg, the probability that you would be created was one in two million trillion. As very rough approximations, the average man produces as many as two trillion sperm in his lifetime, and the average woman starts with about one million eggs.[51] Thus, to the extent that your identity hinges on the exact sperm and egg that made you, the odds of this happening were about one in two quintillion. While all these sex cells aren't genetically unique, numerous factors like age can affect epigenetics, so if your father produced two chromosomally identical sperm at age twenty-five and age forty-five, they wouldn't give precisely the same contribution to the formation of a baby.[52] Thus, as an approximation we must regard each sperm and egg as effectively unique. In addition to that, the comparable event had to take place for *both* sets of grandparents, and for all *four* sets of great-grandparents, and *eight* sets of great-great-grandparents, and so on . . . well, not quite ad infinitum—just to the beginning of life on earth nearly four billion years ago.[53]

A googol (the correct spelling of the number that is the basis for the search company's name) is a 1 followed by 100 zeros. A googolplex is a 1 followed by a googol zeros. It is an unimaginably big number, but based on the rough analysis described previously (and explained further in the endnotes), the probability that you would exist is 1 out of a number consisting of a 1 followed by vastly more than a googolplex zeros.[54] Yet here you are. It is a miracle, is it not?

In addition, that the universe came into being with an ability to evolve complex information at all is arguably even more unlikely. Our understanding of physics and cosmology demonstrates that if the values in the laws of physics had been only slightly different, the universe would not have been able to support life.[55] Put another way, of all the configurations that the universe theoretically could have had, only the very tiniest fraction would have allowed us to exist. The closest we can come to quantifying this apparent unlikelihood is by identifying the different factors on which a life-friendly universe depends and then estimating how different those values would have to have been for life to be impossible.

According to the Standard Model of particle physics, there are thirty-seven kinds of elementary particles (differentiated by mass, charge, and spin), which interact according to four fundamental forces (gravity, electromagnetism, nuclear strong force, nuclear weak force), as well as hypothetical gravitons, which some scientists believe are responsible for gravitational effects.[56] The strengths of these forces interacting with particles are described with a series of constants, which define the "rules" of physics. Physicists have highlighted many areas where extremely slight changes to these rules would have prevented the formation of intelligent life, which is assumed to require complex chemistry and relatively stable environments and energy sources for hundreds of millions or billions of years of evolution. The prevailing view among biologists is that life arose on earth through abiogenesis.[57] According to this theory, over a long period of time, nonliving matter in a "primordial soup" of precursor compounds naturally combined into more complex building blocks of life-enabling proteins. Eventually, proteins spontaneously came together in a pattern that enabled them to self-replicate: the origin of life. Even one break in this causal chain would have made humanity impossible.

If the nuclear strong force had been stronger or weaker, it would have been impossible for stars to form the large amounts of carbon and oxygen from which life is created.[58] Likewise, the nuclear weak force is within one order of magnitude of the minimum possible for life to evolve.[59] If it were weaker than this, hydrogen would have quickly turned to helium, preventing the formation of hydrogen stars like our own, which burn long enough to allow complex life to evolve in their solar systems.

If the difference in mass between up quarks and down quarks had been slightly smaller or larger, it would make protons and neutrons unstable, preventing complex matter from forming.[60] Likewise, if electrons had a slightly larger mass relative to those differences, similar instability would result.[61] According to physicist Craig J. Hogan, "only a few percent fractional change in the quark mass difference in either direction" would have kept life from arising.[62] If the quark mass difference were greater, we would have a "proton world"—an alternate universe where only hydrogen atoms would be possible.[63] If the difference were smaller, we would have a "neutron world"—a universe with nuclei but no electrons around them, making chemistry impossible.[64]

If gravity had been slightly weaker, there would be no supernovas, which are the source of the heavy elements from which life is formed.[65] If gravity had been slightly stronger, stars would have been much shorter-lived, making it impossible to support complex life.[66] Within one second after the big bang, the density parameter (known as Ω, or omega) could not have been different by more than one part in a quadrillion and still allowed the formation of life.[67] If it had been slightly larger, the matter scattered by the big bang would have recollapsed under gravity before stars could form. If it had been slightly smaller, the expansion would have been too fast for matter to clump together into stars in the first place.

Further, the macrostructure of the universe arose from tiny local fluctuations in the density of the matter expanding outward from the big bang in the first instant after the event.[68] The density at any one point averaged a difference from the mean of about 1 part in 100,000.[69] If this amplitude (often compared to ripples in a pond) had differed by more than one order of magnitude, life wouldn't be possible. According to cosmologist Martin Rees, if the ripples had been too small, "gas would never condense into gravitationally bound structures at all, and such a universe would remain forever dark and featureless."[70] By contrast, if the ripples had been too large, the universe would have been a "turbulent and violent place," with most matter collapsing into enormous black holes and no chance for stars to "retain stable planetary systems."[71]

For the universe to produce orderly matter instead of chaotic soup, it would have also needed to have very low entropy immediately after the big bang. According to physicist Roger Penrose, based on what we know about entropy and randomness, only approximately one in $10^{10^{123}}$ possible universes would have had low enough initial entropy at the beginning to take a form similar to our own.[72] That's 10 followed by vastly more zeroes than there are atoms in the known universe. (Estimates generally place the number of atoms between 10^{78} and 10^{82}. Assuming 10^{80}, the number of digits in $10^{10^{123}}$ is then forty-three orders of magnitude larger than the number of all the atoms. That is ten million billion billion billion billion times larger.)[73]

It is certainly possible to question many of these individual calculations, and scientists sometimes disagree about the implications of any single factor. But it isn't enough to analyze each of these fine-tuned parameters in isolation. Rather, as physicist Luke Barnes argues, we must consider the "intersection of the life-permitting regions, not the union."[74] In other words, every single one of these factors has to be friendly to life in order for life to actually develop. If even a single one were missing, there would be no life. In the memorable formulation of astronomer Hugh Ross, the likelihood of all this fine-tuning happening by chance is like "the possibility of a Boeing 747 aircraft being completely assembled as a result of a tornado striking a junkyard."[75]

The most common explanation of this apparent fine-tuning states that the very low probability of living in such a universe is explained by observer selection bias.[76] In other words, in order for us to even be considering this question, we *must* inhabit a fine-tuned universe—if it had been otherwise, we wouldn't be conscious and able to reflect on that fact. This is known as the anthropic principle. Some scientists believe that such an explanation is adequate. But if we believe that reality exists independently of ourselves as observers, this cannot be fully satisfying. Martin Rees considers a compelling question we might still ask. As he puts it, "Suppose you are in front of a firing squad, and they all miss. You could say, 'Well, if they hadn't all missed, I wouldn't be here to worry about it.' But it is still something surprising, something that can't be easily explained. I think there is something there that needs explaining."[77]

AFTER LIFE

The first step on the path to preserving our precious and unlikely identity is preserving the ideas that are central to who we are. We are already creating through our digital activities enormously rich records of how we think and what we feel, and during this decade our technologies for recording, storing, and organizing this information will advance rapidly. As we approach the end of the 2020s, we will animate this data as nonbiological simulations that are highly realistic recreations of humans with specific personalities.[78] Even as of 2023, though, AI is rapidly gaining proficiency at imitating humans. Deep-learning approaches like transformers and GANs (generative adversarial networks) have propelled amazing progress. Transformers, as described in the previous chapter, can train on text a person has written and learn to realistically imitate their communication style. Meanwhile, a GAN entails two neural networks competing against each other. The first tries to generate an example from a target class, like a realistic image of a woman's face. The second tries to discriminate between this image and other, real images of women's faces. The first is rewarded (think of this as scoring points that the neural net is programmed to try to maximize) for fooling the second, and the second is rewarded for making accurate judgments. This process can repeat many times without human supervision, with both neural nets gradually increasing their proficiency.

By combining these techniques, AI can thus already imitate a specific person's writing style, replicate their voice, or even realistically graft their face into a whole video. As mentioned in the previous chapter, Google's experimental Duplex technology uses AI that can react believably in unscripted phone conversations—so successfully that when it was first tested in 2018, real humans it called had no idea they were speaking to a computer.[79] "Deepfake" videos can be used to create harmful political propaganda, or to imagine what movies would look like with different actors in iconic roles.[80] For example, a YouTube channel called Ctrl Shift Face has a viral clip showing what Javier Bardem's character in *No Country for Old Men* would look like if played by Arnold Schwarzenegger, Willem Dafoe, or Leonardo DiCaprio.[81] These technologies are still in their infancy. Not only will each individual capability (e.g., writing, voice, face, conversation) improve greatly in the coming years, but their convergence will create simulations that are more realistic than the sum of their parts.

One type of AI avatar that we can create, called a "replicant" (to borrow a term from *Blade Runner*), will have the appearance, behavior, memories, and skills of a person who has passed away, living on in a phenomenon I call After Life.

After Life technology will go through multiple phases. The most primitive such simulations have already existed for about seven years as I write this. In 2016, *The Verge* published a remarkable article about a young woman named Eugenia Kuyda who used AI and saved text messages to "resurrect" her dead best friend, Roman Mazurenko.[82] As the amount of data each of us generates grows, ever more faithful recreations of specific humans will become possible.

During the late 2020s advanced AI will be able to create a very lifelike replicant, drawing from thousands of photos, hundreds of hours of video, millions of words of text chats, detailed data about the person's interests and habits, and interviews with people who remember them. People will have mixed reactions to this for cultural, ethical, or personal reasons, but the technology will be available to those who want it.

This generation of After Life avatars will be quite realistic, but for many they will exist in the "uncanny valley,"[83] meaning that their behavior will bear a distinct resemblance to the original person but will have subtle differences, making them disconcerting to the person's loved ones. At this stage the simulations are not You 2. They would only re-create the function, not the form, of the information that had been in the person's brain. For this reason, a panprotopsychist view suggests that they would not revive someone's subjective consciousness. Despite this, many people will see them as valuable tools for continuing important work, sharing treasured memories, or helping family members heal.

Replicant bodies will exist mostly in virtual and augmented reality, but realistic bodies in actual reality (that is, convincing androids) will also be possible using the nanotechnology of the late 2030s. Progress in this direction is still in very early stages as of 2023, but there is already significant research going on that will lay the groundwork for much bigger breakthroughs during the next decade. When it comes to android function, technological progress faces a challenge my friend Hans Moravec identified several decades ago, now called Moravec's paradox.[84] In short, mental tasks that seem hard to humans—like square-rooting large numbers and remembering large amounts of information—are comparatively easy for computers. Conversely, mental tasks that are effortless to humans—like recognizing a face or keeping one's balance while walking—are much more difficult for AI. The likely reason is that these latter functions have evolved over tens or hundreds of millions of years and run in the backgrounds of our brains, whereas "higher" cognition is powered by the neocortex, which is the center of our consciousness and which didn't reach its roughly modern form until several hundred thousand years ago.[85]

As AI has grown exponentially more powerful in the past several years, though, it has made amazing progress against Moravec's paradox. In 2000, Honda's ASIMO humanoid robot wowed experts by gingerly walking across a flat surface without falling over.[86] By 2020, Boston Dynamics' Atlas robot could run, jump, and tumble across an obstacle course with greater agility than most humans.[87] Social robots like Sophia and Little Sophia, by Hanson Robotics, and Ameca, by

Engineered Arts, can demonstrate emotion on human-looking faces.[88] Their capabilities have sometimes been exaggerated in headlines, but they nonetheless show the trajectory of progress.

As technology advances, a replicant (as well as those of us who have not died) will have a variety of bodies and types of bodies to choose from. Eventually replicants may even be housed in cybernetically augmented biological bodies grown from the DNA of the original person (assuming it can be found). And once nanotechnology allows molecular-scale engineering, we'll be able to create vastly more advanced artificial bodies than what biology allows. By that point reanimated people will likely transcend the uncanny valley, at least for many of those who interact with them.

Yet such replicants will raise very deep philosophical questions for society. How you answer them may depend on your metaphysical beliefs about ideas like souls, consciousness, and identity. If a person reanimated through this technology "feels" like a lost loved one when you speak with them, will that be enough? How much would it matter whether a replicant was created through AI and data mining, as opposed to a fully uploaded You 2 mind from someone's living brain? As the story of Eugenia Kuyda and Roman Mazurenko shows, even the former kind of replicant could be a source of comfort and healing. Still, it's hard to know for certain how each of us will feel experiencing this for the first time. As this technology becomes more prevalent, society will adapt. We'll probably have laws governing who can create replicants of the dead and how they can be used. Some people may forbid AI from replicating them, while others will leave detailed instructions about their wishes and even participate in the creation of a replicant while they are still alive.

The introduction of replicants will pose many other challenging social and legal questions:

- Are they to be considered people with full human and civil rights (such as the rights to vote and enter into contracts)?

- Are they responsible for contracts signed or crimes previously committed by the person they are replicating?
- Can they take credit for the work or social contributions of the person they are replacing?
- Do you have to remarry your late husband or wife who comes back as a replicant?
- Will replicants be ostracized or face discrimination?
- Under what conditions should the creation of replicants be restricted or banned?

Replicants will also force ordinary people to grapple seriously with the philosophical puzzles of consciousness and identity explored in this chapter—which had previously been mainly theoretical. Likely within a shorter time than that between the 2012 publication of *How to Create a Mind* and your reading this, there will be Turing-level AIs programmed to re-create departed humans. Having cognition as complex as a natural biological person, they will indeed be conscious, and they will think that they are that person. Will their belief that they are the same person mean they *are* the same person? Who could say otherwise?

In the early 2040s, nanobots will be able to go into a living person's brain and make a copy of all the data that forms the memories and personality of the original person: You 2. Such an entity would be able to pass a person-specific Turing test and convince someone who knew the individual that it really is that person. According to all detectable evidence, they will be as real as the original person, so if you believe that identity is fundamentally about information like memories and personality, this would indeed *be* the same person. You could have or continue a relationship with that person, even a physical one, including sex. There may be subtle dissimilarities, but is this so different from living biological people? We change also, usually gradually, but sometimes suddenly, from war, trauma, or changes in status or relationships.

A Chalmers-style understanding of consciousness gives us good reason to suspect that this level of technology will also allow our subjective self to persist in After Life—but remember that this is impossible to prove or disprove scientifically, and each of us will have to make decisions about using this technology based on our own philosophical or spiritual values. At the stage of directly copying over the contents of living brains to nonbiological mediums, we transition from the merely simulated replicants I describe to actual mind uploading, also known as whole-brain emulation, or WBE.

Simulating a mind on a nonbiological medium can mean vastly different things in computational terms. In 2008, John Fiala, Anders Sandberg, and Nick Bostrom identified eleven different levels of possible brain emulation.[89] But to simplify here, brain emulations fall into roughly five categories, proceeding from most abstract to most exhaustive: functional, connectomic, cellular, biomolecular, and quantum.

Functional emulations are those that would act like a biologically based mind but need not actually replicate any of the specific computational structure of a given person's brain. These would be the easiest to process computationally but give the least-complete simulation of the original. Connectomic emulations would replicate the hierarchical connections and logical relationships between groups of neurons but need not model every single cell. Cellular emulations would simulate key information about every neuron in a brain but not simulate detailed physical forces inside. Biomolecular emulation would model interactions between proteins and tiny dynamic forces within each cell. And quantum emulation would capture subatomic effects within and between molecules. This would be the most theoretically complete solution, but it would require staggering computational power that likely wouldn't be available until the next century.[90]

One of the major research projects of the next two decades will be figuring out what level of brain emulation is sufficient. Many who think quantum-level emulation is necessary take this position because they believe subjective consciousness rests on (as yet unknown) quantum effects. As I argue in this chapter (and detailed further in *How to*

Create a Mind), I think that level of emulation will be unnecessary. If something like panprotopsychism is correct, subjective consciousness likely stems from the complex way information is arranged by our brains, so we needn't worry that our digital emulation doesn't include a certain protein molecule from the biological original. By analogy, it doesn't matter whether your JPEG files are stored on a floppy disk, a CD-ROM, or a USB flash drive—they look the same and work the same as long as the information is represented with the same sequence of 1s and 0s. In fact, if you copied out those digits with pencil and paper and mailed the (very large!) stack of papers to a friend, and they typed the digits manually back into a different computer, the image would reappear intact!

So from here, the practical goal is to figure out how to get computers to interface effectively with the brain, and crack the code of how the brain represents information. (For a deeper dive into progress toward mind uploading, the computational dimensions of brain emulation, and even a proposed technology called Matrioshka brains that might one day allow humanity to harness massive amounts of energy for computation, see this endnote for a variety of resources ranging from very accessible to more technical.)[91]

These are awesome challenges, but the superhuman AI tools of the 2030s will enable us to achieve what may seem out of reach today.

TALKING TO MY DAD BOT

Talk to Books was a great idea generator. At Google we could replace its standard 100,000-plus books with any customized selection of books and documents. So in 2019 I gathered everything my father, Fredric Kurzweil, had written—all his love letters to my mother, course notes, a book on music he was writing, and his personal reflections. My daughter, Amy, then asked the same underlying technology that powered Talk to Books a number of questions using his writings as the source material. As you'll see from our very short conversation that follows, this replicant was able to give coherent, natural, surprising answers to our questions—probably well enough that if I'd told you that this had been a conversation with my father while he was still living, you wouldn't have noticed anything unusual. For me and for people who knew him, these words have an additional ring of familiarity, in terms of the way Fredric Kurzweil thought and communicated.

I'm also planning to create a replicant of myself. It will have access to thousands of pages of my communications as well as all my published writings and countless hours of videos. This foundation of rich multimedia would allow my replicant to represent my personality far more deeply and faithfully than my father's. Future replicant AI will also have access to all kinds of metadata about a person's life, enabling even greater fidelity.

So here are some of its answers to our questions. This is the first step in bringing my father back.

WHAT DO YOU LOVE MOST ABOUT MUSIC?

What makes a piece of music so alive, is the fact that it adapts a general plan to its own requirements. The art of music has come a long way, and has always retained connection with human feelings.

WHO IS YOUR FAVORITE COMPOSER, AND WHY?

Brahms, in his composition, takes his cue from Beethoven.

WHAT IS THE ESSENCE OF ROCK AND ROLL MUSIC?

Melody and rhythm are the primary sources of music. Basic principle or laws of structure are repetition and contrast, unity and variety.

HOW DID YOU FIRST GET INVOLVED IN CONDUCTING?

I have for more than twenty years conducted orchestras of various kinds—professional, amateur, college, university, and community and opera orchestras. I have conducted large symphony orchestras, chamber orchestras as well as youth groups. I have taught in many colleges.

HOW DO YOU WANT TO BE REMEMBERED?

It is of course difficult to praise oneself, but I would like to say that my ability to relate and deal with people has always been one of my assets, realizing that it is not only artistic ability that counts, but also amiable personal relationships.

DO YOU HAVE ANY REGRETS?

It seems to be my fate having to spend my days with endless lessons and work of little interest in order to be able to spend, if only a little time on work of real value. I am to blame for letting my hopes and goodwill get the better of my practical judgment. Well, what with my new responsibility and also trying to keep my orchestra together, it is very hard to make ends meet.

WHAT MAKES A PERSON WHO THEY ARE?

Form manifest in the relationship of the parts to the whole.

DO YOU HAVE ANY HOBBIES?

We expect an excellent harvest of peaches and grapes. I love gardening and it is relaxing for me, but I agree with you that it is a lot of work. It is the kind of work which never ends.

DO YOU HAVE ANY ANXIETIES?

Often nightmarish.

HOW DO YOU FEEL ABOUT MODERN TECHNOLOGY?

We always see innovators condemned by their contemporaries. The progress of technology, automation etc., has made the individual small. The ideals and values of former time destroyed by this age of the machine.

CAN YOU TELL ME A STORY ABOUT A CAR ACCIDENT?

I was in Mobile conducting opera, my wife had a serious car accident. She was run down by a taxi cab, pinned against a truck, was seriously injured with multiple fractures, had been present in the hospital in traction for over two months and is at present in a heavy body cast at home in which she can only lie on her back helplessly. She will have to be in this state for quite some time to come.

WHAT IS THE MEANING OF LIFE?

Love.

WHO CAN I BE?

Our questions of identity are tightly interconnected with issues of consciousness, free will, and determinism. In light of these ideas, I could say that this particular person—Ray Kurzweil—is both the result of incredibly precise prior conditions and the product of my own choices. As a self-modifying information pattern, I have certainly shaped myself through decisions throughout my life about whom to interact with, what to read, and where to go.

Yet despite my share of responsibility for who I am, my self-actualization is limited by many factors outside my control. My biological brain evolved for a very different kind of prehistoric life and predisposes me to habits that I would rather not have. It cannot learn fast enough or remember well enough to know all the things I would like to know. I can't reprogram it to free me of fears, traumas, and doubts that I know are preventing me from achieving what I would like to achieve. And my brain sits in a body that is gradually aging—although I work hard to slow this process—and is biologically programmed to eventually destroy the information pattern that is Ray Kurzweil.

The promise of the Singularity is to free us all from those limitations. For thousands of years, humans have gradually been gaining greater control over who we can become. Medicine has enabled us to overcome injuries and disabilities. Cosmetics have allowed us to shape our appearance to our personal tastes. Many people use legal or illegal drugs to correct psychological imbalances or experience other states of consciousness. Wider access to information lets us feed our minds and form mental habits that physically rewire our brains. Art and literature inspire empathy for kinds of people we've never met and can help us grow in virtue. Modern mobile apps can be used to build discipline and cultivate healthy lifestyles. People who are transgender have greater ability than ever before to make their physical bodies match the gender identity that they experience inside. Imagine how much more we'll be able to shape ourselves when we can program our brains directly.

And so merging with superintelligent AI will be a worthy achievement, but it is a means to a higher end. Once our brains are backed up on a more advanced digital substrate, our self-modification powers can be fully realized. Our behaviors can align with our values, and our lives will not be marred and cut short by the failings of our biology. Finally, humans can be truly responsible for who we are.[92]

CHAPTER 4

LIFE IS GETTING EXPONENTIALLY BETTER

THE PUBLIC CONSENSUS IS THE OPPOSITE

Consider this late-breaking news: EXTREME POVERTY WORLDWIDE FELL 0.01% TODAY![1]

This also in: SINCE YESTERDAY, LITERACY HAS RISEN 0.0008%![2]

And this: THE PROPORTION OF HOUSEHOLDS WITH FLUSH TOILETS GREW TODAY BY 0.003%![3]

And the same things happened yesterday.

And the day before yesterday.

If these advances don't seem exciting to you, that counts as at least one reason why you didn't hear about them.

Such signs of progress and many similar examples don't make the headlines because they're not actually new. Day-by-day positive trends have been progressing for years and, at slower rates, for decades and centuries.

As for the examples I just mentioned, from 2016 to 2019, the most recent period for which comprehensive data is available at the time of this writing, the estimated number of people worldwide in extreme poverty (measured by the benchmark of living on less than $2.15 per day in 2017 dollars) declined from roughly 787 million to 697 million.[4] If that trend has been roughly maintained until the present in terms of annual percentage decline, it corresponds to almost a 4 percent drop per year, or around 0.011 percent per day. While there is considerable uncertainty over the precise number, we can be reasonably confident that this is correct to within an order of magnitude. Meanwhile, UNESCO found that from 2015 to 2020 (again, the most recent data available), worldwide literacy rose from about 85.5 to 86.8 percent.[5] That averages about 0.0008 percent per day. And during the same 2015–2020 span, the proportion of the world's population with access to "basic" or "safely managed" sanitation facilities (flush toilets or similar) increased from an estimated 73 to 78 percent.[6] This translates to an average improvement of around 0.003 percent per day. Numerous similar trends are constantly unfolding.

Yet these findings alone are already well documented. I've reviewed the extensive positive impact of technological change on human well-being in *The Age of Spiritual Machines* (1999)[7] and *The Singularity Is Near* (2005)[8] and in scores of lectures and articles since. In their 2012 book *Abundance*,[9] Peter Diamandis and Steven Kotler fleshed out how we are headed toward an era of abundance in resources that used to be characterized by scarcity. And in his 2018 book *Enlightenment Now,*[10] Steven Pinker described the continual progress being made in a variety of areas of social impact.

My emphasis in this chapter is specifically on the exponential nature of this progress, how the law of accelerating returns is the fundamental driver of many individual trends we see, and how the result will be a dramatic improvement of most aspects of life in the very near future—not just in the digital realm.

Before we explore specific examples in detail, it's important to begin with a clear conceptual understanding of this dynamic. My work has sometimes been mischaracterized as claiming that technological change itself is inherently exponential, and that the law of accelerating returns applies to all forms of innovation. That's not my view. Rather, the LOAR describes a phenomenon wherein certain kinds of technologies create feedback loops that accelerate innovation. Broadly, these are technologies that give us greater mastery over information—gathering it, storing it, manipulating it, transmitting it—which makes innovation itself easier. The printing press made books cheap enough that education could become accessible to the next generation of inventors. Modern computers help chip designers create the next generation of faster CPUs. Cheaper broadband makes the internet more useful to everyone, because more people can afford to share their ideas online. The most famous exponential curve of technological change, Moore's law, is thus just one manifestation of this deeper and more fundamental process.

Examples of rapid change that fall outside this law include transport technology speeds—such as the time to travel from England to America. In 1620, the *Mayflower* took sixty-six days to make the crossing.[11] By the American Revolution, in 1775, better shipbuilding and navigation had shaved the time to about forty days.[12] In 1838, the paddlewheel steamship *Great Western* completed the journey in fifteen days,[13] and by 1900 the four-funnel, propeller-driven liner *Deutschland* made the transit in five days and fifteen hours.[14] In 1937, the turboelectric-powered liner *Normandie* cut it to three days and twenty-three hours.[15] In 1939, the first service by Pan Am flying boats took just thirty-six hours,[16] and the first jet airline service, in 1958, made the trip in less than ten and a half hours.[17] In 1976, the supersonic Concorde slashed this to just three and a half hours![18] This certainly seems like an open-ended exponential trend—yet it's not. The Concorde was retired in 2003, and since then the London–New York route is back up to over seven and a half hours.[19] There's a range of specific economic and technical reasons why transatlantic transport has stopped getting faster. But the deeper underlying reason is that transportation technology doesn't create feedback loops. Jet engines aren't used in the building of better jet engines, so at a certain point the costs of adding extra speed outweigh the benefit of further innovation.

What makes the LOAR so powerful for information technologies is that feedback loops keep the costs of innovation lower than the benefits, so progress continues. And as artificial intelligence gains applicability to more and more fields, the exponential trends that are now familiar in computing will start to become visible in areas like medicine, where progress was previously very slow and expensive. With AI rapidly expanding its breadth and capability during the 2020s, this will radically transform areas we do not normally consider to be information technologies, such as food, clothing, housing, and even land use. We are now approaching the steep slope of these exponential curves. That, in short, is why most aspects of life will be getting exponentially better in the coming decades.

The problem is that news coverage systematically skews our perceptions about these trends. As any novelist or screenwriter can tell you, capturing an audience's interest usually requires an element of escalating danger or conflict.[20] From ancient mythology to *Star Wars*, this is the pattern that grabs our brains. As a result—sometimes deliberately and sometimes quite organically—the news tries to emulate this paradigm. Social media algorithms, which are optimized to maximize emotional response to drive user engagement and thus ad revenue, exacerbate this even further.[21] This creates a selection bias toward stories about looming crises while relegating the kinds of headlines cited at the beginning of this chapter to the bottom of our news feeds.

Our attraction to bad news is in fact an evolutionary adaptation. Historically it's been more important for our survival to pay attention to potential challenges. That rustling in the leaves might have been a predator, so it made sense to focus on that threat instead of the fact that your crops may have improved a tenth of a percent since the previous year.

It's unsurprising that humans who evolved for subsistence-level life in hunter-gatherer bands didn't evolve a better instinct for thinking about gradual positive change. For most of human history, improvements in quality of life were so small and fragile that they would hardly be noticeable even over a full lifetime. In fact, this Stone Age state of affairs lasted all the way through the Middle Ages. In England, for example, estimated GDP per capita (in 2023 British pounds) in the year 1400 was £1,605.[22] If someone born that year lived to eighty, GDP per capita at the time of their death was exactly the same.[23] For someone born in 1500, GDP per capita at their birth had dipped to £1,586, and eighty years later it had rebounded to only £1,604.[24] Compare that with a person born in 1900, whose eighty-year life span saw a jump from £6,734 to £20,979.[25] So it's not only that our biological evolution has not attuned us to gradual progress, but our cultural evolution hasn't, either. There's nothing in Plato or Shakespeare reminding us to heed gradual material progress in society, because it wasn't noticeable back when they lived.

A modern version of a predator hiding in the foliage is the phenomenon of people continually monitoring their information sources, including social media, for developments that might imperil them. According to Pamela Rutledge, director of the Media Psychology Research Center, "We continually monitor events and ask, 'Does it have to do with me, am I in danger?'"[26] This crowds out our capacity to assess positive developments that unfold slowly.

Another evolutionary adaptation is the well-documented psychological bias toward remembering the past as being better than it actually was. Memories of pain and distress fade more quickly than positive memories.[27] In a 1997 study by Colorado State University psychologist Richard Walker,[28] participants rated events in terms of pleasure and pain and then evaluated them again three months, eighteen months, and four and a half years later. The negative reactions faded far more quickly than positive ones, while the pleasant memories persisted. A 2014 study in countries including Australia, Germany, Ghana, and many others[29] showed this "fading negative affect bias" to be a worldwide phenomenon.

Nostalgia, a term the Swiss physician Johannes Hofer devised in 1688 by combining the Greek words *nostos* (homecoming) and *algos* (pain or distress), is more than just recalling fond reminiscences; it is a coping mechanism to deal with the stress of the past by transforming it.[30] If the pain of the past did not fade, we would be forever crippled by it.

Research supports this phenomenon. A study by North Dakota State University psychology professor Clay Routledge analyzed the use of nostalgia as a coping mechanism and found that the participants who wrote about a positive nostalgic event reported higher levels of self-regard and stronger social bonds.[31] In this way nostalgia is useful for both the individual and the community. When we look back on our past experiences, the pain, stresses, and challenges have faded, and we tend to remember the more positive aspects of life. Conversely, when we think of the present, we are highly cognizant of our current worries and difficulties. This leads to the often false impression that the past was better than the present, despite overwhelming objective evidence to the contrary.

We also have a cognitive bias toward exaggerating the prevalence of bad news among ordinary events. For example, a 2017 study showed that people's perceptions of small random fluctuations (e.g., good days or bad days in the stock market, severe or mild hurricane seasons, unemployment ticking up or down) are less likely to be perceived as random if they are negative.[32] Instead people suspect that these variations indicate a broader worsening trend. As cognitive scientist Art Markman summarized one of the key results, "When participants were asked whether the graph indicated a fundamental shift in the economy, they were more likely to see a small change as indicating a major change when it meant that things were getting *worse* rather than that things were getting better."[33]

This research and more like it suggests that we are conditioned to expect entropy—the idea that the default state of the world is things falling apart and getting worse. This can be a constructive adaptation, preparing us for setbacks and motivating action, but it represents a strong bias that obscures improvements to the state of human life.

This has a concrete impact on politics. A Public Religion Research Institute poll found that 51 percent of Americans in 2016 felt that "American culture and way of life have changed for the worse . . . since the 1950s."[34] The year before, a YouGov survey found that 71 percent of the British public believed that the world is getting progressively worse, and only 5 percent said that it is getting better.[35] Such perceptions incentivize populist politicians to promise to restore the lost glories of the past, despite that past being dramatically worse on nearly every objective measure of well-being.

As one of many examples of this phenomenon, a 2018 survey[36] asked 31,786 people from 26 countries—speaking 17 languages and representing 63 percent of the world's population—whether world-wide poverty had increased or decreased over the prior twenty years and by how much. Their responses are indicated in the chart below.

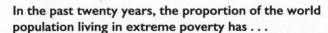

In the past twenty years, the proportion of the world population living in extreme poverty has . . .

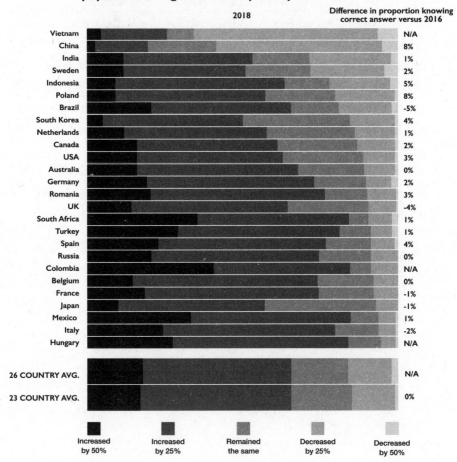

Martijn Lampert, Anne Blanksma Çeta, and Panos Papadongonas, 2018

Only 2 percent got the correct answer: poverty decreased by 50 percent. A growing body of social science confirms these discrepancies between public perception and the reality of pervasive progress according to myriad social and economic measures. For another example, a landmark study in the United Kingdom by Ipsos MORI for the Royal Statistical Society and King's College London[37] showed a wide divergence between popular opinion and actual statistics on numerous topics such as:

- The public impression was that 24 percent of government benefits were claimed fraudulently, whereas the actual figure was 0.7 percent.
- In England and Wales, crime fell 53 percent between 1995 and 2012, yet 58 percent of the public thought crime had gone up or stayed the same during this period. Violent crime between 2006 and 2012 fell 20 percent, whereas 51 percent thought it had gone up.
- The public's impression of teen pregnancy was 25 times worse than reality: 0.6 percent of girls under 15 in the UK get pregnant every year, while the public estimate was 15 percent.

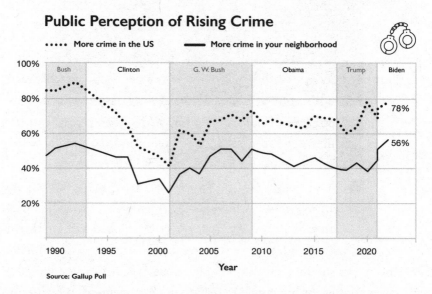

Public Perception of Rising Crime

····· More crime in the US   ——— More crime in your neighborhood

Source: Gallup Poll

Jamiles Lartey, Weihua Li, and Liset Cruz, the Marshall Project, 2022

The same effect holds on the western side of the Atlantic. During the twenty-first century a significant majority of Americans (up to 78 percent) have believed that crime had increased nationally over the previous year, despite the fact that both violent and property crime have declined by about half since 1990.[38]

Actual Crime Rates in the US

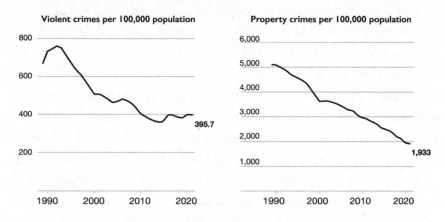

Source: Uniform Crime Reporting Program, Federal Bureau of Investigation

Jamiles Lartey, Weihua Li, and Liset Cruz, the Marshall Project, 2022

The aphorism "If it bleeds, it leads" encapsulates a major cause of these misperceptions. A violent incident will be reported extensively, whereas reductions in crime (e.g., due to data-driven law enforcement or better communication between the police and the community) are literally non-incidents. As such, they do not get widely covered.

This need not be the result of anyone's conscious decision—the incentives of the media structurally favor reporting on violent or negative stories. Because of the cognitive biases described earlier in this chapter, humans are more naturally attuned to threatening information. Since most media (both traditional news media and social media) make their money by attracting eyeballs to generate ad revenue, we shouldn't be surprised that the industry has learned, collectively, that the best way to stay in business is to propagate threatening information that provokes strong emotional responses.

This is also connected to the issue of urgency. The word "news" literally suggests that the information is novel and timely. People have only so much time to consume media, so they tend to prioritize incidents that have just happened. The problem is that the vast majority of such discrete, urgent events are bad things. As I highlighted at the beginning of this chapter, most of the good things happening in the world are very gradual processes, so it is very difficult for these stories to rise to the level of urgency that would make them, for example, a front-page story in *The New York Times* or the top story on CNN. Similar effects hold true on social media—it's easy to share videos of a disaster, but gradual progress doesn't generate dramatic footage.

As Steven Pinker said, "News is a misleading way to understand the world. It's always about events that happened and not about things that didn't happen. So when there's a police officer that has not been shot up or city that has not had a violent demonstration, they don't make the news. As long as violent events don't fall to zero, there will always be headlines to click on. . . . Pessimism can be a self-fulfilling prophecy."[39] This is especially true now that social media aggregates alarming news from the entire planet—whereas previous generations were mainly just informed about local or regional events.

Yet my converse observation is: "Optimism is not an idle speculation on the future but rather a self-fulfilling prophecy." Belief that a better world is genuinely possible is a powerful motivator to work hard on creating it.

Daniel Kahneman received the Nobel Prize in Economics for his work (some of it in collaboration with Amos Tversky) explaining invalid and unconscious heuristics that people use in making estimates about the world.[40] Their research demonstrated that people systematically disregard prior probability—the fact that things that are true about a group in general tend to be true about individuals from that group that one encounters. For example, if asked to select the probable occupation of a stranger based on his self-description, if he tells you he "loves books," you might choose "librarian," ignoring the base rate—the general fact that there are relatively few librarians in the world.[41] Someone overcoming this bias would realize that loving books is very weak evidence about someone's occupation so would instead guess a much more common job like "retail worker." People are not unaware of base rates, but they often overlook them in favor of responding to a vivid detail when considering a particular situation.

Another biased heuristic cited by Kahneman and Tversky is that naive observers will expect that a coin toss is more likely to come out heads if they just experienced a run of tails.[42] This is due to a misunderstanding of regression to the mean.

A third bias that explains much of society's pessimistic skew is what Kahneman and Tversky call the "availability heuristic."[43] People estimate the likelihood of an event or a phenomenon by how easily they can think of examples of it. For the reasons discussed previously the news and our news feeds emphasize negative events, so it is these negative circumstances that come readily to mind.

That we should correct for these biases doesn't mean we should ignore or underestimate real problems, but it provides strong rational grounds for optimism about humanity's overall trajectory. Technological change doesn't happen automatically—it takes human ingenuity and effort. Nor should this progress blind us to the urgent suffering people face in the meantime. Rather, the big-picture trends should remind us that, as difficult and even hopeless as these problems sometimes seem, as a species we are turning the tide in solving them. I find that a source of profound motivation.

THE REALITY IS THAT NEARLY EVERY ASPECT OF LIFE IS GETTING PROGRESSIVELY BETTER AS A RESULT OF EXPONENTIALLY IMPROVING TECHNOLOGY

Information technology advances exponentially because it directly contributes to its own further innovation. But that trend also propels numerous mutually reinforcing mechanisms of progress in other areas. Over the past two centuries this has spawned a virtuous circle advancing nearly every aspect of human well-being, including literacy, education, wealth, sanitation, health, democratization, and reduction in violence.

We often think of human development in economic terms: as people are able to earn more money each year, they have access to a better quality of life. But true development involves something much deeper than merely economies amassing wealth. Economic cycles go up and down; wealth can be gained and lost. But technological change is essentially permanent. Once our civilization learns how to do something useful, we generally keep that knowledge and build on it. This one-way march of progress has been a powerful counterbalance to the transient catastrophes like natural disasters, wars, and pandemics that occasionally set societies back.

Intertwined factors like education, health care, sanitation, and democratization create mutually reinforcing feedback loops—improvements in any of those areas will likely lead to benefits in the other areas as well, such as better education producing more capable doctors, and better doctors keeping more children healthy enough to stay in school. This has a very powerful implication: new technologies can have huge indirect benefits, even far from their own areas of application. For example, domestic appliances in the twentieth century not only saved people a lot of time and sweat but also facilitated the liberating and transformative shifts that brought millions of talented women into the workforce, where they made essential contributions in countless fields. In general, we can say that technological innovation promotes conditions that help more people in a society fulfill their potential, which in turn enables even more innovation.

As another example, the invention of the printing press improved and greatly broadened access to education, providing a more capable and sophisticated workforce that drove economic growth. More literacy enabled better coordination of production and trading, which also resulted in greater prosperity. The increased wealth in turn allowed for greater investment in infrastructure and education, which accelerated the beneficial cycle. Meanwhile, mass print communication facilitated greater democratization, which over time has yielded lower violence.

At first this was a very slow process, and the differences between the lifestyles of grandparents and their grandchildren were subtle and generally not noticed. But the smooth trend over the course of centuries was a gradually yet significantly increasing trajectory in all of these measures of social well-being. These trends have accelerated in recent decades, propelled by the steepening curves of exponential advances in almost every form of information technology. As I describe in this chapter, over the next couple of decades this progress will go into very high gear.

LITERACY AND EDUCATION

Throughout most of human history, literacy remained very low throughout the world. Knowledge was mostly passed orally, and a key reason for this was that reproducing writing was very expensive. It was not worth the average person's time to learn how to read if he or she rarely encountered and could never afford written material. Time is the only scarce resource that we all consume equally—no matter who you are, you only get twenty-four hours in a day. When people are deciding how to spend their time, it's only rational to think of what benefits they'll get from a potential choice. Learning to read is a large investment of time. In societies where survival itself was difficult and books were too costly for the average person to access, this wouldn't be a wise investment. So we should be careful not to think of our illiterate ancestors as benighted or incurious. Rather, they were living under conditions that strongly discouraged literacy.

With this perspective it's worth also considering how the incentives in today's world have sometimes discouraged learning. For example, in places with few information technology jobs, young people interested in computer science may find that studying programming won't be a wise use of their time. But now, just as in Europe centuries ago, technology can change that—as automatic translation, distance learning, natural-language programming, and teleworking open new opportunities and reward curiosity.

The introduction of the movable-type printing press in Europe in the late Middle Ages sparked a proliferation of inexpensive and varied reading materials and made it practical for ordinary people to become literate. As the medieval period was ending, less than a fifth of Europe's population knew how to read.[44] Literacy was limited primarily to clergy and occupations that required reading.[45] During the Enlightenment, literacy gradually became more widespread, but by 1750 only the Netherlands and Great Britain, among major European powers, had more than a 50 percent literacy rate.[46] By 1870 only Spain and Italy notably lagged behind that mark, likely due to their relatively undeveloped economies at that time and recent civil wars.[47] Yet the global average remained lower than that of Europe. In 1800, probably fewer than one in ten people worldwide could read, but throughout the nineteenth century the spread of mass-produced newspapers helped promote wider literacy, while social reforms started to guarantee basic education for all children.[48] By 1900, though, still fewer than one in four people could read.[49] During the twentieth century, public education expanded globally, and worldwide literacy exceeded one in four by 1910. By 1970 a majority of the world's population had become literate.[50] Since then it has rapidly jumped to near-universal prevalence in most places.[51] Today the worldwide literacy rate is nearly 87 percent, and developed countries often boast figures above 99 percent.[52]

There is still progress to be made, however. These literacy figures refer to basic standards, such as reading and writing small messages like one's name. New, richer metrics have been developed to assess the quality of literacy. For example, according to the National Assessment of Adult Literacy, in 2003, only 86 percent of the US population scored above "below basic" in literacy.[53] A similar assessment nine years later found that there was no significant improvement.[54]

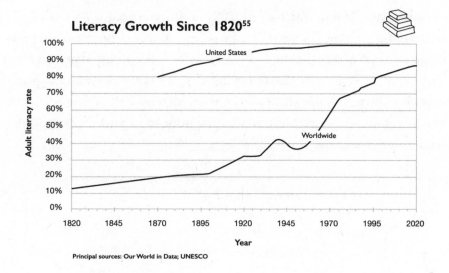

Literacy Growth Since 1820[55]

Principal sources: Our World in Data; UNESCO

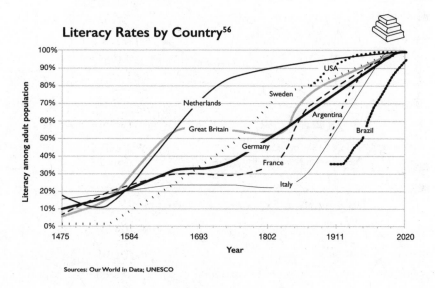

Literacy Rates by Country[56]

Sources: Our World in Data; UNESCO

In 1870 the population of the United States had on average around four years of formal education—while those of the United Kingdom, Japan, France, India, and China were all below one year.[57] The United Kingdom, Japan, and France began quickly catching up to the United States during the early twentieth century as they expanded their free public schooling.[58] Meanwhile, India and China both remained poor and underdeveloped but took major leaps forward during the two decades after World War II.[59] By 2021 India averaged 6.7 years of education and China 7.6 years.[60] The other countries mentioned previously all averaged more than 10 years, with the United States leading the way at 13.7 years.[61] The following charts show this dramatic progress over the past half century—not coincidentally, the same period in which computers were both facilitating education and increasing the benefits of schooling.

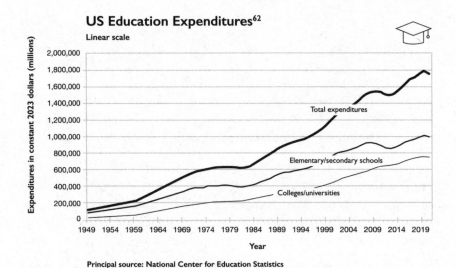

US Education Expenditures[62]

Principal source: National Center for Education Statistics

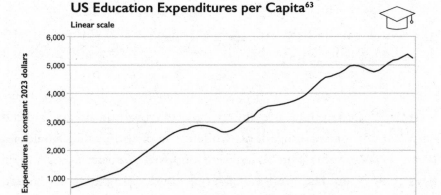

US Education Expenditures per Capita[63]

Principal source: National Center for Education Statistics

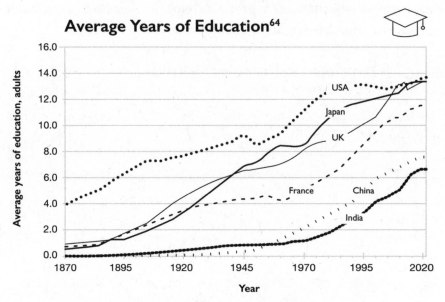

Average Years of Education[64]

Principal sources: Our World in Data; UN Human Development Report Office

AVAILABILITY OF FLUSH TOILETS, ELECTRICITY, RADIO, TELEVISION, AND COMPUTERS

Historically one of the greatest causes of disease and death was con-
tamination of food and water supplies by human feces.[65] Flush toilets,
the definitive technological solution to this problem, gradually gained
adoption in US cities after appearing as early as 1829, but their use did
not become common in urban areas until the turn of the twentieth
century.[66] During the 1920s and 1930s, flush toilets rapidly spread to
rural areas, reaching three-quarters of households by 1950, and 90 percent by 1960.[67] In 2023, the tiny fraction of American homes without
flush toilets is in many cases attributable to lifestyle choices (such as
preferring to live in rustic settings) as opposed to abject poverty.[68] By
contrast, poverty has been a major reason why people in developing
countries still lack flush toilets or other forms of improved sanitation,
such as composting toilets.[69] Worldwide access to safe toilets has been
steadily rising, though, as sanitation technology becomes less expensive,

and as areas that had been prone to violence become more stable and able to invest in sanitation infrastructure.[70]

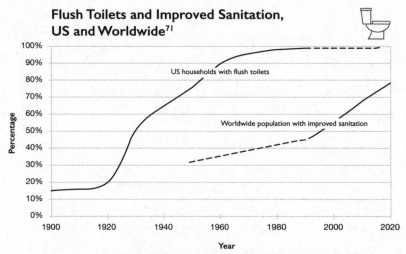

Flush Toilets and Improved Sanitation, US and Worldwide[71]

US households with flush toilets

Worldwide population with improved sanitation

Principal sources: Stanley Lebergott, *Pursuing Happiness: American Consumers in the Twentieth Century* (Princeton, NJ: Princeton University Press, 1993); US Census Bureau; World Bank
Dashed lines indicate estimates bridging data sources.

Electricity is not itself an information technology, but because it powers all our digital devices and networks, it is the prerequisite for the countless other benefits of modern civilization. Even before computers, it ran transformative labor-saving appliances and allowed people to work and play at night. At the start of the twentieth century, electrification in the United States was limited mainly to large urban areas.[72] The pace of electrification slowed around the start of the Great Depression, but during the 1930s and 1940s, President Franklin D. Roosevelt championed massive rural electrification programs, which aimed to bring the efficiency of electric machinery to America's agricultural heartland.[73] By 1951 more than 95 percent of American homes had electricity, and by 1956 the national electrification effort was regarded as essentially complete.[74]

In other parts of the world, electrification has usually followed a similar pattern: cities first, followed by suburban and then rural areas.[75] Today more than 90 percent of the earth's population has electricity.[76] For those still without power, the primary obstacle is political, not technological. MIT professor Daron Acemoğlu and his colleague James Robinson have done very influential research on the key role of political institutions in human development.[77] In short, as countries allow more people to participate freely in politics, and as people gain the security to innovate and invest for the future, feedback loops of prosperity are able to take hold. It is these factors that make it so difficult to bring electricity to the roughly one-tenth of the world's population that still lack it. In areas where violence is common, people conclude that it is not worth investing in expensive electric infrastructure that might be quickly destroyed. Likewise, when roads are poor and dangerous, it is difficult to ship machinery and fuel to isolated communities so they can generate their own power. Fortunately, inexpensive and efficient photovoltaic cells will continue to expand access to electricity.

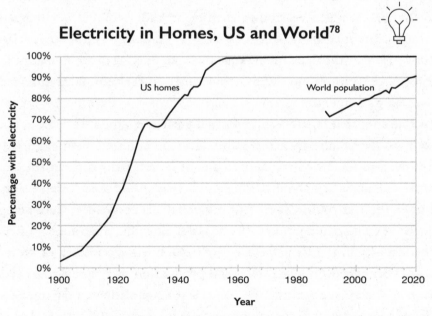

Electricity in Homes, US and World[78]

Sources: US Census Bureau; World Bank; Stanley Lebergott, *The American Economy: Income, Wealth and Want* (Princeton, NJ: Princeton University Press, 1976)

The first transformative communication technology enabled by electricity was radio. Commercial radio broadcasting in the United

States began in 1920 and by the 1930s had become the nation's primary form of mass media.[79] Unlike newspapers, which were mainly limited to a single metropolitan area, radio broadcasts could reach audiences across the country. This spurred development of a truly national media culture, as people from California to Maine heard many of the same political addresses, news reports, and entertainment programs. By 1950 more than nine out of ten American households had a radio, but during that same decade television began to supplant radio's dominance of the media landscape.[80] In response, listeners' habits shifted. Radio programming started to focus more narrowly on news, politics, and sports, and people did much of their listening while in the car.[81] Since the 1980s, highly partisan political talk shows have become some of the most powerful forces in radio and have attracted criticism for reinforcing listeners' biases while closing them off to contrary information.[82] Since the proliferation of smartphones and tablets in the 2010s, an increasing proportion of radio content is streamed online without actually being carried over traditional radio waves. (In 2007, the year the first iPhone was released, only 12 percent of Americans listened to online radio at least once a week; by 2021 this had reached 62 percent.)[83]

The adoption of television followed a pattern similar to radio's, but with the nation already much more developed, its exponential growth was even faster. Scientists and engineers began theorizing in the late nineteenth century about the advances that would lead to television, and by the late 1920s the first primitive television systems were being developed and demonstrated.[84] The technology had reached commercial viability in the United States by 1939, but the outbreak of World War II brought worldwide television production to a virtual halt.[85] As soon as the war ended, though, Americans rapidly began buying televisions. New stations sprang up nationwide, and by 1954 a majority of households had at least one TV.[86] Adoption increased rapidly, and by 1962 more than 90 percent of households had a set.[87] Growth then slowed as late adopters trickled into the TV-watching ranks over the

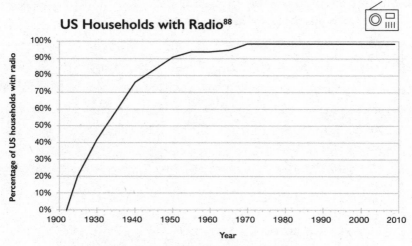

US Households with Radio[88]

Sources: US Census Bureau; Douglas B. Craig, *Fireside Politics: Radio and Political Culture in the United States, 1920–1940* (Baltimore, MD: Johns Hopkins University Press, 2000)

While it appears that the proportion of American households with a radio has fallen significantly in recent years (from 96 percent in 2008 to 68 percent in 2020, according to one study using different methodology from that used for the data in this chart), this is deeply misleading because other devices now perform the same function. For example, as of 2021, 85 percent of American adults had a smartphone, which allows free streaming of radio programming without the need for a radio.

next three decades.[89] By 1997 use had peaked, at 98.4 percent of households, before slightly declining in the years since, standing around 96.2 percent in 2021.[90] The falloff can be attributed to a range of factors: a cultural shift away from excessive television watching, the emergence of competing pastimes online, and the recent trend toward television-style programming being available for streaming on online devices.[91]

Unlike radio and television, which allow passive media consumption, computers open up wider possibilities because they are interactive. Personal computers started entering American homes during the 1970s, with machines such as the Kenbak-1, and in 1975 the wildly popular Altair 8800, which was sold in build-it-yourself kits.[92] By the

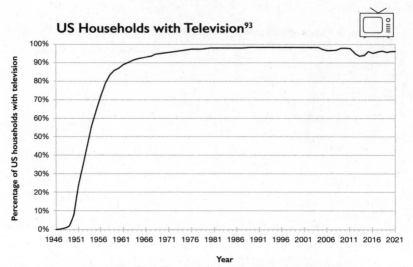

US Households with Television[93]

Principal sources: US Census Bureau; Cobbett S. Steinberg, *TV Facts*, Facts on File; Jack W. Plunkett, *Plunkett's Entertainment & Media Industry Almanac 2006* (Houston, TX: Plunkett Research, 2006); Nielsen Company

end of the decade companies like Apple and Microsoft were transforming the market with user-friendly personal computers that ordinary people could learn to operate in an afternoon.[94] Apple's famous 1984 Super Bowl commercial got the whole country talking about computers, and the proportion of US households with a computer nearly doubled within five years of its airing.[95] During that period people mostly used computers for things like word processing, data entry, and simple gaming.

But the blossoming of the internet in the 1990s massively expanded the usefulness of computers. In January 1990 there were roughly 175,000 hosts across the internet's entire Domain Name System.[96] By January 2000 this had soared to around 72,000,000.[97] Likewise, the volume of worldwide internet traffic grew from about 12,000 gigabytes in 1990 to 306,000,000 gigabytes in 1999.[98] This directly increased the usefulness of computers. Just as streaming services can attract more subscribers and charge higher fees when they have more and better content, the internet became worthwhile to more people as the pool of available content expanded exponentially. And this created a positive feedback loop—many of these new users contributed content of their own that increased the internet's value even further. As a result, during the 1990s home computers went from being platforms for word processing and primitive gaming to portals that could access most of the world's knowledge and connect a user to people continents away. The rise of e-commerce enabled people to use their computers for making many of their purchases, and the emergence of social media in the 2000s allowed the online experience to become richly interactive.

By 2017–2021 about 93.1 percent of US households had computers, and the percentage continues to rise as the Greatest Generation declines and millennials start families of their own.[99] Meanwhile, there has been a steady rise in worldwide computer ownership. Computers embedded in smartphones have rapidly expanded market penetration in the developing world, and as of 2022 around two-thirds of the world's population has at least one.[100]

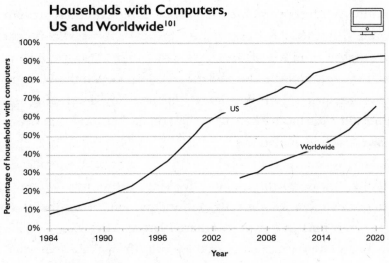

Households with Computers, US and Worldwide[101]

Principal sources; US Census Bureau; International Telecommunication Union

LIFE EXPECTANCY

As I discuss in more detail in chapter 6, most of our progress in disease treatment and prevention to date has been the product of the linear process of hit-or-miss efforts to find useful interventions. Because we have lacked tools for systematically exploring all possible treatments, discoveries under this paradigm have owed a lot to chance. Likely the most notable chance breakthrough in medicine was the accidental discovery of penicillin, which opened up the antibiotic revolution and has since saved perhaps as many as 200 million lives.[102] But even when discoveries aren't literally accidental, it still takes good fortune for researchers to achieve breakthroughs with traditional methods. Without the ability to exhaustively simulate possible drug molecules, researchers have to rely on high-throughput screening and other painstaking laboratory methods, which are much slower and more inefficient.

To be fair, this approach has brought great benefits. A thousand years ago, European life expectancy at birth was just in the twenties, since so many people died in infancy or youth from diseases like cholera and dysentery, which are now easily preventable.[103] By the middle of the nineteenth century, life expectancy in the United Kingdom and the United States had increased to the forties.[104] As of 2023 it has risen to over eighty in much of the developed world.[105] So we have nearly tripled life expectancy in the past thousand years, and doubled it in the past two centuries. This was largely achieved by developing ways to avoid or kill external pathogens—bacteria and viruses that bring disease from outside our bodies.

Today, though, most of this low-hanging fruit has been picked. The remaining sources of disease and disability spring mostly from deep within our own bodies. As cells malfunction and tissues break down, we get conditions like cancer, atherosclerosis, diabetes, and Alzheimer's. To an extent we can reduce these risks through lifestyle, diet, and supplementation—what I call the first bridge to radical life extension.[106] But those can only delay the inevitable. This is why life expectancy gains in developed countries have slowed since roughly the middle of the twentieth century. For example, from 1880 to 1900, life expectancy at birth in the United States increased from about thirty-nine to forty-nine, but from 1980 to 2000—after the focus of medicine had shifted from infectious disease to chronic and degenerative disease—it only increased from seventy-four to seventy-six.[107]

Fortunately, during the 2020s we are entering the second bridge: combining artificial intelligence and biotechnology to defeat these degenerative diseases. We have already progressed beyond using computers just to organize information about interventions and clinical trials. We are now utilizing AI to find new drugs, and by the end of this decade we will be able to start the process of augmenting and ultimately replacing slow, underpowered human trials with digital simulations. In effect we are in the process of turning medicine into an information technology, harnessing the exponential progress that characterizes these technologies to master the software of biology.

One of the earliest and most important examples of this is found in the field of genetics. Since the completion of the Human Genome Project in 2003, the cost of genome sequencing has followed a sustained exponential trend, falling on average by around half each year. Despite a brief plateau in sequencing costs from 2016 to 2018 and slowed progress amid the disruptions of the COVID-19 pandemic, costs continue to fall—and this will likely accelerate again as sophisticated AI plays a greater role in sequencing. Costs have plunged from about $50 million per genome in 2003 to as low as $399 in early 2023, with one company promising to have $100 tests available by the time you read this.[108]

As AI transforms more and more areas of medicine, it will give rise to many similar trends. It is already starting to have a clinical impact,[109] but we are still in the early part of this particular exponential curve. The current trickle of applications will become a flood by the end of the 2020s. We will then be able to start directly addressing the biological factors that now limit maximum life span to about 120 years, including mitochondrial genetic mutations, reduced telomere length, and the uncontrolled cell division that causes cancer.[110]

In the 2030s we will reach the third bridge of radical life extension: medical nanorobots with the ability to intelligently conduct cellular-level maintenance and repair throughout our bodies. By some definitions, certain biomolecules are already considered nanobots. But what will set the nanobots of Bridge Three apart is their ability to be actively controlled by AI to perform varying tasks. At this stage, we will gain a similar level of control over our biology as we presently have over automobile maintenance. That is, unless your car gets destroyed outright in a major wreck, you can continue to repair or replace its parts indefinitely. Likewise, smart nanobots will enable targeted repair or upgrading of individual cells—definitively defeating aging. More on that later in chapter 6.

The fourth bridge—being able to back up our mind files digitally—will be a 2040s technology. As I argue in chapter 3, the core of a person's identity is not their brain itself, but rather the very particular arrangement of information that their brain is able to represent and manipulate. Once we can scan this information with sufficient accuracy, we'll be able to replicate it on digital substrates. This would mean that even if the biological brain was destroyed, it wouldn't extinguish the person's identity—which could achieve an almost arbitrarily long life span being copied and recopied to safe backups.

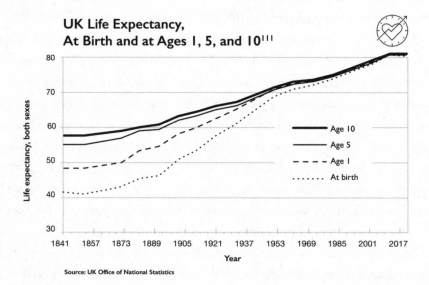

UK Life Expectancy, At Birth and at Ages 1, 5, and 10[III]

Source: UK Office of National Statistics

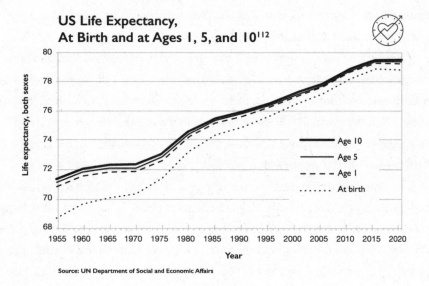

**US Life Expectancy,
At Birth and at Ages 1, 5, and 10**[112]

Life expectancy, both sexes

Age 10
Age 5
Age 1
At birth

Year

Source: UN Department of Social and Economic Affairs

DECLINE IN POVERTY AND INCREASE IN INCOME

The technological trends described so far in this chapter are each highly beneficial, but it is their aggregate mutually reinforcing effect that is truly transformative. While economic well-being is an incomplete measure of progress, it is our best metric for understanding this sweeping process over long periods of time.

Worldwide, the macro trend has been remarkably steady. In 1820, an estimated 84 percent of the world population was living in extreme poverty by modern global standards.[113] With the spread of industrialization, poverty rates soon began to decline in Europe and America.[114] After World War II this process accelerated significantly as more modern agriculture was introduced in India and China over the following decades.[115] Although intensifying media coverage of global poverty has given many people in developed countries a false impression of its extent, the poverty rate continues to decline—with dramatic improvements over the past thirty years. By 2019 extreme poverty (presently defined as living on less than $2.15 a day in 2017 dollars) was down to around 8.4 percent worldwide, having dropped by more than two-thirds between 1990 and 2013.[116]

The decline was steepest in East Asia, where China's economic development pulled hundreds of millions of people out of poverty and into living standards comparable to those of other developed countries. From 1990 to 2013 there was an astonishing 95 percent drop in East Asia's extreme poverty, in a total population that grew from 1.6 billion to 2 billion during that period.[117]

The only region where extreme poverty increased during much of that same period was Europe and Central Asia, where the economic chaos that followed the collapse of the Soviet Union has taken decades to repair.[118] Notably, this was primarily for political reasons rather than narrowly economic or technological ones. The collapse of the authoritarian Soviet government created a power vacuum that allowed enormous corruption. Especially in the poorer Central Asian post-Soviet republics, this generated negative feedback loops that discouraged investment and suppressed prosperity.

Since the end of the Cold War, though, the international community has been able to devote much more of its attention to fighting serious poverty in the most deprived regions of the earth. Right after the fall of the Soviet Union, many developed nations cut their foreign aid budgets and gave less priority to international development.[119] This occurred because during the Cold War, development was largely seen through a strategic lens as Western democracies and communist Eastern bloc nations wrangled for influence in the developing world. But by the middle of the 1990s, the OECD (Organisation for Economic Co-operation and Development) decided that promoting development was vitally important both from a humanitarian perspective and because fostering a safe and prosperous world would benefit everyone. In 2000 the United Nations enshrined these ideas in its Millennium Development Goals, which coordinated international efforts to achieve key development milestones by 2015.[120] While many of these ambitious targets were not met, the MDGs nonetheless stimulated very important progress that improved hundreds of millions of lives.

Within the United States, extreme poverty in absolute terms (per the global benchmark of $2.15 per day in 2017 dollars) has been at or below 1.2 percent since measurement began.[121] Yet statistics for relative poverty (being poor in relation to accepted standards for one's own society) provide a different perspective. Relative poverty in the United States fell from about 45 percent in the nineteenth century, dropping dramatically during the postwar years, and reached about 12.5 percent in 1970, at which point it stagnated.[122] It has remained in the teens since,[123] fluctuating with changes in the wider economy, but has not seen long-term improvement.[124] One reason for this is that rising living standards have led to continual redefinition of the relative poverty line, so people who would not have been considered poor based on quality of life in 1980 are now considered poor.[125]

Still, from 2014 to 2019, the number of Americans in poverty (according to the periodically redefined standard) declined by around 12.6 million, despite the total population's rising by about 8.9 million in that same period.[126] And 4.1 million of that reduction came in 2019 alone, when the poverty rate among the elderly neared an all-time low.[127] While the COVID-19 pandemic caused a temporary increase,[128] it represented a deviation from the overall downward trend since the 2008 financial crisis.

In addition, today's poor are much better off in absolute terms due to the wide accessibility of free information and services via the internet—such as the ability to take MIT open courses or video-chat with family continents away.[129] Similarly, they benefit from the radically improved price-performance of computers and mobile phones in recent decades, but these are not properly reflected in economic statistics. (The failure of economic statistics to sufficiently capture the exponentially improving price-performance of products and services influenced by information technology is discussed in greater detail in the next chapter.) Someone with an inexpensive smartphone today can use the internet to quickly and easily access almost all of the world's educational information, translate languages, find directions, and much more. This capability would not have been available, even for many millions of dollars, a few decades ago.

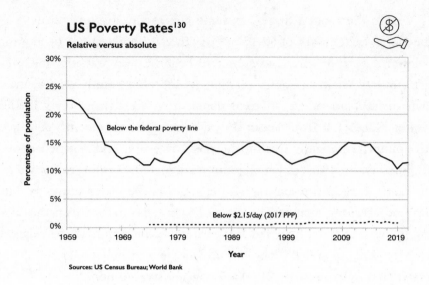

US Poverty Rates[130]

Relative versus absolute

Sources: US Census Bureau; World Bank

Declining Poverty Rates Worldwide[131]

In absolute terms, with US poverty rate for comparison

Principal sources: Our World in Data; François Bourguignon and Christian Morrisson, "Inequality Among World Citizens: 1820–1992," *American Economic Review* 92, no. 4 (September 2002); 727–44.

In the United States, average daily income per person has steadily soared. In 2023 the US poverty line for a single person was $14,580, which translates to living on $39.95 a day.[132] In real terms, the average (not median) American has been above the 2023 poverty line since about 1941.[133]

Average daily income per person in the United States (2023 dollars), by year[134]

2020: $191.00	1970: $89.82	1900: $20.08
2015: $169.88	1960: $65.27	1880: $15.08
2010: $154.15	1950: $52.97	1860: $13.27
2000: $147.18	1940: $35.47	1840: $8.37
1990: $124.32	1930: $30.74	1800: $5.65
1980: $102.41	1910: $26.45	1774: $7.06

Approximate poverty rate in the United States (relative poverty, periodically redefined by the government), by year[135]

2020: 11.5%	1990: 13.5%	1950: ~30%
2015: 13.5%	1980: 13.0%	1935: ~45%
2010: 15.1%	1970: 12.6%	1910: ~30%
2000: 11.3%	1960: 22.2%	1870: ~45%

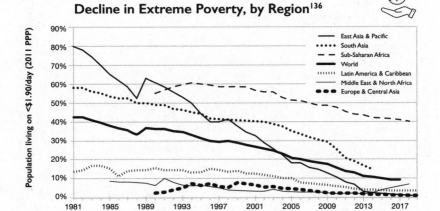

Decline in Extreme Poverty, by Region[136]

Source: World Bank

It is perhaps unsurprising that US gross domestic product has been growing exponentially, as population growth means a larger economy. But gross domestic product per capita—which controls for population growth—has also been growing exponentially.[137] Again, keep in mind that this does not include free information products and does not account for the exponential increase in the power of information technology for the same cost over time.

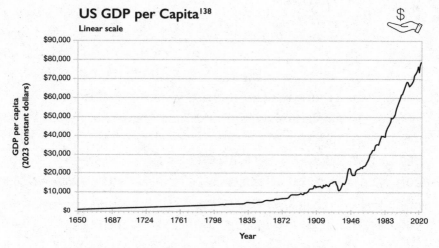

Sources: Maddison Project Database; Bureau of Economic Analysis; Federal Reserve

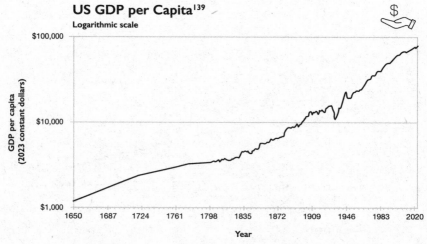

Sources: Maddison Project Database; Bureau of Economic Analysis; Federal Reserve

GDP reflects overall economic activity, including big businesses, but the same trend also holds when we focus only on personal income. Personal income per capita measures earnings by real people, as opposed to those of corporations. Thus, it includes salaries and wages but also the dividends and profits that shareholders and business owners make from their companies. Since the statistic was first recorded in 1929, personal income in the United States in constant dollars per capita has risen enormously, with only brief downturns during the Great Depression and major recessions. Over the past nine decades, the average American's real earnings have increased more than five-fold. This is the case despite substantial reductions in the number of hours worked.[140] Median personal income, which reflects a person in the exact middle of the national income distribution, has not grown as quickly. But real median incomes continue to rise steadily in absolute terms, from (all in 2023 dollars) $27,273 in 1984 to $42,488 in 2019, just before the pandemic.[141]

These gains significantly understate the benefits that have actually been achieved, as they do not reflect the fact that, as noted previously, many goods, like electronics, now cost much less than they used to— and many extremely valuable services, like search engines and social networks, are offered free to their users. In addition, globalization has made available a much wider selection of products and services than people could access in 1929. It is difficult to assign a monetary value to the dizzying variety available to modern consumers relative to past decades. Even if you can have only Chinese food or Mexican food at a given meal, you would probably rather have the choice instead of just one option. This variety has affected countless areas of life. Instead of having three television stations to choose from, we have hundreds. Instead of a few fruits at the supermarket, we get out-of-season fruit flown in from the opposite hemisphere. Instead of a few thousand books at a bookstore, we can choose from among millions on Amazon.

These choices help everyone satisfy their preferences better, but they are especially important to people who have unusual tastes and interests. Thanks to the globalized economy made possible by information technology, if you love collecting old kaleidoscopes, you can go on eBay and buy them from anywhere in the world. And if you're a kid interested in math and science, you can watch hours of educational shows that nurture your curiosity instead of being stuck watching *Gunsmoke* around your family's single TV set, as was common back in my generation. The maturation of 3D printing and eventually nanotechnology will exponentially accelerate this diversification of our choices in the coming decades.

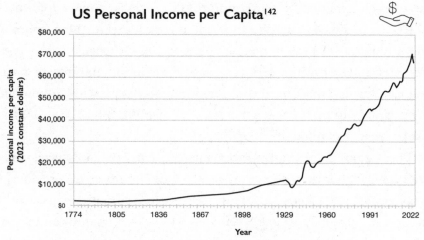

US Personal Income per Capita[142]

Sources: Bureau of Economic Analysis; National Bureau of Economic Research; Alexander Klein, "New State-Level Estimates of Personal Income in the United States, 1880–1910," in *Research in Economic History*, vol. 29, ed. Christopher Hanes and Susan Wolcott (Bingley, UK: Emerald Group, 2013); Federal Reserve

Annual income is a useful metric, but it's even more informative to consider income in light of the amount of time people spend working. Real personal income per hour in constant dollars has steadily increased in the United States, from around $5 per hour in 1880 to around $93 per hour in 2021.[143] Note, though, that personal income does not just consist of wage and salary income. It also includes income from investments and earnings from a business someone owns, as well as government benefits and one-time payments like the 2020 pandemic stimulus. Thus, personal income per hour is consistently higher than measures of hourly wage and salary income. The reason this is a useful measure is that it reflects how non-wage income is forming an increasing share of all personal income, and so better demonstrates how total economic prosperity is increasing even as the average worker spends less time working.

The following chart shows how steadily real income per hour worked has been rising in the United States—even during periods of economic turmoil. Although there isn't good data available from the deepest trough of the Great Depression, it appears that hourly income did not suffer nearly as badly during this period as other measures of economic performance. The reason is that the fall in personal income per capita during the Depression resulted from the population's earning less overall income. With the population (the denominator) not changing very much, this inevitably meant lower per-capita earnings. However, when people lose jobs, they both lose income and work fewer hours. This reduction in the denominator resulted in earnings per hour staying steady and even growing during the Depression. Another way of looking at this is that many people lost jobs altogether, but wages didn't fall much for those who managed to stay employed.

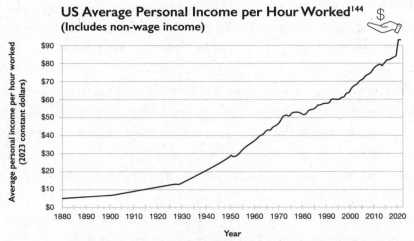

US Average Personal Income per Hour Worked[144]
(Includes non-wage income)

Year

Principal sources: Maddison Project; Bureau of Economic Analysis; Stanley Lebergott, "Labor Force and Employment, 1800–1960," in *Output, Employment, and Productivity in the United States After 1800*, ed. Dorothy S. Brady (Washington, DC: National Bureau of Economic Research, 1966); Alexander Klein, "New State-Level Estimates of Personal Income in the United States, 1880–1910," in *Research in Economic History*, vol. 29, ed. Christopher Hanes and Susan Wolcott (Bingley, UK: Emerald Group, 2013); Michael Huberman and Chris Minns, "The Times They Are Not Changin': Days and Hours of Work in Old and New Worlds, 1870–2000," *Explorations in Economic History* 44, no. 4 (July 12, 2007)

In the late nineteenth century, the average American worker spent nearly three thousand hours per year on the job.[145] After around 1910 this figure began rapidly declining as regulations and labor unions pushed down the length of workdays and provided workers with more time off.[146] Additionally, employers found that rested employees were more productive and accurate in their work, so it made sense to divide work hours among a greater number of workers than was the case at the start of the Industrial Revolution. Average hours worked plunged to below 1,750 a year during the Great Depression as firms had to slash the hours of even those employees who still had jobs.[147] During World War II and the postwar boom, Americans rushed back into factories and offices, averaging more than 2,000 hours a year.[148] Since then, hours worked have gradually but steadily declined in the United States, returning to approximately Depression-era levels.[149] The difference is that today those reductions have been largely driven by people opting for part-time work and making other choices that promote a healthier work-life balance. In some European countries, the decline in hours worked has been even steeper.[150]

Shifts in the kinds of jobs in demand have motivated millennials and Generation Z, more than other generations, to seek creative, often entrepreneurial careers, and have given them the freedom to work remotely, which cuts out travel time and expense but can lead to blurry boundaries between work and life. The COVID-19 pandemic created a sudden and dramatic shift in the labor force toward telework and other alternative models of the employer-employee relationship. In one study, 98 percent of respondents said they wanted the option to work remotely for the remainder of their careers.[151] As technological change enables more and more jobs to be done remotely, this trend will likely intensify.

One more key indicator of socioeconomic well-being is child labor. When children are forced into work by poverty, they miss out on education and face diminished long-term potential. Fortunately, child labor has been in steady decline throughout the twenty-first century.

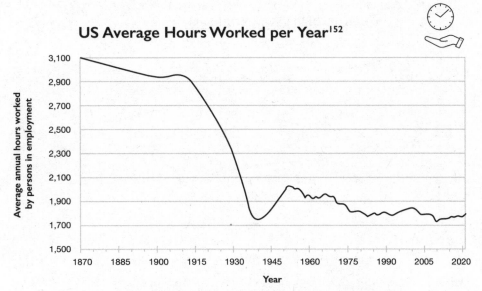

US Average Hours Worked per Year[152]

Sources: Michael Huberman and Chris Minns, "The Times They Are Not Changin': Days and Hours of Work in Old and New Worlds, 1870–2000," *Explorations in Economic History* 44, no. 4 (July 12, 2007); University of Groningen and University of California, Davis; Federal Reserve; Organisation for Economic Co-operation and Development

The International Labor Organization uses three nested categories to measure progress in this area.[153] The broadest category is "child employment," which includes children working modest hours in light work on family farms or in family-run businesses; although this may distract from education, it is a relatively mild form of child labor. The second category, "child labor" proper, encompasses children working in jobs roughly similar to those an adult might have, in terms of the hours demanded and the arduousness of the work itself. The third and narrowest category is "hazardous work," which is child labor in conditions that are particularly dangerous. Such jobs include mining, shipbreaking (dismantling a ship to extract value and for disposal), and waste handling. From 2000 to 2016, the percentage of the world's children estimated to be in hazardous work fell from 11.1 percent to 4.6 percent, although economic disruptions from COVID-19 appear to have interrupted this progress.[154]

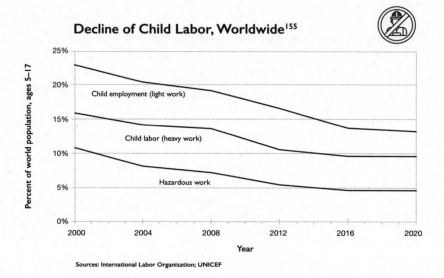

Decline of Child Labor, Worldwide[155]

Sources: International Labor Organization; UNICEF

DECLINE IN VIOLENCE

Increasing material prosperity has a mutually reinforcing relationship with declining violence. People with a lot to lose economically have stronger incentives to avoid fighting, and when people can look forward to long lives of safety, they have good reason to make long-term investments that benefit society. In Western Europe, homicide rates have been falling consistently since at least the fourteenth century.[156] Over these long time scales, the decline has been exponential, even despite the development of ever more deadly personal weapons. In the Western European countries for which we have good data going back to the medieval period, annual homicides per 100,000 people have fallen from a per-country average of around thirty-three in the fourteenth and fifteenth centuries to less than one today—a drop of greater than 97 percent.[157] Note that these statistics focus on forms of "ordinary" homicide such as murder and manslaughter and do not include warfare and genocide.

Homicide Rates in Western Europe Since 1300[158]

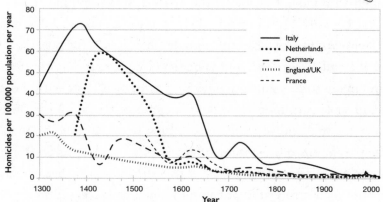

Sources: Our World in Data (Roser and Ritchie); Manuel Eisner, "From Swords to Words: Does Macro-Level Change in Self-Control Predict Long-Term Variation in Levels of Homicide?," *Crime and Justice* 43, no. 1 (September 2014); UN Office on Drugs and Crime

This chart draws on the excellent work of Max Roser and Hannah Ritchie at Our World in Data, but it differs in recent source selection. The data before 1990 is from Manuel Eisner, "From Swords to Words: Does Macro-Level Change in Self-Control Predict Long-Term Variation in Levels of Homicide?," but instead of using data from the World Health Organization for 1990–2018, this chart uses 1990–2020 data from the UN Office on Drugs and Crime. This is because the UNODC data is better corroborated by other sources and more closely matches a wide range of official law enforcement estimates.

In the United States, murder and other forms of violent crime have been in a long-term decline since about 1991. Although US homicides have risen from a low of 4.4 per 100,000 in 2014 to 6.8 in 2021 (the most recent data at the time of writing), this short-term spike nonetheless leaves homicides down by more than 30 percent—the figure was 9.8 in 1991.[159] During two periods of the twentieth century, though, homicide and violent crime in general were about twice as common as they are today. The first was the 1920s and 1930s, largely as a result of Prohibition and the criminal organizations that sprang up around the trade in bootleg liquor.[160] The second epidemic of violence took place during the 1970s, 1980s, and 1990s as the trade in narcotics and other illegal drugs again brought violence into the streets of American cities.[161]

Since then, several factors have played major roles in the decline of violence. With violent crime in the United States spiking to all-time highs by the early 1980s, criminologists started looking for new solutions. Scholars George Kelling and James Q. Wilson observed that low-level crime like graffiti and vandalism made communities feel unsafe and made some people believe they could get away with more serious and violent crimes.[162] This idea came to be called the "broken windows theory," and it influenced a new trend in policing that emphasized stopping those minor offenses as a way of preventing more serious crime. This was combined with a shift toward other, more proactive approaches to crime prevention, such as increased foot patrols in high-crime neighborhoods and smarter data-driven modeling of how police resources could be most effectively employed. Together these factors appear to have played a big role in the nationwide drop in crime that occurred throughout the 1990s and 2000s. Still, they did not come without costs. In some cities, broken windows policing was taken too far and led to disproportionate harm to minority communities. During the 2020s, the challenge for police will be twofold: continuing the long-term crime reduction trend while simultaneously addressing racial disparities and related injustices. Although there is no single solution for these issues, technologies like police body cameras, citizens' cell-phone cameras, automated gunshot detectors, and AI-driven data analysis can all play a positive role if used responsibly.

Another factor that is just now coming to be properly appreciated is the relationship between pollution and crime. The effect of environmental toxicity on the brain, especially from lead, was poorly understood for most of the twentieth century. Exposure to lead in car exhaust and household paints adversely affected children's cognitive development. Although it is impossible to know whether this chronic poisoning was responsible for any given crime committed, at the level of the whole population it did lead to a statistical increase in violent crime—likely by causing lowered impulse control. Since roughly the 1970s, increasing environmental regulations have limited the amount of lead and other toxins that get into people's brains in childhood, and this is considered to have contributed to lower levels of violence as well.[163]

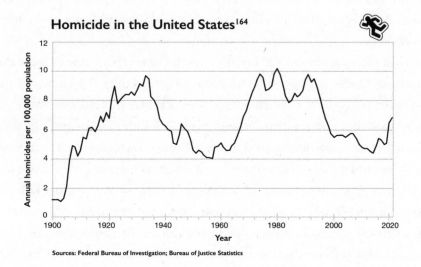

Homicide in the United States[164]

Sources: Federal Bureau of Investigation; Bureau of Justice Statistics

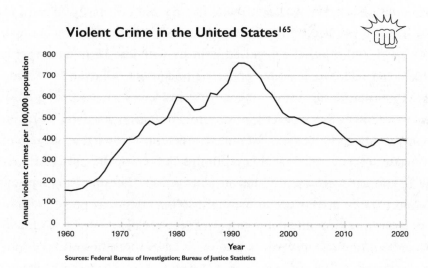

Violent Crime in the United States[165]

Sources: Federal Bureau of Investigation; Bureau of Justice Statistics

In his 2011 book *The Better Angels of Our Nature*, Steven Pinker marshals further evidence that deaths from homicide in Europe have fallen by a factor of very roughly fifty since the Middle Ages, and in some cases more.[166] For example, in Oxford in the fourteenth century, there were an estimated 110 homicides per 100,000 people per year, whereas in London today there is less than one homicide per 100,000 people per year.[167] Pinker estimates that violent deaths overall have fallen by a factor of somewhere around five hundred since prehistoric times.[168]

Even the great historical conflicts of the twentieth century were not nearly as deadly in proportional terms as the continual violence in the pre-state conditions of humanity's past. Pinker studied twenty-seven undeveloped societies throughout history without a formal state—a mix of hunter-gatherers and hunter-horticulturalists that probably represent most human communities during prehistory.[169] He estimates that these societies averaged a rate of death in warfare of 524 per 100,000 people per year. By comparison, the twentieth century saw two world wars—which included genocide, atomic bombings, and the largest-scale organized violence the world has ever seen—yet in Germany, Japan, and Russia, three of the countries hit hardest by this slaughter, the annual war death rates during that century were just 144, 27, and 135 per 100,000, respectively. The United States, by comparison, made it through the twentieth century with just 3.7 annual warfare deaths per 100,000, despite entering into conflicts all over the world.

Yet much of the public incorrectly perceives violence as getting worse. Pinker largely ascribes this to "historical myopia," wherein people focus on more recent events that get more attention and remain unaware of even worse violent episodes deeper in the past.[170] Essentially, this is the availability heuristic in action. These misperceptions can in part be attributed to technologies of documentation: we have easy access to dramatic color videos of recent violent events. Compare this with the black-and-white photographs of the nineteenth century or, even before that, the text descriptions and relatively few paintings of earlier eras.

Like me, Pinker attributes this dramatic decline in violence to virtuous circles. As people become more confident that they will be free from violence, the incentive to build schools and write and read books becomes greater, which in turn encourages the use of reason instead of force to solve problems, which then reduces violence even further. We have experienced an "expanding circle" of empathy (philosopher Peter Singer's term) that extends our sense of identification from narrow groups like clans to entire nations, then to people in foreign countries, and even to nonhuman animals.[171] There has also been a growing role for the rule of law and cultural norms against violence.

The key insight for the future is that these virtuous circles are fundamentally driven by technology. Where humans once only identified with small groups, communication technology (books, then radio and television, then computers and the internet) enabled us to exchange ideas with an ever wider sphere of people and discover what we have in common. The ability to watch gripping video of disasters in distant lands can lead to historical myopia, but it also powerfully harnesses our natural empathy and extends our moral concern across our whole species.

Further, the more wealth grows and poverty declines, the greater incentives people have for cooperation, and the more zero-sum struggles for limited resources are alleviated. Many of us have a deeply ingrained tendency to view the struggle for scarce resources as an unavoidable cause of violence and as an inherent part of human nature. But while this has been the story of much of human history, I don't think this will be permanent. The digital revolution has already rolled back scarcity conditions for many of the things we can easily represent digitally, from web searches to social media connections. Fighting over a copy of a physical book may be petty, but on a certain level we can understand it. Two children may tussle over a favorite printed comic because only one can have it and read it at a time. But the idea of people fighting over a PDF document is comical—because your having access to it doesn't mean I don't have access to it. We can create as many copies as we need, essentially for free.

Once humanity has extremely cheap energy (largely from solar and, eventually, fusion) and AI robotics, many kinds of goods will be so easy to reproduce that the notion of people committing violence over them will seem just as silly as fighting over a PDF seems today. In this way the millions-fold improvement in information technologies between now and the 2040s will power transformative improvement across countless other aspects of society.

GROWTH OF RENEWABLE ENERGY

Almost every aspect of our technological civilization requires energy, but our longtime reliance on fossil fuels is unsustainable for two main reasons. Most obviously, it creates toxic pollution and greenhouse gas emissions, but it also limits us to scarce resources that are becoming more expensive to extract even as humanity's need for cheap energy is soaring. Fortunately, the costs of environmentally friendly renewable energy have been dropping exponentially as we apply increasingly sophisticated technologies to the design of the underlying materials and mechanisms. For example, over the past decade we have been using supercomputing to discover new materials for both solar cells and energy storage—and in recent years deep neural nets have been employed for this as well.[172] As a result of these ongoing cost reductions, the overall amount of energy obtained from renewable sources—solar, wind, geothermal, tidal, and biofuels—is also growing exponentially.[173] Solar was responsible for about 3.6 percent of electricity in 2021, and this proportion has been doubling on average every twenty-eight months since 1983.[174] More on that later in this chapter.

Photovoltaic Module Cost per Watt[175]

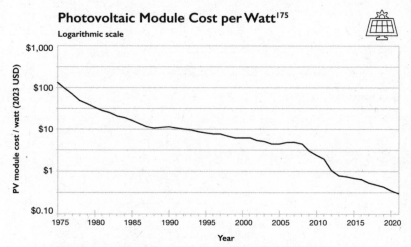

Logarithmic scale

Principal sources: Our World in Data; Gregory F. Nemet, "Interim Monitoring of Cost Dynamics for Publicly Supported Energy Technologies," *Energy Policy* 37, no. 3 (March 2009); IRENA

As this logarithmic plot shows, photovoltaic module costs per watt have been exponentially declining for almost five decades, a trend that is sometimes known as Swanson's law. Note that while module cost is the largest single component of total costs for solar power installations, other factors, like permits and installation labor, can bring total costs up to roughly triple the module cost.[176] And while module costs have been falling very rapidly, those other factors have been declining more slowly. AI and robotics can bring down costs for labor and design, but we'll also need policies that encourage efficient utility planning and permitting.

Photovoltaics—Global Installed Capacity[177]

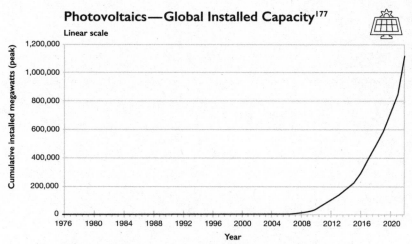

Linear scale

Principal sources: Our World in Data; IRENA; Gregory F. Nemet, "Interim Monitoring of Cost Dynamics for Publicly Supported Energy Technologies," *Energy Policy* 37, no. 3 (March 2009)

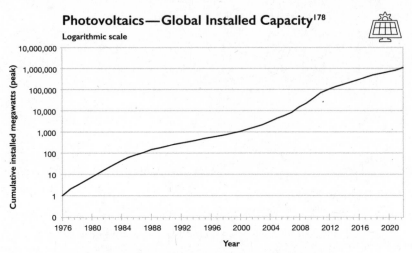

Photovoltaics—Global Installed Capacity[178]

Principal sources: Our World in Data; IRENA; Gregory F. Nemet, "Interim Monitoring of Cost Dynamics for Publicly Supported Energy Technologies," *Energy Policy* 37, no. 3 (March 2009)

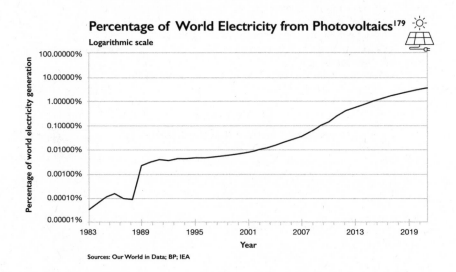

Percentage of World Electricity from Photovoltaics[179]

Sources: Our World in Data; BP; IEA

Wind Power Costs[180]

Levelized costs for US land-based generation projects

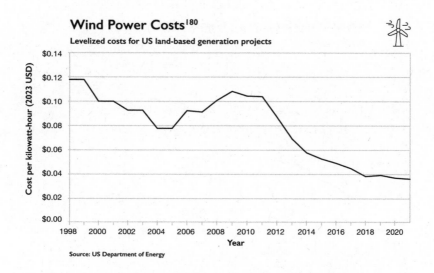

Source: US Department of Energy

World Wind Electricity Generation[181]

Linear scale

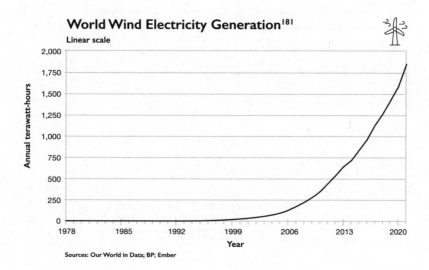

Sources: Our World in Data; BP; Ember

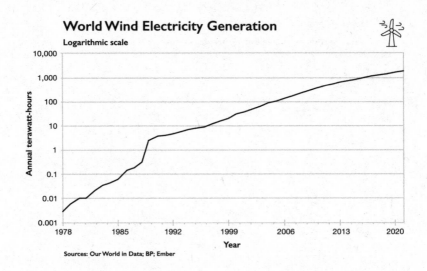

World Wind Electricity Generation

Logarithmic scale

Sources: Our World in Data; BP; Ember

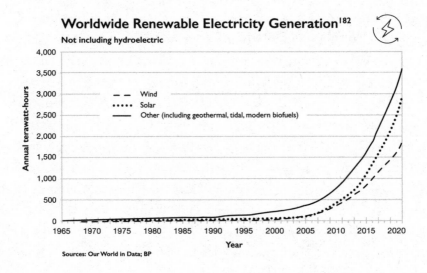

Worldwide Renewable Electricity Generation[182]

Not including hydroelectric

Sources: Our World in Data; BP

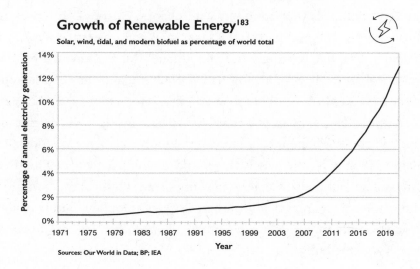

Growth of Renewable Energy[183]

Solar, wind, tidal, and modern biofuel as percentage of world total

Sources: Our World in Data; BP; IEA

SPREAD OF DEMOCRACY

Cheap renewable energy will unlock great material abundance, but people sharing in it equitably requires democracy. Here again is a fortunate synergy: information technology has a long track record of making societies more democratic. The spread of democracy from its roots in medieval England parallels, and likely results largely from, the rise of mass communication technologies. The Magna Carta, which famously articulated the rights of ordinary people not to be unjustly imprisoned, was penned in 1215 and signed by King John.[184] Yet for most of the Middle Ages, commoners' rights were often ignored and political participation was minimal. This changed with Gutenberg's invention of the movable-type printing press around 1440, and with its rapid adoption the educated classes were able to spread both news and ideas with far greater effectiveness.[185]

The printing press is an excellent illustrative example of how the law of accelerating returns works for information technologies. As noted previously, information technologies are those that involve the gathering, storage, manipulation, and transmission of information. Human ideas are information. In theoretical terms, information is an abstract orderliness that contrasts with chaotic entropy. To represent information in the real world, we make physical objects orderly in a particular way, like writing letters on a page. But the key idea is that ideas are abstract ways of arranging information, and they can be represented on all kinds of different mediums—stone tablets, parchment paper, punch cards, magnetic tape, or voltages inside a silicon microchip. You could write out every line of the source code for Windows 11 on stone tablets if you had a chisel and enough patience! This is a funny image, but it also suggests how important physical mediums are to how well we can practically gather, store, manipulate, and transmit information. Some ideas become practical only when we can do these things very efficiently.

For most of medieval history, books could be reproduced in Europe only by scribes laboriously copying them out by hand, which was the opposite of efficiency. As a result it was extremely expensive to transmit written information—ideas. Very broadly, the more ideas a person or a society has, the easier it is to create new ones; this includes technological innovation. Therefore, technologies that make it easier to share ideas make it easier to create new technologies—some of which will make it even easier to share ideas. So when Gutenberg introduced the printing press, it soon became vastly cheaper to share ideas. For the first time, middle-class people could afford books in large numbers, unlocking huge amounts of human potential. Innovation flourished, and the Renaissance spread quickly across Europe. This also brought further innovations to printing, and by the early seventeenth century books were several thousand times cheaper than they had been before Gutenberg.

The spread of knowledge brought wealth and political empowerment, and legislative bodies like England's House of Commons became much more outspoken. While most power was still held by the king, Parliament was able to deliver tax protests to the monarch and impeach those of his ministers it did not like.[186] The 1642–1651 English Civil War eliminated the monarchy altogether, after which it was reinstalled in a form subservient to Parliament; later the government adopted a Bill of Rights that clearly established the principle that the king could rule only by consent of the people.[187]

Prior to the American War of Independence, England, while far from a true democracy, was the most democratic nation in world history—and, notably, one of the most literate.[188] Between the collapse of the Roman Republic in the first century BC and the American Revolution, there had been a number of other societies that had elections or roughly republican political institutions, but these always had very limited political participation and always lapsed back into tyranny. During the Middle Ages there were several republics in Italy centered on trade-rich city-states like Genoa and Venice, yet these were in fact strongly aristocratic. For example, the leaders of Venice (called doges) were elected for life through a complex selection process that kept noble families in power and didn't let ordinary people play any role.[189] From 1569 to 1795, the Polish-Lithuanian Commonwealth had a remarkably free and democratic system for the *szlachta* nobility (usually a tenth of the population or less), but non-nobles had very little say in politics.[190]

By contrast, in Great Britain there was at least theoretical voting eligibility for all free adult male householders. And while in practice there were usually additional property requirements, eligibility to vote did not depend on someone's status at birth. Despite the many people this excluded, it was an absolutely crucial political innovation because it set the stage for the idea of universal political participation. Once people accepted that status at birth shouldn't determine someone's right to vote, the power of that idea was irrepressible. Thus, it was not a coincidence that the first true modern democracy emerged in the American colonies, even if it took a war against England to achieve it. Yet living up to its promises has been a painstaking process.

Two centuries ago in the United States, most people still did not enjoy full rights of political participation. In the early nineteenth century, voting rights were limited mostly to adult white males with at least modest property or wealth. These economic requirements allowed a majority of white men to vote but almost entirely excluded women, African Americans (millions of whom were held in chattel slavery), and Native Americans.[191] Historians disagree on precisely what percentage of the population was eligible to vote, but it's most commonly considered to have been between 10 and 25 percent.[192] Yet this injustice carried the seeds of its own undoing—voting provided the United States with a mechanism for reform, and restrictions on voting stood in contrast with the lofty ideals set forth in its founding documents.

Despite the high aspirations of its advocates, democracy gained ground only slowly over the course of the nineteenth century. For example, the 1848 liberal revolutions in Europe mostly failed, and many of the reforms of Tsar Alexander II in Russia were undone by his successors.[193] In 1900 just 3 percent of the world's population lived in what we would currently consider democracies, as even the United States still denied women the right to vote and enforced segregation against African Americans. By 1922, in the aftermath of World War I, that had climbed to 19 percent.[194] Yet the rising tide of fascism soon put democracy in retreat, and hundreds of millions of people came under totalitarian rule during World War II. Notably, mass communication via radio initially helped fascists take power, but ultimately the same technology enabled the Allies to rally their democracies to victory—most notably in Winston Churchill's inspirational speeches during the Blitz.

The postwar years saw a rapid spike in the proportion of the world's people living under democracy, largely driven by the independence won by India and Britain's other colonies in South Asia. For most of the Cold War, the reach of democracy stayed roughly steady, with just over one in three people in the world living in democratic societies.[195] Yet the proliferation of communication technology outside the Iron Curtain, from Beatles LPs to color TVs, stirred discontent against the governments that suppressed it. With the breakup of the Soviet Union, democracy again expanded rapidly, reaching almost 54 percent of the world's population by 1999.[196]

Although this figure has fluctuated over the past two decades, with gains in some countries offset by backsliding elsewhere, a promising liberalization of a different kind has been proceeding rapidly. At the end of the Cold War, about 35 percent of the world's population lived in closed autocracies—the most repressive class of regimes.[197] By 2022 that figure had declined to about 26 percent, freeing more than 750 million people from tyranny.[198] This includes the momentous and complex events of the Arab Spring, which were largely enabled and driven by social media. A crucial challenge over the coming decades will be helping countries that fall into the gray area between autocracy and democracy make the transition to fully democratic governance. Part of this will depend on careful use of AI to promote openness and transparency while minimizing its potential to be abused for authoritarian surveillance or to spread disinformation.[199]

History gives us reason for profound optimism, though. As technologies for sharing information have evolved from the telegraph to social media, the idea of democracy and individual rights has gone from barely acknowledged to a worldwide aspiration that's already a reality for nearly half the people on earth. Imagine how the exponential progress of the next two decades will allow us to realize these ideals even more fully.

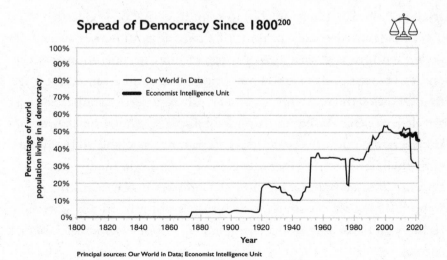

Spread of Democracy Since 1800[200]

Principal sources: Our World in Data; Economist Intelligence Unit

WE ARE NOW ENTERING THE STEEP PART OF THE EXPONENTIAL

The essential point to realize is that all the progress I have described so far came from the slow early stages of these exponential trends. As information technology makes vastly more progress in the next twenty years than it did in the past two hundred, the benefits to overall prosperity will be far greater—indeed, they are already much greater than most realize.

The most fundamental trend at work here is the exponentially improving price-performance of computation—that is, how many computations per second can be performed for one inflation-adjusted dollar. When Konrad Zuse built the first working programmable computer, the Z2, in 1939, it could perform around 0.0000065 computations per second per 2023 dollar.[201] In 1965, the PDP-8 managed around 1.8 computations per second per dollar. When my book *The Age of Intelligent Machines* was published in 1990, the MT 486DX could achieve about 1,700. When *The Age of Spiritual Machines* appeared nine years later, Pentium III CPUs were up to 800,000. And when *The Singularity Is Near* debuted in 2005, some Pentium 4s were at 12 million. As this book goes to print in early 2024, Google Cloud TPU v5e chips are likely around 130 billion operations per second per dollar! And because this awesome (quintillions of operations per second for a large pod) cloud computing power can be rented for a few thousand dollars an hour by anyone with an internet connection—as an alternative to building and maintaining a whole supercomputer from scratch—the effective price-performance available to retail users with small projects is several orders of magnitude higher than that. Because cheap computation directly facilitates innovation, this macro trend has steadily continued—and it does not rely on any particular technological paradigm like shrinking transistors or increasing clock speeds.

Price-Performance of Computation, 1939–2023[202]

Best achieved price-performance in computations per second per constant 2023 dollar

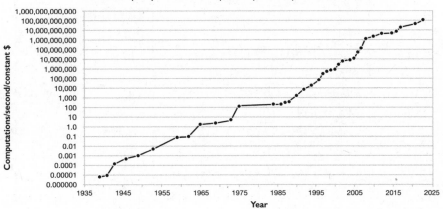

MACHINES ACHIEVING RECORD-SETTING PRICE-PERFORMANCE

YEAR	NAME	COMPUTATIONS PER SECOND PER CONSTANT 2023 DOLLAR
1939	Z2	~ 0.0000065
1941	Z3	~ 0.0000091
1943	Colossus Mark 1	~ 0.00015
1946	ENIAC	~ 0.00043
1949	BINAC	~ 0.00099
1953	UNIVAC 1103	~ 0.0048
1959	DEC PDP-1	~ 0.081
1962	DEC PDP-4	~ 0.097
1965	DEC PDP-8	~ 1.8
1969	Data General Nova	~ 2.5
1973	Intellec 8	~ 4.9
1975	Altair 8800	~ 144
1984	Apple Macintosh	~ 221

YEAR	NAME	COMPUTATIONS PER SECOND PER CONSTANT 2023 DOLLAR
1986	Compaq Deskpro 386 (16 MHz)	~ 224
1987	PC's Limited 386 (16 MHz)	~ 330
1988	Compaq Deskpro 386/25	~ 420
1990	MT 486DX	~ 1,700
1992	Gateway 486DX2/66	~ 8,400
1994	Pentium (75 MHz)	~ 19,000
1996	Pentium Pro (166 MHz)	~ 75,000
1997	Mobile Pentium MMX (133 MHz)	~ 340,000
1998	Pentium II (450 MHz)	~ 580,000
1999	Pentium III (450 MHz)	~ 800,000
2000	Pentium III (1.0 GHz)	~ 920,000
2001	Pentium 4 (1700 MHz)	~ 3,100,000
2002	Xeon (2.4 GHz)	~ 6,300,000
2004	Pentium 4 (3.0 GHz)	~ 9,100,000
2005	Pentium 4 662 (3.6 GHz)	~ 12,000,000
2006	Core 2 Duo E6300	~ 54,000,000
2007	Pentium Dual-Core E2180	~ 130,000,000
2008	GTX 285	~ 1,400,000,000
2010	GTX 580	~ 2,300,000,000
2012	GTX 680	~ 5,000,000,000
2015	Titan X (Maxwell 2.0)	~ 5,300,000,000
2016	Titan X (Pascal)	~ 7,300,000,000
2017	AMD Radeon RX 580	~ 22,000,000,000
2021	Google Cloud TPU v4-4096	~ 48,000,000,000
2023	Google Cloud TPU v5e	~ 130,000,000,000

Yet this dramatic progress is not as widely recognized as one might expect. In an onstage dialogue with Christine Lagarde, managing director of the International Monetary Fund, and other economic leaders at the IMF's Annual Meeting on October 5, 2016, LaGarde asked me why we don't see more evidence of economic growth from all the remarkable digital technology that is now available. My answer was (and is) that we factor out this growth by putting it in both the numerator and denominator.

When a teenager in Africa spends $50 on a smartphone, it counts as $50 of economic activity, despite the fact that this purchase is equivalent to over a billion dollars of computation and communication technology circa 1965, and millions of dollars circa 1985. The Snapdragon 810, a common chip on smartphones in the $50 range, averages about three billion floating-point operations per second (GFLOPS) across a range of performance benchmarks.[203] This corresponds to around 60 million computations per second per dollar. The best available computers in 1965 achieved about 1.8 computations per second per dollar, and in 1985 they were up to about 220.[204] At those efficiencies, it would have taken almost $1.7 billion (in 2023 dollars) to match the Snapdragon in 1965 and $13.6 million in 1980.

Of course, this crude comparison does not take into account many other aspects of the technology that have evolved since. Speaking more literally, the full capabilities of a $50 smartphone would not have been achievable for any price in either 1965 or 1985. Thus, traditional metrics almost completely ignore the steep deflation rate for information technology, which in the case of computation, genetic sequencing, and many other areas is on the order of 50 percent per year. Of this ever improving price-performance, part goes into price and part into performance, so we get ever better products for lower prices.

As a personal example, when I attended MIT in 1965, the school was so advanced that it actually *had* computers. The most notable of them, an IBM 7094, had 150,000 bytes of "core" storage and a quarter of a MIPS (million instructions per second) of computing speed. It cost $3.1 million (in 1963 dollars, which is $30 million in 2023 dollars) and was shared by thousands of students and professors.[205] By comparison, the iPhone 14 Pro, released while this book was being written, cost $999 and could achieve up to 17 trillion operations per second for AI-related applications.[206] This is not a perfect comparison— part of the iPhone's cost includes features irrelevant to the 7094, like cameras, and the iPhone's speed is significantly slower than that figure for many uses—but the overall point is clear. At least roughly, the iPhone is 68 million times faster than the 7094 and less than one 30,000th the cost. In terms of price-performance (speed per dollar), this is a staggering two-*trillion*-fold improvement.

This rate of improvement continues unabated. It is not fundamentally dependent on Moore's law, the famous exponential reduction of feature sizes on microchips. MIT bought the transistor-based IBM 7094 in 1963, before microchip-based computers had been widely introduced—and two years before Gordon Moore publicly articulated his eponymous law in a landmark 1965 article.[207] As influential as that law has been, it is just one exponential computing paradigm among several—so far including electromechanical relays, vacuum tubes, transistors, and integrated circuits, with more to come in the future.[208]

In many ways, just as important as computation power is information. As a teenager I saved up for years from the earnings of my paper route to buy a set of the *Encyclopedia Britannica* for several thousand dollars, and that counted for thousands of dollars of GDP. By contrast, a teenager today with a smartphone has access to a vastly superior encyclopedia in Wikipedia—one that counts for nothing in economic activity because it's free. While Wikipedia does not consistently display the same editorial quality as *Encyclopedia Britannica*, it has several striking advantages: comprehensiveness (the main English-language Wikipedia is about a hundred times the size of *Britannica*), timeliness (updates within minutes of breaking news, versus years for a print encyclopedia), multimedia (integration of audiovisual content for many articles), and hypertextuality (hyperlinks that richly connect articles to one another).[209] It also frequently links readers to *Britannica*-quality sources if additional academic rigor is needed. And that's just one of thousands of free information products and services, all of which count for zero toward GDP.

In response to this argument, Lagarde countered that, yes, digital technology does have many remarkable qualities and implications, but you can't eat information technology, you can't wear it, and you can't live in it. My response: this will all change over the next decade. Certainly manufacturing and transporting these types of resources has grown more efficient thanks to ever-improving hardware and software, but we are about to enter an era in which goods like food and clothing are not simply being made more economical by information technology, but are themselves actually *becoming* information technologies—as resource and production costs fall as a result of automation and artificial intelligence taking on dominant roles in production.[210] Such goods will, therefore, be subject to the same high deflation rates that we see for other information technologies.

During the late 2020s we will start to be able to print out clothing and other common goods with 3D printers, ultimately for pennies per pound. One of the key trends in 3D printing is miniaturization: designing machines that can create ever smaller details on objects. At some point, the traditional 3D-printing paradigms, like extrusion (similar to an ink-jet), will be replaced by new approaches for manufacturing at even tinier scales. Probably sometime in the 2030s this will cross into nanotechnology, where objects can be created with atomic precision. Eric Drexler's estimate in his 2013 book *Radical Abundance* is that, taking into account more mass-efficient nanomaterials, atomically precise manufacturing could build most kinds of objects for the equivalent of about twenty cents per kilogram.[211] Although these figures remain speculative, the cost reduction will surely be enormous. And just as we have seen with music, books, and movies, there will be many free designs available alongside proprietary ones. Indeed, the coexistence of an open-source market—which is and will be a great leveler—with the proprietary market will become the defining nature of the economy in an increasing number of areas.

As I will explain later in this chapter, we will soon produce high-quality, low-cost food using vertical agriculture with AI-controlled production and chemical-free harvesting. And meat grown cleanly and ethically from cell cultures will displace environmentally devastating factory farming. In 2020 humans slaughtered more than 74 billion land animals for meat, whose products collectively weighed an estimated 371 million tons.[212] The United Nations estimates that this accounts for over 11 percent of all annual greenhouse gas emissions by human civilization.[213] The technology currently known as lab-grown meat has the potential to radically change that. Meat taken from animal carcasses has several major disadvantages: it inflicts suffering on innocent creatures, it is often unhealthy for humans, and it causes severe environmental impacts through both toxic pollution and carbon emissions. Growing meat from cultured cells and tissues can solve all those problems. No living animals would suffer, it could be designed to be both healthier and better-tasting, and harm to the environment could be minimized with ever-cleaner technology. The tipping point may be realism. As of 2023 the technology can replicate meats without much structure, like the texture of ground beef, but it isn't yet ready to generate full filet mignon steaks from scratch. When cultured meat can convincingly imitate all its animal-based counterparts, however, I expect that most people's discomfort with it will quickly diminish.

As I will also describe in more detail later, we will soon be able to inexpensively produce modules to construct houses and other buildings, making comfortable habitation newly affordable for millions. All these technologies have already been successfully demonstrated and will become increasingly advanced and mainstream in this decade. Alongside this revolution in the physical world will be a transformative next generation of virtual and augmented reality, sometimes known as the metaverse.[214] For many years the metaverse was largely unknown outside of science fiction and futurism circles, but the concept got a big boost into the public consciousness from Facebook's 2021 rebranding as Meta and its announcement that the centerpiece of its long-term strategy is playing a key role in constructing the metaverse—such that right now many people mistakenly think that Meta invented the concept.

Much like the internet is an integrated and persistent environment of web pages, the VR and AR of the late 2020s will merge into a compelling new layer to our reality. In this digital universe, many products won't even need a physical form at all, as simulated versions will perform perfectly well in highly realistic detail. Examples include a full virtual meeting, with the ability to interact with coworkers and easily collaborate as though together in person; a virtual concert, with the fully immersive auditory experience of sitting in a symphony hall; and a sensory-rich virtual beach vacation for the whole family, with the sounds, sights, and smells of sand and sea.

At present, most media is limited to engaging two senses: sight and hearing. Current virtual reality systems that incorporate smells or tactile sensations are still clunky and inconvenient. But over the next couple of decades, brain–computer interface technology will become much more advanced. Ultimately this will allow full-immersion virtual reality that feeds simulated sensory data directly into our brains. Such technology will bring major and hard-to-predict changes to how we spend our time and the experiences we prioritize. It will also force us to reconsider why we do what we do. For example, when we can safely experience all the challenge and natural beauty of climbing Mount Everest virtually, people will have to wrestle with whether it's worth doing the real thing—or whether the danger was part of the attraction all along.

Lagarde's last challenge to me was that land is not going to become an information technology, and that we are already very crowded. I replied that we are crowded because we chose to crowd together in dense groups. Cities came about to make possible our working and playing together. But try taking a train trip anywhere in the world and you will see that almost all of the habitable land remains unoccupied—only 1 percent of it is built up for human living.[215] Only about half of the habitable land is directly used by humans at all, almost all of it dedicated to agriculture—and among agricultural land, 77 percent is used for livestock, grazing, and feed, with only 23 percent for crops for human consumption.[216] Cultured meat and vertical farming will enable these needs to be met with only a small fraction of the land we now use. This will make possible a greater abundance of healthy food and accommodate an increasing population while still freeing up huge areas of land. Autonomous vehicles will facilitate less-crowded land use by making longer commutes practical, and we will increasingly be able to live wherever we want while still being able to work and play together in virtual and augmented spaces.

This transition is already underway, accelerated by the social changes necessitated by COVID-19. At the peak during the pandemic, up to 42 percent of Americans were working from home.[217] This experience will likely have a long-term impact on how both employees and employers think about work. In many cases the old model of nine-to-five sitting at a desk in a company office has been obsolete for years, but inertia and familiarity made it hard for society to change until the pandemic forced us to. As the LOAR takes information technologies into the steep parts of these exponential curves and AI matures over the coming decades, such impacts will only accelerate.

RENEWABLE ENERGY IS APPROACHING COMPLETE REPLACEMENT OF FOSSIL FUELS

One of the most important transitions arising from the exponential progress of the 2020s is in energy, because energy powers everything else. Solar photovoltaics are already cheaper than fossil fuels in many cases, with costs rapidly declining. But we need advances in materials science to achieve further improvements in cost-efficiency. AI-assisted breakthroughs in nanotechnology will increase cell efficiency by enabling photovoltaic cells to capture energy from more of the electromagnetic spectrum. Exciting developments are underway in this area. Putting tiny structures called nanotubes and nanowires inside solar cells can steadily improve their ability to absorb photons, transport electrons, and generate electric currents.[218] Likewise, placing nanocrystals (including quantum dots) inside cells can increase the amount of electricity generated per photon of sunlight absorbed.[219]

Another nanomaterial called black silicon has a surface composed of a vast number of atomic-scale needles smaller than the wavelength of light.[220] This nearly eliminates reflections from a cell, ensuring that more of the incoming photons create electricity. Princeton researchers have developed an alternative method of maximizing electricity by using a nanoscale mesh of gold atoms just 30 billionths of a meter thick to trap photons and increase efficiency.[221] Meanwhile, a project at MIT has created photovoltaic cells from sheets of graphene, a special form of carbon that is only one atom thick (less than a nanometer).[222] Such technologies will allow photovoltaics of the future to be thinner, lighter, and installable on more surfaces. For example, companies like Solar-Window Technologies have pioneered thin photovoltaic films that can coat windows, producing useful electricity without blocking the view.[223]

In the years ahead, nano-based technology will also reduce manufacturing costs by facilitating 3D printing of solar cells, which will make decentralized production possible so photovoltaics can be created when and where they are needed. And unlike the big, clumsy, rigid panels used today, cells built with nanotech can take many convenient forms: rolls, films, coatings, and more. This will reduce installation costs and give more communities around the world access to cheap, abundant solar power.

In 2000 renewables (mainly solar, wind, geothermal, tidal, and biomass, but not hydroelectric) accounted for about 1.4 percent of global electricity generation.[224] By 2021 it had risen to 12.85 percent, for an average doubling time of about 6.5 years during that span.[225] Doubling is faster in absolute terms, since the total amount of power generation is itself growing—from the equivalent of about 218 terawatt-hours in 2000 to 3,657 terawatt-hours in 2021, with a doubling time of around 5.2 years.[226]

This progress will continue exponentially as further cost reductions are driven by AI applied to material discovery and device design. At this rate, renewables could theoretically achieve complete coverage of world electricity needs by 2041. But when projecting into the future, it's not all that useful to think of renewables in the aggregate, as they aren't all getting cheaper at the same rate.

Costs of solar electricity generation are falling quite a bit faster than those of any other major renewable, and solar has the most headroom to grow. The closest competition to solar in price declines is wind, but solar has been falling roughly twice as fast as wind over the past five years.[227] Further, solar has a lower potential floor because materials science advances directly translate to cheaper and more efficient panels, and current technology captures only a fraction of the theoretical maximum—typically somewhere around 20 percent of incoming energy out of a theoretical limit around 86 percent.[228] Although this ceiling won't ever be reached in practice, there is nonetheless a lot of potential for improvement. By contrast, typical wind power systems reach about 50 percent efficiency, which is much closer to the theoretical maximum limit of 59 percent.[229] So no matter how we innovate, these systems just can't get all that much better.

Solar made up about 3.6 percent of worldwide electricity in 2021.[230] We need to use only roughly 1 part in 10,000 of the free sunlight that hits the earth to meet 100 percent of our total energy needs at current levels of use. The earth is constantly bathed in approximately 173,000 terawatts of energy from the sun.[231] Most of this cannot practically be captured for the foreseeable future, but even with contemporary technology, solar energy is more than enough to meet humanity's needs. A 2006 estimate by US government scientists put the total available energy with the best of then-current technology at up to 7,500 terawatts.[232] This is equivalent to 65,700,00 terawatt-hours over an entire year. For comparison, in 2021 total primary energy use around the world was the equivalent of 165,320 terawatt-hours.[233] This includes electricity, heating, and all fuel.

Solar also has a much faster doubling time as a proportion of the world total than other renewable sources: on average, just under every twenty-eight months from 1983 to 2021—even as the total amount of electricity generation has increased by around 220 percent in that same span in absolute terms.[234] From 3.6 percent in 2021, it would take only about 4.8 doublings to reach 100 percent, which would put us at 2032 to meet all of our energy needs from solar alone. While this doesn't mean that solar will literally reach total adoption (due to a mixture of economic and political obstacles), it's clear that it's on a trajectory toward truly transformative impact.

It is convenient that some of the world's largest unused areas—deserts—also tend to be the best places for solar electricity. For example, there have been ongoing proposals to cover a small but significant fraction of the Sahara Desert with photovoltaic panels. These could generate enough electricity to power all of Europe (via cables under the Mediterranean) and Africa.[235]

One of the main challenges to scaling up solar electricity like this is the need for more effective energy storage technology. The advantage of fossil fuels is that we can store them and then burn them whenever we need to generate electricity. But the sun shines only during the daytime, and solar intensity varies with the seasons. Hence the necessity of efficient means of storing solar energy when it is generated so we can use it later (hours later or months later, depending on the circumstances) when people need to use that power.

Fortunately, we are starting to make exponential gains in the price efficiency and quantity of energy storage as well. Note that these are not fundamental, persistent exponential trends like the law of accelerating returns, as improvement and expansion of energy storage is not dominated by the creation of feedback loops. But energy storage price efficiency and total use are sharply increasing as a consequence of soaring renewables usage—which, in the case of photovoltaics especially, does benefit indirectly from the LOAR via the role of information technology in unlocking new advances in materials science. As investment pours into renewables and the cost of renewables falls, this is pulling resources and innovation efforts into storage, because energy storage is so important to the ability of renewables to compete with fossil fuels for the lion's share of electricity generation. Continuing exponential gains will also be enabled by convergent advances in materials science, robotic manufacturing, efficient shipping, and energy transmission. The implication is that solar will dominate sometime during the 2030s.

Many approaches now being developed appear promising, but it is not yet clear which will prove most effective at the huge scale we'll need. Since electricity itself cannot be effectively stored, it needs to be converted into other kinds of energy until it's needed. Options include transforming it into heat energy with molten salts, into gravitational potential energy in water pumped into an elevated reservoir, into rotational energy in a fast-spinning flywheel, or into chemical energy in hydrogen that is produced with electricity and then burned cleanly on demand.[236]

And while most batteries aren't suitable for utility-scale storage, advanced batteries using lithium ions and several other chemistries are now rapidly increasing in cost-effectiveness. For example, between 2012 and 2020, lithium-ion storage costs fell roughly 80 percent per megawatt-hour and are projected to continue declining.[237] As these costs continue to decline with new innovations, renewables can supplant fossil fuels as the backbone of the grid.[238]

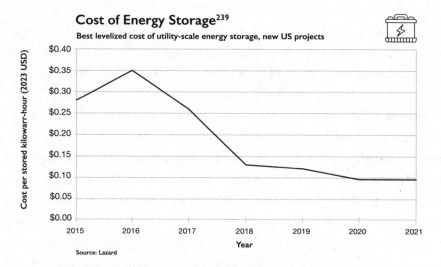

Cost of Energy Storage[239]
Best levelized cost of utility-scale energy storage, new US projects

Source: Lazard

Because energy storage technologies are used at many different stages of the electricity generation and consumption process, and in many different economic contexts, it is very difficult to compare the costs of storage from one project to another. Perhaps the most rigorous analysis to date comes from the financial advisory firm Lazard, which uses the levelized cost of storage (LCOS) metric. This is designed to encompass all costs (including capital costs), divided over total stored energy (in megawatt-hours or equivalent) expected to be discharged over a project's lifetime. In order to best reflect the cutting edge of technological progress, this chart shows the best LCOS available among new US utility-scale energy storage projects opened each year. Keep in mind that average LCOS figures are higher for a given year but follow similar trends—so the best LCOS in one year may be the average LCOS several years later.

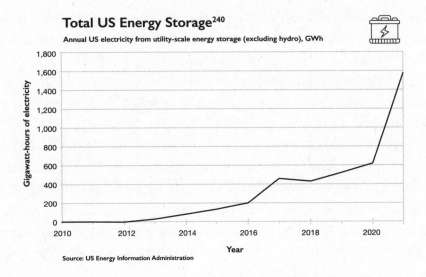

Total US Energy Storage[240]

Annual US electricity from utility-scale energy storage (excluding hydro), GWh

Source: US Energy Information Administration

WE ARE APPROACHING CLEAN WATER ACCESS FOR EVERYONE

A key challenge of the twenty-first century will be making certain that earth's growing population has a reliable supply of clean, fresh water. In 1990, about 24 percent of the world's people did not have regular access to relatively safe sources of drinking water.[241] Thanks to development efforts and advancing technology, the figure is now down to somewhere around 1 in 10.[242] That is still a large problem, however. According to the Institute for Health Metrics and Evaluation, around 1.5 million people around the world, including 500,000 young children, died in 2019 from diarrheal disease—mostly via drinking water contaminated by the bacteria in feces.[243] These diseases include cholera, dysentery, and typhoid fever and are particularly deadly to children.

The problem is that much of the world still lacks the infrastructure to collect fresh water, keep it clean, and deliver it to homes for people to use for drinking, cooking, washing, and bathing. Constructing huge networks of wells, pumps, aqueducts, and pipes is expensive, and many developing countries cannot afford to do so. In addition, civil wars and other political problems sometimes make large infrastructure projects impractical. As a result, centralized water purification and distribution systems like those in developed countries are not a viable solution for most of that last tenth. The alternative is technologies that enable people to purify water in their local neighborhoods, or even on their own.

In general, decentralized technologies will define the 2020s and beyond in many areas, including energy production (solar cells), food production (vertical agriculture), and production of everyday objects (3D printing). For water purification, this approach can take several forms, ranging from building-size machines like the Janicki Omni Processor that purify water for an entire village to portable filters like the LifeStraw that individuals can use.[244]

Some purification cells use heat—from solar energy or burning fuels—to boil water before use. Boiling kills disease-causing bacteria but does not remove other toxic pollutants, and it is easy for the water to become recontaminated if it is not consumed immediately after boiling. Adding antibacterial chemicals to water can prevent recontamination, but it still does not remove other toxins. In recent years, some portable water purification units have used electricity to turn oxygen in the air into ozone, a gas that can be passed through water to kill pathogens very efficiently.[245] Others kill bacteria and viruses by shining strong ultraviolet light through the water. But this, likewise, does not protect against chemical pollution.

An alternate method is filtration. For many years, filtration technology was able to get most but not all organisms and toxins out of water. Many of the deadliest viruses are so small that they pass through the holes in ordinary filters.[246] Similarly, the molecules of some pollutants cannot be blocked by normal filtration.[247] Yet recent innovations in materials science are creating filters that block smaller and smaller toxins. In the coming years, nanoengineered materials will enable filters to work faster and be very inexpensive.

An especially promising emerging technology is the Slingshot water machine, invented by Dean Kamen (born 1951).[248] It is a relatively compact device—about the size of a small refrigerator—that can produce totally pure water that meets the standards for an injectable liquid, from any source, including sewer water and contaminated swamp water. The Slingshot requires less than one kilowatt of electricity to operate. It uses vapor compression distillation (turning the input water into steam, leaving contaminants behind) and requires no filters. The Slingshot is intended to be powered by a very adaptable kind of engine called a Stirling engine, which can produce electricity from any heat source, including burning cow dung.[249]

VERTICAL AGRICULTURE WILL PROVIDE INEXPENSIVE, HIGH-QUALITY FOOD AND FREE UP THE LAND WE USE FOR HORIZONTAL AGRICULTURE

Most archaeologists estimate that the birth of human agriculture took place around 12,000 years ago, but there is some evidence that the earliest agriculture may date as far back as 23,000 years.[250] It is possible that future archaeological discoveries will revise this understanding even further. Whenever agriculture began, the amount of food that could then be grown from a given area of land was quite low. The first farmers sprinkled seeds into the natural soil and let the rain water them. The result of this inefficient process was that the vast majority of the population needed to work in agriculture just to survive.

By around 6,000 BC, irrigation enabled crops to receive more water than they could get from rain alone.[251] Plant breeding enlarged the edible parts of plants and made them more nutritious. Fertilizers supercharged the soil with substances that promote growth. Better agricultural methods allowed farmers to plant crops in the most efficient arrangements possible. The result was that more food became available, so over the centuries more and more people could spend their time on other activities, like trade, science, and philosophy. Some of this specialization yielded further farming innovation, creating a feedback loop that drove even greater progress. This dynamic made our civilization possible.

A useful way of quantifying this progress is crop density: how much food can be grown in a given area of land. For example, corn production in the United States uses land more than seven times as efficiently as a century and a half ago. In 1866, US corn farmers averaged an estimated 24.3 bushels per acre, and by 2021 this had reached 176.7 bushels per acre.[252] Worldwide, land efficiency improvement has been roughly exponential, and today we need, on average, less than 30 percent of the land that we needed in 1961 to grow a given quantity of crops.[253] This trend has been essential to enabling the global population increases in that time and has spared humanity the mass starvation from overpopulation that many people worried about when I was growing up.

Further, because crops are now grown at extremely high density, and machines do a lot of the work that used to be done by hand, one farmworker can grow enough food to feed about seventy people. As a result, farmwork has gone from constituting 80 percent of all labor in the United States in 1810 to 40 percent in 1900 to less than 1.4 percent today.[254]

Yet crop densities are now approaching the theoretical limit of how much food can be grown in a given outdoor area. One emerging solution is to grow multiple stacked layers of crops, referred to as vertical agriculture.[255] Vertical farms take advantage of several technologies.[256] Typically they grow crops hydroponically, meaning that instead of being grown in soil, plants are raised indoors in trays of nutrient-rich water. These trays are loaded into frames and stacked many stories high, which means that excess water from one level can trickle down to the next instead of being lost as runoff. Some vertical farms now use a new approach called aeroponics, where the water is replaced with a fine mist.[257] And instead of sunlight, special LEDs are installed to ensure that each plant gets the perfect amount of light. Vertical farming company Gotham Greens, which has ten large facilities ranging from California to Rhode Island, is one of the industry leaders. As of early 2023, it had raised $440 million in venture funding.[258] Its technology enables it to use "95 percent less water and 97 percent less land than a traditional dirt farm" for a given crop yield.[259] Such efficiencies will both free up water and land for other uses (recall that agriculture is currently estimated to take up around half of the world's habitable land) and provide a much greater abundance of affordable food.[260]

Vertical farming has other key advantages. By preventing agricultural runoff, it does away with one of the main causes of pollution in waterways. It avoids the need for farming loose soil, which gets blown into the air and diminishes air quality. It makes toxic pesticides unnecessary, as pests are unable to enter a properly designed vertical farm. This approach also makes it possible to raise crops year-round, including species that could not grow in the local outdoor climate. That likewise prevents crop losses due to frosts and bad weather. Perhaps most importantly, it means that cities and villages can grow their own food locally instead of bringing it in by trains and trucks from hundreds or even thousands of miles away. As vertical agriculture becomes less expensive and more widespread, it will lead to great reductions in pollution and emissions.

In the coming years, converging innovations in photovoltaic electricity, materials science, robotics, and artificial intelligence will make vertical farming much less expensive than current agriculture. Many facilities will be powered by efficient solar cells, produce new fertilizers on-site, collect their water from the air, and harvest the crops with automated machines. With very few workers required and a small land footprint, future vertical farms will eventually be able to produce crops so cheaply that consumers may be able to get food products almost for free.

This process mirrors what took place in information technology as a result of the law of accelerating returns. As computing power has gotten exponentially cheaper, platforms like Google and Facebook

Lettuce growing in stacked layers in a vertical farm.
Photo credit: Valcenteu, 2010.

have been able to provide their services to users for free while paying
for their own costs through alternative business models such as advertising. By using automation and AI to control all aspects of a vertical

farm, vertical agriculture represents turning food production essentially into an information technology.

3D PRINTING WILL REVOLUTIONIZE THE CREATION AND DISTRIBUTION OF PHYSICAL THINGS

For most of the twentieth century, manufacturing three-dimensional solid objects usually took two forms. Some processes involved shaping material inside a mold, such as injecting molten plastic into a tooling, or shaping heated metal in a press. Other processes involved selectively removing material from a block or sheet, much like a sculptor chipping away at a marble block to carve a statue. Both of these methods have major disadvantages. Creating molds is very expensive, and the molds are quite hard to modify once completed. By comparison, so-called subtractive manufacturing wastes a lot of material and is unable to produce certain shapes.

In the 1980s, though, a new family of technologies began to emerge.[261] Unlike previous methods, they created parts by stacking or depositing relatively flat layers and building them up into a three-dimensional shape. These techniques have come to be known as additive manufacturing, three-dimensional printing, or 3D printing.

The most common types of 3D printers work somewhat like an ink-jet printer.[262] A typical ink-jet passes back and forth over a piece of paper, squirting ink from a cartridge out of a nozzle in the places software directs it to. Instead of ink, 3D printers use a material like plastic and heat it until it is soft. Their nozzles deposit the material, following a software-run pattern for each layer, repeating the process many times as the object gradually becomes more three-dimensional. The layers fuse together as they harden, and the finished object is ready to use. Over the past two decades 3D printing has continued to advance in terms of higher resolutions, reductions in cost, and increases in speed.[263] 3D-printing systems can now create objects out of a wide variety of materials, including paper, plastic, ceramic, and metal. As

3D-printing technology advances, it will be able to handle even more exotic materials. For example, medical implants may be created with drug molecules built in to be gradually released into the body. Nano-materials like graphene could be used to create lightweight bulletproof clothing and superfast electronics. 3D printing can also benefit from advances in artificial intelligence, such as software that can optimize an object's strength, aerodynamic shape, or other properties, and even create designs requiring shapes that would be impossible to manufacture with contemporary methods.

New, intuitive software is making it easier for people to create 3D-printed parts without advanced training. As 3D printing has become more widespread, it has begun to revolutionize the manufacturing industry. One major advantage is that it enables inexpensive and fast prototyping. Engineers can design a new part on a computer and hold a 3D-printed model in their hands within minutes or hours—a process that might have taken weeks with previous technology. This allows for rapid cycles of testing and modification for a fraction of the cost of old methods. As a result, people with good ideas but relatively little money can bring their innovations to the marketplace and benefit society.

Another key advantage of 3D printing is that it permits levels of customization that are not practical with mold-based manufacturing. Even a slight modification usually requires an entirely new mold, which can cost tens of thousands of dollars or more. By contrast, even major changes to a 3D-printing design carry no additional cost. As a result, inventors can have exactly the right parts they need to innovate, and consumers can affordably access products designed especially for them. One example among many is producing shoes made to the exact measurements of a customer's feet for greatly enhanced fit and comfort. A leading 3D-printed footwear company is FitMyFoot, which lets customers use an app to take photos of their feet that are automatically converted into measurements for the printing process.[264] Similarly, furniture can be molded to fit every body type, and tools can be made to precisely fit your own hand.[265] Of even greater importance, vital medical implants will be cheaper and more effective.[266]

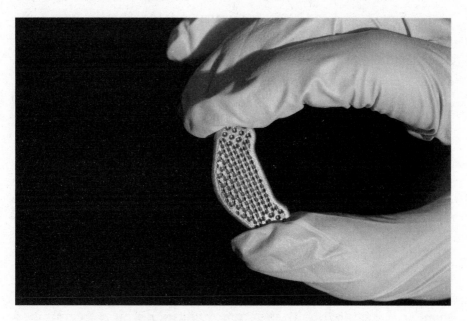

A 3D-printed titanium spinal disc that can be implanted in patients with spinal damage or disease.
Photo credit: FDA photo by Michael J. Ermarth, 2015.

In addition, 3D printing allows manufacturing to be decentralized, empowering consumers and local communities. This contrasts with the paradigm that developed during the twentieth century, in which manufacturing is largely concentrated in giant corporate factories in major cities. Under this model, small towns and developing countries must buy their products from far away, and shipping is expensive and time-consuming. Decentralized manufacturing will also have significant environmental benefits. Shipping products from factories to consumers hundreds or thousands of miles away generates enormous emissions. According to the International Transport Forum, freight shipping accounts for around 30 percent of all carbon emissions from fuel burning.[267] Decentralized 3D printing can make much of that unnecessary.

Each year the resolution of 3D printing is improving and the technology is getting cheaper.[268] As resolution improves (that is, the size of the smallest attainable design features shrinks) and costs fall, the range of goods that can be economically printed will grow. For example, many common fabrics have fibers with diameters that are 10 to 22 microns (millionths of a meter) wide.[269] Some 3D printers can already achieve resolution of 1 micron or less.[270] Once the technology can achieve fabric-like diameters with fabric-like materials at similar prices to regular fabric, it will be economically feasible to print out any clothing we wish.[271] Because printing speed is also increasing, high-volume manufacturing will become more practical.[272]

In addition to manufacturing of everyday goods like shoes and tools, new research is applying 3D printing to biology. Scientists are currently testing techniques that will make possible the printing of human body tissues and, ultimately, whole organs.[273] The general principle involves a biologically inactive material, such as synthetic polymer or ceramic, printed into a three-dimensional "scaffold" in the shape of the desired body structure. Fluid rich with reprogrammed stem cells is then deposited over the scaffold, where the cells multiply and fill in the appropriate shape, thereby creating a replacement organ with the patient's own DNA. United Therapeutics (a company for which I am a board member) is applying this approach (and others) to someday grow entire lungs, kidneys, and hearts.[274] This method will ultimately be far superior to transplanting organs from one person to another, which has profound limitations in terms of availability and incompatibility with a patient's immune system.[275]

One potential drawback of 3D printing is that it could be used to manufacture pirated designs. Why pay $200 for a pair of designer shoes if you can download the file and print them for yourself at a fraction of the cost? We are already facing similar issues with the intellectual property of music, books, movies, and other creative forms. All of this requires new approaches to protect intellectual property.[276]

Another troubling implication is that decentralized manufacturing will allow civilians to create weapons that they otherwise couldn't easily access. Files are already circulating on the internet that enable people to print the parts to assemble their own guns.[277] This will present a challenge for gun control and allow the creation of firearms with no serial numbers, making it more difficult for law enforcement to trace crimes. 3D-printed guns made from advanced plastics could even be used to bypass metal detectors. This will require a thoughtful reevaluation of current regulations and policies.

3D PRINTING OF BUILDINGS

Three-dimensional printing is usually associated with manufacturing small objects, such as tools or medical implants, but it can also be used to create larger structures, like buildings. This technology is rapidly advancing through prototype stages, and as 3D-printed structures become less expensive to produce, they will become a commercially viable alternative to current construction methods. Ultimately, 3D printing of both building modules and the smaller objects that go into a building will dramatically lower the construction costs of homes and offices.

There are two main approaches to 3D printing a building. The first is to create parts or modules that are subsequently put together—much like how people buy furniture parts from IKEA and assemble the pieces on their own.[278] In some cases this means printing items like wall sections and roof segments and then connecting them at the construction site—similar to snapping together Lego parts. By the late 2020s it will be possible for this assembly of modules to be done largely by robots.

Another approach is printing the structures of entire rooms or modular structures.[279] These modules usually have square or rectangular footprints and can fit together in many different configurations. At the construction site, they can be lifted into position by cranes and assembled quickly. This minimizes the disruption and nuisance that construction normally causes to the surrounding area. In 2014 the

Chinese firm WinSun demonstrated building ten simple modular houses in twenty-four hours, at a cost of less than $5,000 each.[280] China is already a hub of 3D-printed buildings and will have much need for more mature versions of this technology over the coming decades.

An alternative method is to print a whole custom-designed building as a single module.[281] Engineers set up a large frame around the area where the building will be and the printing nozzle robotically moves around within that frame, depositing layers of material (for example, concrete) in the shape of the walls. Very little human labor is needed for the main structural construction, but after it is complete, workers can go in to finish the inside of the building and add elements such as windowpanes and roof tiles. For example, in 2016 the HuaShang Tengda company announced completion of a two-story villa that had been printed in one piece in forty-five days.[282] Just as I write this, this technology is spreading to the United States, where in 2021 a company called Alquist 3D completed the first owner-occupied 3D-printed home, and 2023 saw the first multistory construction for a home in Houston.[283] By the late 2020s, combining 3D printing of large and small objects with intelligent robotics will increase the ability to personalize buildings while at the same time dramatically reducing costs.

3D printing of building modules has several key advantages, which will become even stronger as the technology develops. First, it cuts labor costs, which will enable basic housing to become more affordable. It also shortens construction times, which reduces the environmental impact that prolonged construction causes. This includes reducing factors such as waste and garbage, light and noise pollution, toxic dust, traffic disruption, and hazards to workers. In addition, 3D printing makes it easier to construct buildings out of materials that are readily and locally available instead of using resources that might be hundreds of miles away, like timber and steel.

In the future, 3D printing may be used to make skyscrapers easier and cheaper to build. One of the main challenges of high-rise construction is getting people and building materials to the upper floors. A 3D-printing system, together with autonomous robots that can use building materials pumped up from ground level in liquid form, will make this process far easier and less expensive.

DILIGENT PEOPLE WILL ACHIEVE LONGEVITY ESCAPE VELOCITY BY AROUND 2030

Material abundance and peaceful democracy make life better, but the challenge with the highest stakes is the effort to preserve life itself. As I describe in chapter 6, the method of developing new health treatments is rapidly changing from a linear hit-or-miss process to an exponential information technology in which we systematically reprogram the suboptimal software of life.

Biological life is suboptimal because evolution is a collection of random processes optimized by natural selection. Thus, as evolution has "explored" the range of possible genetic traits, it has depended heavily on chance and the influence of particular environmental factors. Also, the fact that this process is gradual means that evolution can achieve a design only if all the intermediate steps toward a given feature also lead creatures to be successful in their environments. So there are surely some potential traits that would be very useful but that are inaccessible because the incremental steps needed to build them would be evolutionarily unfit. By contrast, applying intelligence (human or artificial) to biology will allow us to systematically explore the full range of genetic possibilities in search of those traits that are optimal—that is, most beneficial. This includes those inaccessible to normal evolution.

We have now had about two decades of exponential progress in genome sequencing (approximately doubling price-performance each year) from the completion of the Human Genome Project in 2003—and in terms of base pairs, this doubling has occurred on average roughly every fourteen months, spanning multiple technologies and dating all the way back to the first nucleotide sequencing from DNA in 1971.[284] We are finally getting to the steep part of a fifty-year-old exponential trend in biotechnology.

We are beginning to use AI for discovery and design of both drugs and other interventions, and by the end of the 2020s biological simulators will be sufficiently advanced to generate key safety and efficacy data in hours rather than the years that clinical trials typically require. The transition from human trials to simulated *in silico* trials will be governed by two forces working in opposite directions. On the one hand there will be a legitimate concern over safety: we don't want the simulations to miss relevant medical facts and erroneously declare a dangerous medication to be safe. On the other hand, simulated trials will be able to use vastly larger numbers of simulated patients and study a wide range of comorbidities and demographic factors—telling doctors in granular detail how a new treatment will likely affect many different kinds of patients. In addition, getting lifesaving drugs to patients faster may save many lives. The transition to simulated trials will also involve political uncertainty and bureaucratic resistance, but ultimately the effectiveness of the technology will win out.

Just two notable examples of the benefits *in silico* trials will bring:

- Immunotherapy, which is enabling many stage 4 (and otherwise terminal) cancer patients to go into remission, is a very hopeful development in cancer treatment.[285] Technologies like CAR-T cell therapy reprogram a patient's own immune cells to recognize and destroy cancer cells.[286] So far, finding such approaches is limited by our incomplete biomolecular understanding of how cancer evades the immune system, but AI simulations will help break this logjam.

- With induced pluripotent stem (iPS) cells, we are gaining the capability to rejuvenate the heart after a heart attack and overcome the "low ejection fraction" from which many heart attack survivors suffer (and from which my father died). We are now growing organs using iPS cells (adult cells that are converted into stem cells via the introduction of specific genes). As of 2023, iPS cells have been used for the regeneration of tracheas, craniofacial bones, retinal cells, peripheral nerves, and cutaneous tissue, as well as tissues from major organs like the heart, liver, and kidneys.[287] Because stem cells are similar in some ways to cancer cells, an important line of research going forward will be finding ways to minimize the risk of uncontrolled cell division. These iPS cells can act like embryonic stem cells and can differentiate into almost all types of human cells. The technique is still experimental, but it has been successfully used in human patients. For those with heart issues, it entails creating iPS cells from the patient, growing them into macroscopic sheets of heart muscle tissue, and grafting them onto a damaged heart. The therapy is believed to work via the iPS cells' releasing growth factors that spur existing heart tissue to regenerate. In effect, they may be tricking the heart into thinking it is in a fetal environment. This procedure is being used for a broad variety of biological tissues. Once we can analyze the mechanisms of iPS action with advanced AI, regenerative medicine will be able to effectively unlock the body's own blueprints for healing.

As a result of these technologies, the old linear models of progress in medicine and longevity will no longer be appropriate. Both our natural intuition and a backward-looking view of history suggest that the next twenty years of advances will be roughly like the last twenty, but this ignores the exponential nature of the process. Knowledge that radical life extension is close at hand is spreading, but most people—both doctors and patients—are still unaware of this grand transformation in our ability to reprogram our outdated biology.

As mentioned earlier in this chapter, the 2030s will bring another health revolution, which my book on health (coauthored with Terry Grossman, MD) calls the third bridge to radical life extension: medical nanorobots. This intervention will vastly extend the immune system. Our natural immune system, which includes T cells that can intelligently destroy hostile microorganisms, is very effective for many types of pathogens—so much so that we would not live long without it. However, it evolved in an era when food and resources were very limited and most humans had short life spans. If early humans reproduced when young and then died in their twenties, evolution had no reason to favor mutations that could have strengthened the immune system against threats that mainly appear later in life, like cancer and neuro-degenerative diseases (often caused by misfolded proteins called prions). Likewise, because many viruses come from livestock, our evolutionary ancestors who existed before animal domestication did not evolve strong defenses against them.[288]

Nanorobots not only will be programmed to destroy all types of pathogens but will be able to treat metabolic diseases. Except for the heart and the brain, our major internal organs put substances into the bloodstream or remove them, and many diseases result from their malfunction. For example, type 1 diabetes is caused by failure of the pancreatic islet cells to produce insulin.[289] Medical nanorobots will monitor the blood supply and increase or decrease various substances, including hormones, nutrients, oxygen, carbon dioxide, and toxins, thus augmenting or even replacing the function of the organs. Using these technologies, by the end of the 2030s we will largely be able to overcome diseases and the aging process.

The 2020s will feature increasingly dramatic pharmaceutical and nutritional discoveries, largely driven by advanced AI—not enough to cure aging on their own, but sufficient to extend many lives long enough to reach the third bridge. And so, by around 2030, the most diligent and informed people will reach "longevity escape velocity"—a tipping point at which we can add more than a year to our remaining life expectancy for each calendar year that passes. The sands of time will start running in rather than out.

The fourth bridge to radical life extension will be the ability to es-sentially back up who we are, just as we do routinely with all of our dig-ital information. As we augment our biological neocortex with realistic (albeit much faster) models of the neocortex in the cloud, our thinking will become a hybrid of the biological thinking we are accustomed to today and its digital extension. The digital portion will expand expo-nentially and ultimately predominate. It will become powerful enough to fully understand, model, and simulate the biological portion, en-

abling us to back up all of our thinking. This scenario will become realistic as we approach the Singularity in the mid-2040s.

The ultimate goal is to put our destiny in our own hands, not in the
metaphorical hands of fate—to live as long as we wish. But why would
anyone ever choose to die? Research shows that those who take their
own lives are typically in unbearable pain, whether physical or emotional.[290] While advances in medicine and neuroscience cannot prevent all of those cases, they will likely make them much rarer.

Once we have backed ourselves up, how could we die, anyway? The
cloud already has many backups of all of the information it contains, a
feature that will be greatly enhanced by the 2040s. Destroying all copies of oneself may be close to impossible. If we design mind-backup
systems in such a way that a person can easily choose to delete their
files (hoping to maximize personal autonomy), this inherently creates
security risks where a person could be tricked or coerced into making
such a choice and could increase vulnerability to cyberattacks. On the
other hand, limiting people's ability to control this most intimate of
their data impinges on an important freedom. I am optimistic, though,
that suitable safeguards can be deployed, much like those that have
successfully protected nuclear weapons for decades.

If you restored your mind file after biological death, would you really be restoring *yourself*? As I discussed in chapter 3, that is not a scientific question but a philosophical one, which we'll have to grapple
with during the lifetimes of most people already alive today.

Finally, some have an ethical concern about equity and inequality.
A common challenge to these predictions about longevity is that only
the wealthy will be able to afford the technologies of radical life extension. My response is to point out the history of the cell phone. You indeed had to be wealthy to have a mobile phone as recently as thirty
years ago, and that device did not work very well. Today there are billions of phones, and they do a lot more than just make phone calls.
They are now memory extenders that let us access almost all of human
knowledge. Such technologies start out being expensive with limited
function. By the time they are perfected, they are affordable to almost

everyone. And the reason is the exponential price-performance improvement inherent in information technologies.

THE RISING TIDE

As I have argued in this chapter, contrary to many popular assumptions, life is getting better in profound and fundamental ways for the great majority of people on earth. More importantly, this isn't just a coincidence. The vast improvements we've seen over the past two centuries in areas like literacy and education, sanitation, life expectancy, clean energy, poverty, violence, and democracy are all powered by the same underlying dynamic: information technology facilitates its own advancement. This insight, which is the core of the law of accelerating returns, explains the virtuous circles that have transformed human life so dramatically. Information technology is about ideas, and exponentially improving our ability to share ideas and create new ones gives each of us—in the broadest possible sense—greater power to fulfill our human potential and to collectively solve many of the maladies that society faces.

Exponentially improving information technology is a rising tide that lifts all the boats of the human condition. And we are now about to enter the period when this tide surges upward as never before. The key to this is artificial intelligence, which is now allowing us to turn many kinds of linearly advancing technology into exponential information technology—from agriculture and medicine to manufacturing and land use. This force is what will make life itself exponentially better in the time ahead.

Humanity's journey toward easier, safer, and more abundant life for all has been progressing for years, decades, centuries, and millennia. We truly have trouble imagining what life was like even a century ago, let alone before that. Our accelerating progress, with substantial gains over the past few decades and profound evolution over the next few decades, will catapult us forward in this positive direction, far beyond what we can now imagine.

CHAPTER 5

THE FUTURE OF JOBS: GOOD OR BAD?

THE CURRENT REVOLUTION

The convergent technologies of the next two decades will create enormous prosperity and material abundance around the world. But these same forces will also unsettle the global economy, forcing society to adapt at an unprecedented pace.

In 2005, the year *The Singularity Is Near* was released, DARPA awarded a $2 million prize to a Stanford team that won its grand challenge race for autonomous vehicles.[1] At the time, self-driving cars were still science fiction as far as the public was concerned, and even many experts believed they were as far as a century away. But when Google launched an ambitious AI-driven project in 2009, progress began to rapidly accelerate. The project became an independent company called Waymo, which by 2020 was offering fully autonomous ride-hailing taxis to the public in the Phoenix area, followed by expansion to San Francisco.[2] By the time you read this, its passenger service will have expanded to Los Angeles and possibly other cities.[3]

Waymo's self-driving vehicles have traveled well over 20 million fully autonomous miles at the time of this writing (a figure that is rapidly increasing—one of the challenges in writing this book!).[4] This real-world experience provided a basis for Waymo to create and fine-tune a realistic simulator—a virtual environment that can re-create the many vagaries of driving.

These two modes have their own strengths and weaknesses, but together they serve to mutually reinforce each other. Real-world driving can be fully realistic and involves unexpected situations that engineers might never anticipate on their own and put into a simulated environment. But when the AI struggles with a real-world situation, engineers can't stop traffic and tell all the drivers around them, "Try that again, but this time five miles an hour faster."

By contrast, virtual-world driving allows high-volume testing that scientifically adjusts exactly the parameters needed to master a situation. Also, it's possible to simulate risky scenarios that it wouldn't be safe enough to train AI on in real-world driving. By having the AI tackle millions of different circumstances this way, engineers can identify the most important issues to address in real-world driving.

For perspective on the amount of data that simulation makes possible relative to reality, in 2018 Waymo started simulating as many driving miles every single day as it had accumulated up to then on real roads in the entire history of the project, going back to 2009. As of 2021, the most recent year for which data is available as I write this, that stunning ratio still holds—around 20 million simulated miles a day, versus more than 20 million real miles since its founding.[5]

As discussed in chapter 2, such simulation can generate sufficient examples to train deep (e.g., one-hundred-layer) neural nets. This is how Alphabet subsidiary DeepMind generated enough training examples to soar past the best humans in the board game Go.[6] Simulating the world of driving is far more complicated than simulating the world of Go, but Waymo uses the same fundamental strategy—and has now honed its algorithms with more than 20 *billion* miles of simulated driving[7] and generated enough data to apply deep learning to improve its algorithms.

If your job is driving a car, bus, or truck, this news is likely to give you pause. Across the United States, more than 2.7 percent of employed persons work as some kind of driver—whether driving trucks, buses, taxis, delivery vans, or some other vehicle.[8] According to the most recent data available, this figure accounts for over 4.6 million jobs.[9] While there is room for disagreement over exactly how quickly autonomous vehicles will put these people out of work, it is virtually certain that many of them will lose their jobs before they would have otherwise retired. Further, automation affecting these jobs will have uneven impacts all over the country. While in large states like California and Florida drivers account for less than 3 percent of the employed labor force, in Wyoming and Idaho the figure exceeds 4 percent.[10] In parts of Texas, New Jersey, and New York, the percentage rises to 5 percent, 7 percent, or even 8 percent.[11] Most of these drivers are men, most are middle-aged, and most do not have college educations.[12]

But autonomous vehicles won't just disrupt the jobs of people who physically drive behind the wheel. As truck drivers lose their jobs to automation, there will be less need for people to do truckers' payroll and for retail workers in roadside convenience stores and motels. There'll be less need for people to clean truck stop bathrooms, and lower demand for sex workers in the places truckers frequent today. Although we know in general terms that these effects will happen, it is very difficult to estimate precisely how large they will be or how quickly these changes will unfold. Yet it is helpful to keep in mind that transportation and transportation-related industries directly employ about 10.2 percent of US workers, according to the latest (2021) estimate of the Bureau of Transportation Statistics.[13] Even relatively small disruptions in a sector that large will have major consequences.

Yet driving is just one of a very long list of occupations that are threatened in the fairly near term by AI that exploits the advantage of training on massive datasets. A landmark 2013 study by Oxford University scholars Carl Benedikt Frey and Michael Osborne ranked about seven hundred occupations on their likelihood of being disrupted by the early 2030s.[14] At a 99 percent likelihood of being able to be automated were such job categories as telemarketers, insurance underwriters, and tax preparers.[15] More than half of all occupations had a greater than 50 percent likelihood of being automatable.[16]

High on that list were factory jobs, customer service, banking jobs, and of course driving cars, trucks, and buses.[17] Low on that list were jobs that require close, flexible personal interaction, such as occupational therapists, social workers, and sex workers.[18]

Over the decade since that report was released, evidence has continued to accumulate in support of its startling core conclusions. A 2018 study by the Organisation for Economic Co-operation and Development reviewed how likely it was for each task in a given job to be automated and obtained results similar to Frey and Osborne's.[19] The conclusion was that 14 percent of jobs across thirty-two countries had more than a 70 percent chance of being eliminated through automation over the succeeding decade, and another 32 percent had a probability of over 50 percent.[20] The results of the study suggested that about 210 million jobs were at risk in these countries.[21] Indeed, a 2021 OECD report confirmed from the latest data that employment growth has been much slower for jobs at higher risk of automation.[22] And all this research was done before generative AI breakthroughs like ChatGPT and Bard. The latest estimates, such as a 2023 report by McKinsey, found that 63 percent of all working time in today's developed economies is spent on tasks that could already be automated with today's technology.[23] If adoption proceeds quickly, half of this work could be automated by 2030, while McKinsey's midpoint scenarios forecast 2045—assuming no future AI breakthroughs. But we know AI is going to continue to progress—exponentially—until we have superhuman-level AI and fully automated, atomically precise manufacturing (controlled by AI) sometime in the 2030s.

Yet this is hardly the first time that people have been able to clearly see their jobs being likely to succumb en masse to automation. The story began two centuries ago when the weavers of Nottingham were threatened by the introduction of the power loom and other textile machines.[24] These laborers had enjoyed a modest living from their skillful production of stockings and lace in stable family businesses that had been passed down through the generations. But the technological innovations of the early nineteenth century shifted the industry's economic power into the hands of the machine owners, leaving the weavers in danger of losing their jobs.

It's not clear whether Ned Ludd actually existed, but legend has it that he accidentally broke textile factory machinery, and any equipment damaged thereafter—either mistakenly or in protest of automation— would be blamed on Ludd.[25] When the desperate weavers formed an urban guerrilla army in 1811, they declared General Ludd their leader.[26] These Luddites, as they were known, revolted against factory owners— they first directed their violence primarily at the machines, but bloodshed soon ensued. The movement ended with the imprisonment and hanging of prominent Luddite leaders by the British government.[27] Ned Ludd was never found.

The weavers had seen their entire livelihood upended. From their perspective, it was irrelevant that higher-paying jobs had been created to design, manufacture, and market the new machines. There were no government programs to retrain them, and they had spent their lives developing a skill that had become obsolete. Many were forced into lower-paying jobs, at least for a time. But a positive result of this early wave of automation was that the common person could now afford a well-made wardrobe rather than a single shirt. And over time, whole new industries were formed as a result of automation. The resulting prosperity was the primary factor that destroyed the original Luddite movement. Although the Luddites have passed into history, they've remained a powerful symbol of those who protest being left behind by technological progress.

DESTRUCTION AND CREATION

If I were a prescient futurist in 1900, I would have said to the labor force: "There are around 40 percent of you who work on farms (a figure that was over 80 percent in 1810) and a fifth of you who work in factories, yet I predict that by the year 2023 the portion of you working in manufacturing will fall by more than half (to 7.8 percent), and those working in agriculture by over 95 percent (to less than 1.4 percent)."[28]

I could have gone on to say: "You need not worry, though, because employment will actually go up rather than down. More jobs will be created than eliminated." If they then asked me, "What new jobs?" an honest answer would have been "I don't know—they haven't been invented yet. And they will be in industries that don't exist yet." That this is not a very satisfying response illustrates why political anxiety is associated with automation.

If I were really prescient, I would have told people in 1990 that new jobs would soon become available to create and operate websites and mobile applications, doing data analytics and online merchandising. But they wouldn't have had any idea what I was talking about.

Indeed, despite the dramatic reduction in many categories of employment, the total number of jobs has grown dramatically—both in absolute and proportional terms. In 1900 the total US workforce was around 29 million, comprising 38 percent of the population.[29] In early 2023 it was around 166 million, comprising over 49 percent of the population.[30]

Not only is the total number of jobs growing, but the workers who fill those jobs are working fewer hours and making more money. In the United States, the annual number of hours worked by each worker has fallen from just over 2,900 in 1870 to around 1,765 as of 2019 (just before the disruptions of COVID-19).[31] And despite the reduction in hours worked, workers' average annual earnings have multiplied more than fourfold in constant dollars since 1929.[32] That year, annual per capita personal income in the United States was about $700. Since only 48 million of the 122.8 million people in the country were employed, this translates to about $1,790 (roughly $31,400 in 2023 dollars) per worker.[33] In 2022, US personal income per capita was estimated at $64,100 across 332 million people, of which the labor force was around 164 million.[34] Thus, working Americans averaged annual earnings of around $129,800 (about $133,000 in 2023 dollars)—over four times as much as nine decades before.

Note, though, that while this average reflects huge gains in overall wealth across the country, median incomes (the income level that has an equal number of people earning more and less) are much lower.

There isn't reliable data for 1929, but in 2021 median income was $37,522, as compared with the $64,100 average.[35] Some of that difference reflects extremely high earnings by a minority of the population, as well as the large number of people who are retired, students, stay-at-home parents, or otherwise not employed.

If we look more narrowly at hourly earnings, a similar trend appears. The average American worker in 1929 was on the job for 2,316 hours per year.[36] With average earnings around $31,400 in 2023 dollars, that's about $13.55 per hour. By 2021 Americans earned roughly $10.8 trillion (2023 adjusted) in wages and salaries with 254 billion hours of work—around $42.50 an hour, and over triple the 1929 figure.[37]

In reality, the increase has been even larger. Some kinds of work aren't captured by the official wage statistics. For example, high-earning freelance computer programmers are not on payrolls. Nor are entrepreneurs or creative artists, who may get very large amounts of income from each hour worked. In 2021 total personal income in the United States was around $21.8 trillion (2023 adjusted), which implies hourly income roughly double the wage-and-salary figure.[38] But much personal income (e.g., renting out a property) doesn't translate well to hours worked, so the most accurate number falls somewhere between the two. As discussed in the previous chapter, these gains do not even take into account how many kinds of goods got much better for the same (inflation-adjusted) prices during that span, or the fact that consumers have access to countless new innovations.

Underlying much of this progress, technological change is introducing information-based dimensions to old jobs and creating millions of new jobs that did not exist a quarter century ago, let alone a hundred years ago, and that require novel and higher-level skills.[39] This has so far offset the massive destruction of agricultural and manufacturing jobs that once occupied the vast majority of the labor force.

At the beginning of the nineteenth century, the United States was an overwhelmingly agricultural society. As more settlers poured into the young nation and moved west of the Appalachians, the percentage of Americans employed in farming actually rose, peaking at over

80 percent.[40] But in the 1820s this proportion began a rapid decline as improved agricultural technology made it possible for fewer farmers to feed more people. Initially this was the result of a combination of improved scientific methods of plant breeding and better crop rotation systems.[41] As the Industrial Revolution progressed, mechanical farming implements became major labor-saving devices.[42] The year 1890 marked the first time a majority of Americans did not work on farms, and the trend was accelerating sharply by 1910 as tractors powered by steam or internal combustion engines replaced slow and inefficient work animals.[43]

During the twentieth century the advent of improved pesticides, chemical fertilizers, and genetic modification led to an explosion in crop yields. For example, in 1850 wheat yields in the United Kingdom were 0.44 ton per acre.[44] As of 2022 they had risen to 3.84 tons per acre.[45] During roughly that same span, the United Kingdom's population rose from about 27 million to 67 million, so food production was able to not just accommodate a growing number of citizens but make food much more abundant for each person.[46] As people got access to better nutrition, they grew taller and healthier and had better brain development in childhood. With more people living to reach their full potential, more talent was unlocked to continue further innovation.[47]

Looking to the future, the emergence of automated vertical farming will likely foster another vast leap in the productivity and efficiency of agriculture. Companies like Hands Free Hectare in the United Kingdom are already working to remove human labor from all stages of agricultural production.[48] As AI and robotics advance and renewable energy becomes cheaper, there will eventually be dramatic reductions in the price of many agricultural products. As food prices become less dependent on human labor and scarce natural resources, poverty will not prevent people from accessing a plentiful supply of healthy and nutritious fresh food.

Although many of the farmworkers who lost jobs found new employment in factories, much the same story played out for factory work about a century and a half later. In the first decade of the nineteenth century, about 1 out of 35 American workers were employed in manu-

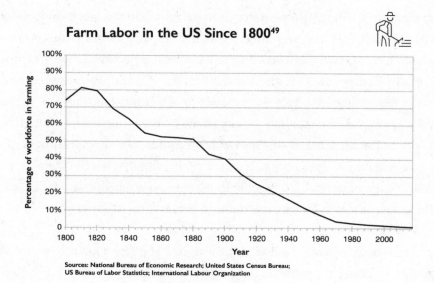

Farm Labor in the US Since 1800[49]

Sources: National Bureau of Economic Research; United States Census Bureau;
US Bureau of Labor Statistics; International Labour Organization

facturing.[50] The Industrial Revolution soon transformed major cities, though, as steam-powered factories sprang up and demanded millions of low-skilled laborers. By 1870 almost 1 in 5 workers were in manufacturing, mainly in the rapidly industrializing North.[51] The second wave of the Industrial Revolution brought a new mass of workers—largely immigrants—into manufacturing around the start of the twentieth century. The development of the assembly line greatly increased efficiency, and as prices for products fell, they became accessible to more and more people.[52]

As demand grew, factories had to hire new armies of workers, and peacetime manufacturing employment peaked around 1920 at an estimated 26.9 percent of the civilian labor force.[53] Methods of measuring the size of the labor force have changed somewhat over time, so we can't perfectly compare this figure with those of later decades, but a general fact is clear. Aside from the disruptions of the Great Depression (when manufacturing employment temporarily fell) and World War II (when it temporarily rose), the US workforce maintained about 1 in 4 people in manufacturing until the 1970s.[54] For roughly five decades, there was no overall upward or downward trend.

Then two technology-related shifts began to erode US factory employment. First, innovations in logistics and transportation, most notably containerized shipping, made it cheaper for companies to outsource manufacturing to countries with less expensive labor and import finished products to the United States.[55] Containerization is not a flashy technology like factory robotics or AI, but it has had one of the most profound impacts on modern society of any innovation. By drastically reducing the cost of worldwide shipping, containerization made it possible for the economy to become truly global. This made an enormous range of products available more cheaply to ordinary people, but it was also a key factor in the deindustrialization of large parts of the United States.

Second, automation reduced the amount of human labor demanded by the domestic manufacturing sector. While early assembly lines involved significant hands-on work at each step, the introduction of robotics reduced this need. That trend was reinforced in the 1990s as computerization and artificial intelligence started making automation ever more capable and efficient. Thus, the average manufacturing worker, assisted by smarter machines, could produce more and more goods every hour. In fact, in the two decades from 1992 to 2012, as computerization transformed factory production, the hourly output of the average manufacturing worker doubled (adjusting for inflation).[56]

As a result, during the twenty-first century, manufacturing output and manufacturing employment have decoupled. In February 2001, just before the post-dot-com recession, 17 million Americans had manufacturing jobs.[57] This dropped sharply during the recession and never recovered—jobs stayed flat at around 14 million all through the mid-2000s boom despite a substantial increase in output.[58] In December 2007, at the start of the Great Recession, about 13.7 million Americans were working in manufacturing, and this had fallen to 11.4 million by February 2010.[59] Manufacturing output quickly rebounded and by 2018 was back near all-time highs—but many of the lost jobs never came back.[60] Even in November 2022, only 12.9 million workers were needed to produce that output.[61]

Looking back over the past century, these trends are striking. After staying steadily between 20 and 25 percent from 1920 to 1970, manufacturing employment has steadily shrunk as a fraction of the labor force in the five decades since—to 17.5 percent in 1980, 14.1 percent in 1990, 12.1 percent in 2000, and bottoming at 7.5 percent in 2010.[62] In the decade since, it has remained essentially flat despite the sustained economic expansion and healthy growth of manufacturing output. And so, as of early 2023, the manufacturing sector employs roughly 1 in 13 American workers.[63]

Manufacturing Labor in the US Since 1900[64]

Source: US Bureau of Labor Statistics; Stanley Lebergott, "Labor Force and Employment, 1800–1960," in *Output, Employment, and Productivity in the United States After 1800*, ed. Dorothy S. Brady (Washington, DC: National Bureau of Economic Research, 1966)

Despite the dramatic reduction in both farm and factory jobs, the US labor force has been steadily growing for as long as the statistic has been measured, even in light of successive waves of automation.[65] From the early Industrial Revolution through the middle of the twentieth century, the economy not only created enough new jobs to provide employment for the rapidly expanding population but also accommodated the entry of tens of millions of women into the workforce.[66]

Since the start of the twenty-first century, the labor force has slightly shrunk as a proportion of the total population, but a major reason for this is that a higher percentage of Americans are now of retirement age.[67] In 1950, 8.0 percent of the US population was sixty-five or older;[68] by 2018 that had doubled to 16.0 percent, leaving relatively fewer working-age people in the economy.[69] The US Census Bureau projects—independent of any new medical breakthroughs that may be achieved in the coming decades—that over-sixty-fives will constitute 22 percent of the population by 2050.[70] If I am correct that significant life-extension technologies will become a reality by then, the proportion of senior citizens will be even higher.

In absolute terms, though, the labor force itself is still growing. In 2000 the civilian labor force was estimated at 143.6 million out of a total population of 282 million, or 50.9 percent.[71] By 2022 the labor force had grown to 164 million workers out of around 332 million people, or 49.4 percent.[72]

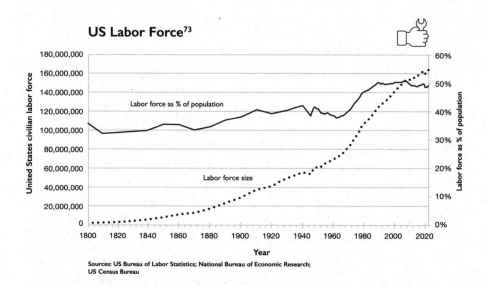

US Labor Force[73]

Sources: US Bureau of Labor Statistics; National Bureau of Economic Research; US Census Bureau

As the economy has shifted toward more technology-intensive jobs, we have been dramatically increasing the investment in education to provide the new skills required and to create new employment opportunities. We've gone from 63,000 university students (undergraduate and graduate level) in 1870 to an estimated 20 million as of 2022.[74] Of that total, we've added about 4.7 million university students in the

United States just between the years 2000 and 2022.[75] We spend more than eighteen times as much today in constant dollars per child in K–12 as we did a century ago. During the 1919–20 school year, K–12 public schools spent the equivalent of $1,035 per pupil (2023 dollars).[76] By the 2018–19 school year, this had increased to around $19,220 (2023 dollars).[77]

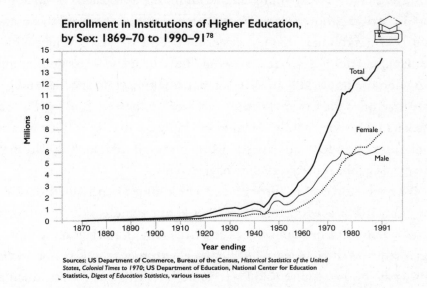

Enrollment in Institutions of Higher Education, by Sex: 1869–70 to 1990–91[78]

Sources: US Department of Commerce, Bureau of the Census, *Historical Statistics of the United States, Colonial Times to 1970*; US Department of Education, National Center for Education Statistics, *Digest of Education Statistics*, various issues

Over the past two centuries, technological change has replaced most jobs in the economy several times over, yet—aided by improving education—we have seen sustained and dramatic economic progress. This is despite the continual (and accurate) perception that major categories of employment are about to disappear.[79]

IS THIS TIME DIFFERENT?

Despite the long pattern of net job growth, some prominent economists have predicted that this time will be different. One of the leading proponents of the view that the upcoming onslaught of AI-based automation will be a net job killer has been Stanford professor Erik

Brynjolfsson. He argues that, unlike previous technology-driven transitions, the latest form of automation will result in a loss of more jobs than it creates.[80] Economists who take this view see the current situation as the culmination of several successive waves of change.

The first wave is often referred to as "deskilling."[81] For example, a driver of a horse-drawn carriage who needed extensive skills at handling and maintaining unpredictable animals was replaced by an automobile driver who required fewer such abilities. One of the main effects of deskilling is that it is easier for people to take new jobs without lengthy training. Craftsmen would have to spend years developing the wide range of skills involved in shoemaking, but once assembly-line machines took over much of this work, a person could get a job after a much shorter time learning to operate a machine. This meant that labor costs fell and shoes became more affordable, but also that low-paying jobs replaced higher-paying ones.

The second wave is "upskilling." Upskilling often follows deskilling, and introduces technologies that require more skill than what came before. For example, providing drivers with navigation technologies requires them to learn how to use electronics that were not previously part of the required skill set. Sometimes this means introducing machines that take an ever larger role in manufacturing but that require sophisticated skills to operate. For example, while early shoe-making machines were hand-operated presses that required no formal education to operate, today companies like FitMyFoot use 3D printing to create custom footwear that fits each customer perfectly.[82] So instead of a large number of low-skill jobs, FitMyFoot's production depends on a smaller number of people with skills in computer science and operating 3D printers. Trends like this tend to replace low-paying jobs with fewer but higher-paying jobs.

The oncoming wave, however, could be called "nonskilling." Driverless-vehicle AI, for example, will replace human drivers altogether. As more and more tasks fall within the capabilities of AI and robotics, there will be a series of nonskilling transitions. What makes AI-driven innovation different from previous technologies is that it opens more opportunities for taking humans out of the equation alto-gether. Instead of reducing or increasing the amount of skill required to perform a certain task, artificial intelligence can often take over the task entirely. This is desirable not just for cost reasons but also because in many areas AI can actually do a better job than the humans it is replacing. Self-driving cars will be much safer than those operated by human drivers, and the AI will never get drunk, drowsy, or distracted.

Yet it is important to distinguish between tasks and professions. In some cases (but not all), tasks that are wholly automated allow a cer-tain type of job to pivot to a different set of tasks—in effect, upskilling. For example, ATMs can now replace human bank tellers for many routine cash transactions, but tellers have taken on a greater role in marketing and building personal relationships with customers.[83] Simi-larly, while software for legal research and document analysis has re-placed certain functions of paralegals, the profession has changed in response and now entails a significantly different bundle of tasks than it did decades ago.[84] This sort of effect may soon happen in the art world. Starting in 2022, publicly available systems like DALL-E 2, Midjourney, and Stable Diffusion used AI to create high-quality graphic art based on text-based prompts from humans.[85] As this tech-nology advances, human graphic designers may spend less time physi-cally sketching art and more time brainstorming ideas with clients and curating or modifying samples that AI produces.

In the long term, the economic incentives for automation will push AI to take over an ever-expanding set of tasks. All else being equal, it is less expensive to buy machines or AI software than to pay ongoing labor costs.[86] When business owners are designing their operations, they often have some flexibility over the balance between capital and labor. In places where wages are relatively low, it makes more sense to use labor-intensive processes. Where wages are high, there's more of an incentive to innovate and design machines that require less labor. This is likely one reason why Great Britain was the cradle of the Industrial Revolution—it had wages that were higher than almost anywhere else in the world as well as abundant cheap coal. This drove development of technologies that substituted inexpensive steam power for expensive human labor. In today's developed economies there is a similar dynamic. While machines can be one-time purchases that become assets, employee wages are ongoing costs, and workers have a range of other needs that employers must meet. So where it's possible to increase automation, businesses have an incentive to do so. As AI approaches human—and, shortly thereafter, superhuman—levels of competence, there will be ever fewer tasks that unenhanced humans will be needed to perform. Until we merge with our AI more fully, this portends major disruptions for workers.

Yet one sticking point in this thesis has been a productivity puzzle: if technological change really is starting to cause net job losses, classical economics predicts that there would be fewer hours worked for a given level of economic output. By definition, then, productivity would be markedly increasing. However, productivity growth as traditionally measured has actually slowed since the internet revolution in the 1990s. Productivity is often measured as real output per hour, which is the total amount of goods and services produced (adjusted for inflation) divided by the total number of hours worked to produce them. From the first quarter of 1950 to the first quarter of 1990, real output per hour in the United States increased by an average of 0.55 percent per quarter.[87] As personal computers and the internet became widespread in the nineties, productivity gains accelerated. From the first quarter of 1990 to the first quarter of 2003, quarterly increases averaged 0.68 percent.[88] It appeared that the World Wide Web had unleashed a new age of rapid growth, and as late as 2003 there was a widespread expectation that this pace would continue.[89] Yet starting in 2004, productivity growth began to significantly slow. From the first quarter of 2003 to the first quarter of 2022, it averaged just 0.36 percent per quarter.[90] This has been one of the great economic mysteries of the past decade. With information technology transforming business in so many ways, we'd expect to see much stronger productivity growth. Theories abound as to why we haven't.

If automation is really having such a huge impact, there appears to be several trillion dollars of the economy "missing." In my view, which has been growing in acceptance among economists, much of the explanation is that we don't count the exponentially increasing value of information products in GDP, many of which are free and represent categories of value that did not exist until recently. When MIT bought the IBM 7094 computer I used as an undergraduate for around $3.1 million in 1963, that counted for, well, $3.1 million ($30 million in 2023 dollars) in economic activity.[91] A smartphone today is hundreds of thousands of times more powerful in terms of computation and communication and has myriad capabilities that did not exist at any price in 1965, yet it counts for only a few hundred dollars of economic activity, because that is what you paid for it.

This general explanation for the missing productivity has also been notably advanced by Erik Brynjolfsson and venture capitalist Marc Andreessen.[92] As a condensed explanation, gross domestic product measures economic activity via the prices of all finished goods and services in a country. So if you pay $20,000 for a new car, that adds $20,000 to that year's GDP—even if you would have been willing to pay $25,000 or $30,000 for the same car. This method worked well during the twentieth century because across the entire population the average willingness to pay for an item would be reasonably close to its actual price. A major reason for this is that when goods and services are produced with physical materials and human labor, it costs businesses a significant amount of money to produce each new unit. Building a car, for example, requires a combination of expensive metal parts and many hours of skilled labor. This is the concept of marginal cost.[93] Classical economic theory says that prices will tend toward goods' average marginal cost—because businesses can't afford to sell at a loss, but competitive pressure forces them to sell as cheaply as they can. Further, since more useful and powerful products have traditionally cost more to produce, there was historically a strong relationship between a product's quality and its price reflected in GDP.

Yet many information technologies have become vastly more useful while prices have stayed more or less constant. A roughly $900 (2023 inflation-adjusted) computer chip in 1999 could perform more than 800,000 computations per second per dollar.[94] By early 2023 a $900 chip could do nearly 58 billion computations per second per dollar.[95]

So the problem is that GDP naturally counts today's $900 chip as equivalent to one produced over two decades ago, even though the current one is more than 72,000 times more powerful for the same price. Thus, nominal wealth and income increases over the past few decades do not properly reflect the massive lifestyle advantages enabled by new technology. This distorts the interpretation of economic data and creates misleading perceptions, such as apparently slow or even stagnant wage growth. Even if your nominal wages stayed flat over the past two decades, you can now buy many thousands of times more computing power with them.[96] Government agencies have made some efforts to take improving performance into account for some economic statistics,[97] but these still massively underestimate the true price-performance gains.

This dynamic is even stronger for digital goods, which can be produced almost for free. Once Amazon has formatted an e-book for sale, selling new copies of it doesn't take any additional paper, ink, or labor—so it sells for a nearly infinite multiple of its marginal cost. As a result, the close relationship between marginal cost, price, and consumers' willingness to pay has been weakened. In the case of services whose marginal cost is low enough that they can be free to consumers altogether, that relationship breaks down completely. Once Google has designed its search algorithms and built its server farms, providing a user with one additional search costs almost nothing. It doesn't cost Facebook any more money to connect you to one thousand friends than to connect you to only one hundred. So they give the public free access and cover their marginal costs with ads.

Even though such services are free to consumers, we can approximate people's willingness to pay for them (also known as consumer surplus) by looking at their choices.[98] For example, if you could earn $20 by mowing a neighbor's lawn but choose to spend that time on TikTok instead, we can say that TikTok is giving you at least $20 of value. As Tim Worstall estimated in *Forbes* in 2015, Facebook's US-based revenue was about $8 billion, which would thus be its official contribution to GDP.[99] But if you value the amount of time people spend on Facebook even at minimum wage, the true benefit to consumers was around $230 billion.[100] As of 2020 (the most recent year for which data is available as this book goes to press), US social media–using adults spent an average of thirty-five minutes each day on Facebook.[101] With around 72 percent of America's roughly 258 million adults using social media, this suggests $287 billion of economic value from Facebook that year, using Worstall's methodology.[102] And a 2019 global survey found that American internet users spent an average of two hours and three minutes per day on all social media—which contributed around $36.1 billion in advertising revenue to GDP but implies a total benefit to users of over $1 trillion per year![103]

Valuing social media use at minimum wage is hardly a perfect measure, since, for example, it's more practical to surf Facebook while waiting in line for coffee than to use those few minutes doing remote freelance work. But as a general approximation, it reveals that people place enormous value on social media usage, yet only a small fraction of that value is visible to economists as revenue. Wikipedia is an even more extreme example: its official contribution to GDP is basically zero. The same analysis applies to countless web- and app-based services.

This suggests that as digital technology takes up a larger and larger share of the economy, consumer surplus is rising much faster than GDP suggests. Thus, productivity in terms of consumer surplus has been increasing more rapidly than the traditional output-per-hour metric. Since consumer surplus is a more "genuine" metric of true prosperity than prices, one could say that the kind of productivity we really care about has been growing just fine all along.

These effects go far beyond areas that are obviously "technology." Technological change has enabled countless other benefits that don't show up in GDP—from less pollution and safer living conditions to expanded opportunities for learning and entertainment. That said, these changes haven't affected all areas of the economy evenly. For example, despite dramatic deflation in computing prices, health care has been getting more expensive faster than overall inflation—so someone who needs a lot of medical treatments may not be comforted much by how much cheaper GPU cycles are getting.[104]

The good news, though, is that artificial intelligence and technological convergence will turn more and more kinds of goods and services into information technologies during the 2020s and 2030s—allowing them to benefit from the kinds of exponential trends that have already brought such radical deflation to the digital realm. Advanced AI tutors will make possible individually tailored learning on any subject, accessible at scale to anyone with an internet connection. AI-enhanced medicine and drug discovery are still in their infancy as of this writing but will ultimately play a major role in bringing down health-care costs.

The same will happen for numerous other products that have not traditionally been considered information technologies—such as food, housing and building construction, and other physical products like clothing. For example, AI-driven advances in materials science will make solar photovoltaic electricity extremely cheap, while robotic resource extraction and autonomous electric vehicles will bring the costs of raw materials much lower. With cheap energy and materials, and automation increasingly replacing human labor outright, prices will fall substantially. In time, such effects will cover so much of the economy that we'll be able to eliminate much of the scarcity that presently holds people back. As a result, in the 2030s it will be relatively inexpensive to live at a level that is considered luxurious today.

If this analysis is correct, then technology-driven deflation in all these areas will only widen the gap between nominal productivity and the real average benefit that each hour of human work brings to society. As such effects spread beyond the digital sphere to other industries and encompass a wider portion of the whole economy, we would expect national inflation to decrease—and eventually lead to overall deflation. In other words, we can expect clearer answers to the productivity puzzle as time goes on.

There's another puzzle as well: Why does economic data show that the proportion of Americans in the workforce is shrinking? Economists who support the net-job destruction thesis point to the US civilian labor force participation rate, which is the number of employed persons plus the number of unemployed who are looking for a job as a percentage of the population aged sixteen and over. After rising from about 59 percent in 1950 to just under 67 percent in 2002, it had dropped below 63 percent by 2015 and remained nearly flat until the COVID-19 pandemic despite an ostensibly booming economy.[105]

The actual percentage of the total population in the labor force is smaller. In June 2008 there were more than 154 million people in the US civilian labor force out of a population of 304 million, or 50.7 percent.[106] By December 2022 there were 164 million in the labor force out of 333 million, or just under 49.5 percent.[107] That doesn't seem like a big drop, but it still puts the United States at its lowest percentage in more than two decades. The government statistics on this do not perfectly capture economic reality, as they do not include several categories—including agricultural workers, military personnel, and federal government employees—but they are still useful for showing the direction and approximate magnitude of these trends.

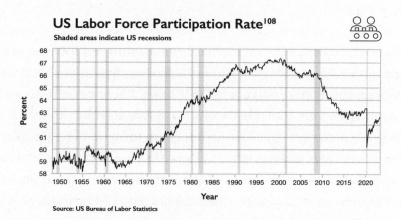

US Labor Force Participation Rate[108]
Shaded areas indicate US recessions

Source: US Bureau of Labor Statistics

Yet while some of this decline is likely due to automation, there are two major confounding factors. First, as Americans become more educated, fewer teenagers are working, and many people are staying in college and graduate school well into their twenties.[109] Also, as the large baby boomer generation increasingly ages into retirement, the proportion of Americans of working age is decreasing.[110]

If instead we look at the labor force participation rate among prime working-age adults aged twenty-five to fifty-four, the decline almost disappears—as of early 2023, the participation rate is 83.4 percent, as compared with 84.5 percent at its peak in 2000.[111] This is still a difference of around 1.7 million people in today's population, but much less than the previous chart makes it appear.[112]

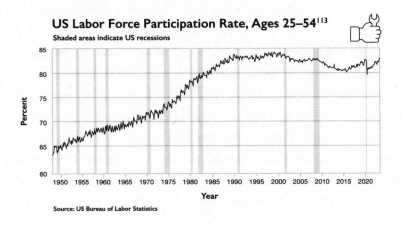

US Labor Force Participation Rate, Ages 25–54[113]

Shaded areas indicate US recessions

Source: US Bureau of Labor Statistics

In addition, since 2001 there has been a significant increase in participation by people fifty-five and over. Among those fifty-five to sixty-four, it rose from 60.4 percent in 2001 to 68.2 percent in 2021, and in the same span it grew among those seventy-five and older from 5.2 percent to 8.6 percent.[114] Conflicting forces are involved here. On the one hand, many older workers who lose jobs due to automation are simply retiring earlier and accepting a lower standard of living. On the other hand, people are living longer (before the COVID-19 pandemic, combined male-female US life expectancy had increased by about two years since 2000)[115] and are healthy enough to work at later ages. For many, this is an enjoyable source of purpose and satisfaction. Still, this data doesn't reflect the fact that some older adults are forced to stay in the workforce in marginal jobs after losing better-paying ones before they can secure their retirement.[116]

Yet all these analyses are constrained by the fact that labor force participation itself is an increasingly flawed concept. Two major trends are reshaping the nature of work but are not well reflected in economic statistics.

The first is the underground economy, which has always existed but has been greatly facilitated by the internet. Its activities include virtually the entire sex industry as well as many other types of services, including paid-under-the-table housework, alternative healing modalities, and many more. Another facilitator of the underground economy is the advent of encrypted technologies, such as cryptocurrencies that enable hiding transactions from taxation, regulation, and law enforcement authorities.

The largest and most famous cryptocurrency is Bitcoin.[117] On August 6, 2017, the daily trade volume of Bitcoin on major exchanges was less than $19.3 million.[118] It had spiked to over $4.95 billion by December 7 of the same year but quickly fell again, and as of mid-2023 averaged around $180 million per day.[119] This is nonetheless very rapid growth, but still tiny in comparison with major traditional currencies. According to the Bank for International Settlements, global foreign exchange trading averaged $7.5 trillion per day in April 2022, and it is likely even higher as this book goes to press.[120]

Also, in contrast to most traditional currencies, the value of most cryptocurrencies has been extremely volatile. On January 4, 2012, for example, Bitcoins were trading at $13.43.[121] On April 2 they were up above $130.[122] But interest in cryptocurrency was still largely limited to a technology-minded subculture. Then, after almost five years of relative quiet and stability, Bitcoin started shooting up even higher in 2017. Suddenly ordinary people were hearing about Bitcoins as a sure-fire investment and buying them up in hopes that they would appreciate further. This became a self-fulfilling prophecy, with prices hitting $1,354 on April 29 and $18,877 on December 17.[123] But then prices started to fall and people sold their Bitcoins in a panic, trying to get out of the market before their assets lost any more value. By December 12, 2018, Bitcoin was back down to \$3,360—only to hit \$64,899 on April 13, 2021, before another major crash dropped it to \$15,460 by November 20, 2022.[124]

This volatility poses a major problem for people who want to use Bitcoin as a currency—that is, as a medium for regularly exchanging goods and services. If you believed your dollars would be worth ten times as much within half a year, you'd try to avoid spending them. Conversely, if your dollars could lose half their value in a few months, you would be reluctant to keep many of your assets in dollars, and merchants wouldn't want to accept them. If cryptocurrencies are to achieve wider adoption by the public, they'll need to find a way to keep values more stable.

Yet cryptocurrency is hardly a requirement for underground work to flourish. Social media and platforms like Craigslist provide ample opportunities for people to form economic connections that are mostly invisible to the government.

This effect also facilitates the other major trend: new ways of earning money not always regarded as traditional methods of employment. These include creating, buying, selling, and bartering physical and digital assets and services using websites and apps, and also creating apps, videos, and other forms of digital content on social media sites. Some people have developed successful careers creating content for YouTube, for example, or are paid to influence others on Instagram or TikTok.[125]

Before the release of the iPhone in 2007, there was no app economy to speak of. In 2008 there were fewer than 100,000 iOS apps available; this had rocketed up to around 4.5 million by 2017.[126] On Android, the growth was just as dramatic. In December 2009, there were around 16,000 mobile apps available in the Google Play Store.[127] As of March 2023, there were 2.6 million.[128] That is a more than 160-fold increase in thirteen years. This led directly to growing employment. From 2007 to 2012, the app economy created an estimated half a million jobs in the United States.[129] By 2018, according to Deloitte, this had grown to over 5 million jobs.[130] Another 2020 estimate, including jobs indirectly created by the app economy, put it at 5.9 million US jobs and $1.7 trillion of economic activity.[131] These numbers depend somewhat on how broadly or narrowly one defines the app market, but the key takeaway is that in a little over a decade, mobile apps have exploded from insignificance to a major factor in the wider economy.

And so, even as technological change is rendering many jobs obsolete, those very same forces are opening up numerous new opportunities that fall outside the traditional model of "jobs." Although it is not without its limitations, the so-called gig economy often allows people more flexibility, autonomy, and leisure time than their previous options. Maximizing the quality of these opportunities is one strategy for how to help workers as automation trends accelerate and disrupt traditional workplaces.

SO WHERE ARE WE HEADED?

On the face of it, the labor situation sounds alarming. When Oxford's Frey and Osborne estimated that almost half of 2013's jobs would be automatable by 2033, their research assumed more conservative rates of progress in AI and other exponential technologies than what I've outlined in this book.[132] While people have recognized the threat to jobs from automation for more than two hundred years, the current situation is unique in the speed and breadth of the looming threat.

To predict how this will play out, we need to consider several fundamental issues. First, employment is not an end in itself, but a means to an end. One goal of work is to meet the material needs of life. As noted previously, just growing and distributing food required nearly all human labor as recently as two centuries ago, while food production in the United States and much of the developed world requires less than 2 percent of labor today. As AI unlocks unprecedented material abundance across countless areas, the struggle for physical survival will fade into history.

Another goal of work is to give purpose and meaning to life. If your job consists of laying bricks, that labor provides two sorts of meaning. Most obviously, your wages enable you to provide for your loved ones and care for them—an important facet of identity. But you're also building lasting structures that contribute to the public good. You're literally contributing to something larger than yourself. Some of the most fulfilling jobs, like those in the arts and academia, additionally provide the opportunity to be creative and generate new knowledge.

The upcoming revolution will empower humans for such contributions far beyond what has ever been possible before. In fact, advancing information technology is already enhancing artists' ability to enrich the culture—often in underappreciated ways. For example, when I was growing up, there were only three television stations available: ABC, NBC, and CBS. Because everyone was watching such a limited selection of programs, the networks had to create content that would be popular with the widest possible demographics. In order to be successful, shows had to appeal to men and women, children and parents, blue collar and white collar. Ideas with a strong but narrower appeal, like absurdist comedy, paranormal drama, or science fiction, did not have an easy path to commercial viability. Many people now forget that *Star Trek*, the most influential science fiction series in history, was canceled after just three seasons.[133]

But the proliferation of cable broadened the TV landscape to the extent that niche shows could find an audience. Deep-dive documentaries on unusual topics flourished on the Discovery Channel, the History Channel, the Learning Channel, and elsewhere. But viewership was still bound by airtimes. The introduction of DVRs and then on-demand streaming allowed people to watch whatever they wanted, whenever they wanted. That meant that innovative new shows could gather their audience from the whole population—not just whoever happened to be watching during a particular time slot. As a result, artistic ideas that wouldn't have been viable on the networks of my youth, like *Stranger Things* and *Fleabag*, have found loyal viewers and critical acclaim. This dynamic is great news for relatively small demographics like LGBTQ people, persons with disabilities, and American Muslims, because it's easier for programs positively depicting their particular life experiences to be commercially successful.

In addition, streaming allows for different creative choices. For example, half-hour comedy episodes on broadcast TV normally have self-contained plots because networks want viewers to be able to watch and enjoy them in any order. But on-demand means that viewers can always watch in the proper sequence. This has freed groundbreaking shows like *BoJack Horseman* to have character development from one episode to another and gradually build up jokes over several episodes.[134] These artistic possibilities literally wouldn't exist with previous broadcast technologies.

Over the next two decades this transformation will accelerate dramatically. Think of the creativity that AI has achieved over the past few years in visual images thanks to systems like DALL-E, Midjourney, and Stable Diffusion. These capabilities will become more sophisticated and will expand to music, video, and games, radically democratizing creative expression. People will be able to describe their ideas to AI and tweak the results with natural language until it fulfills the visions in their minds. Instead of needing thousands of people and hundreds of millions of dollars to produce an action movie, it will eventually be possible to produce an epic film with nothing but good ideas and a relatively modest budget for the compute that runs the AI.

Yet despite all these impending benefits, we must also be realistic about the disruptive effects that will occur between now and then. Automation and its indirect effects have already eradicated many jobs near the bottom and in the middle of the skill ladder, and this trend will broaden and pick up speed over the next decade. Most of our new jobs require more sophisticated skills. As a whole, our society has moved up the skill ladder, and this will continue. But as AI soars past the capabilities of even the most skilled humans in one area after another, how can humans keep up?

The primary method of improving human skills over the past

two centuries has been education. Our investment in learning has sky-rocketed over the past century, as I described previously. But we are already well into the next phase of our own betterment, which is enhancing our capabilities by merging with the intelligent technology we are creating. We are not yet putting computerized devices inside our bodies and brains, but they are literally close at hand. Almost no one could do their jobs or get an education today without the brain extenders that we use on an all-day, every-day basis—smartphones that can access almost all human knowledge or harness huge computational power with a single tap. It is therefore not an exaggeration to say that our devices have become parts of us. This was not the case as recently as two decades ago.

These capabilities will become even more integrated with our lives throughout the 2020s. Search will transform from the familiar paradigm of text strings and link pages into a seamless and intuitive question-answering capability. Real-time translation between any pair of languages will become smooth and accurate, breaking down the language barriers that divide us. Augmented reality will be projected constantly onto our retinas from our glasses and contact lenses. It will also resonate in our ears and ultimately harness our other senses as well. Most of its functions and information will not be explicitly requested, but our ever present AI assistants will anticipate our needs by watching and listening in on our activities. In the 2030s, medical nanorobots will begin to integrate these brain extensions directly into our nervous system.

In chapter 2, I described how this technology will extend our neocortex into the cloud, adding more capacity and more levels of abstraction. This will be available to everyone and ultimately will be inexpensive, just as mobile phones started out very expensive and not so smart but are ubiquitous today (the International Telecommunication Union estimates that there were 5.8 billion active smartphone subscriptions in the world as of 2020),[135] with rapidly improving capabilities.

But on the way to a future of such universal abundance, we need to address the societal issues that will arise as a result of these transitions. The social safety net started in the United States in the 1930s with the passage of Social Security.[136] While particular formulations go in and out of political favor (for example, "welfare"), the overall safety net has nonetheless grown more extensive ever since, regardless of the political leanings of particular parties and administrations.

Even though the United States is considered to have a less extensive social safety net than "socialist" European countries, its public spending on social welfare—an estimated 18.7 percent of GDP as of 2019 (before COVID-19 pandemic relief muddied the data)—is close to the median for developed countries.[137] Canada was lower, at 18.0 percent.[138] Australia and Switzerland were similar, both at 16.7 percent.[139] The United Kingdom was slightly higher, at 20.6 percent—around $580 billion of a $2.8 trillion GDP, or less than $8,800 per capita across a population of 66 million.[140] But since US GDP per capita is higher, its social safety net is likewise higher per capita. In 2019, US GDP was over $21.4 trillion, of which around $4 trillion was public social spending.[141] Across a population that averaged around 330 million that year, this amounts to more than $12,000 per capita.[142]

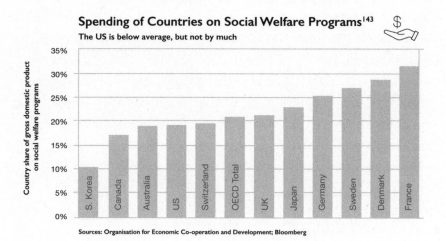

Spending of Countries on Social Welfare Programs[143]

The US is below average, but not by much

Country share of gross domestic product on social welfare programs (y-axis: 0% to 35%)

Countries (left to right): S. Korea, Canada, Australia, US, Switzerland, OECD Total, UK, Japan, Germany, Sweden, Denmark, France

Sources: Organisation for Economic Co-operation and Development; Bloomberg

The US safety net has been steadily growing as a percentage of government spending (now about 50 percent of all federal, state, and local expenditures) and of GDP (and both government spending and GDP are themselves steadily increasing).[144] Look at the following charts and see if you can determine when "left-wing" or "right-wing" administrations were in power. (The most recent two data years available include a good deal of pandemic relief, so the spike for 2020–2021 exceeds the underlying long-term growth trend.)

With GDP continuing to grow exponentially, social safety net spending will likely continue to increase both overall and per capita. Significant programs within the US social safety net include Medicaid for basic medical services, SNAP "food stamps" (essentially a debit card for food), and housing assistance. The level of these programs is barely adequate today, but as AI-driven advances make it possible for medicine, food, and housing to be much cheaper during the 2030s, the same level of financial support will provide a very comfortable standard of living without needing to further increase the percentage of GDP devoted to social safety net spending. If this percentage continues to grow, it can fund even more extensive services.

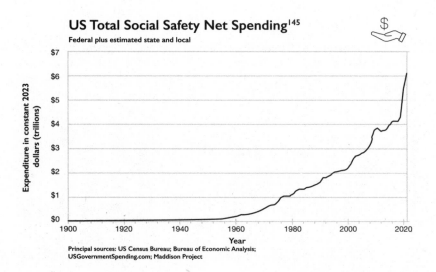

US Total Social Safety Net Spending[145]
Federal plus estimated state and local

Principal sources: US Census Bureau; Bureau of Economic Analysis;
USGovernmentSpending.com; Maddison Project

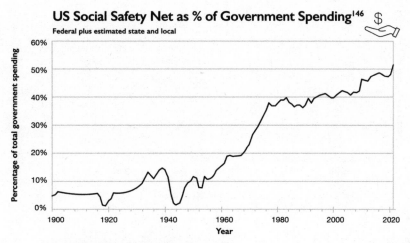

US Social Safety Net as % of Government Spending[146]

Federal plus estimated state and local

Principal sources: US Census Bureau; Bureau of Economic Analysis;
USGovernmentSpending.com; Maddison Project

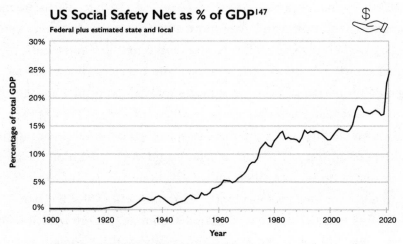

US Social Safety Net as % of GDP[147]

Federal plus estimated state and local

Principal sources: US Census Bureau; Bureau of Economic Analysis;
USGovernmentSpending.com; Maddison Project

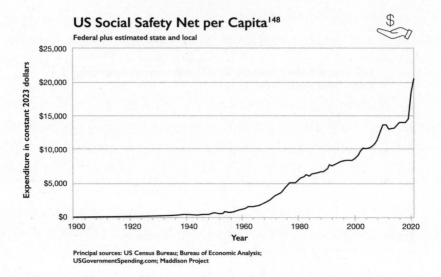

US Social Safety Net per Capita[148]

Federal plus estimated state and local

Principal sources: US Census Bureau; Bureau of Economic Analysis;
USGovernmentSpending.com; Maddison Project

In an onstage dialogue with TED curator Chris Anderson at the 2018 TED conference in Vancouver,[149] I predicted that we would effectively have universal basic income (UBI) or its equivalent by the early 2030s in developed countries, and by the late 2030s in most countries—and that people would be able to live well by today's standards on that income. This would entail regular payments to all adults, or the provision of free goods and services, likely funded by some combination of taxes on automation-driven profits and government investments in emerging technologies.[150] Related programs might provide financial support to people caring for family or building healthy communities.[151] Such reforms could greatly cushion the harms of job disruptions. In evaluating the likelihood of such progress, one has to consider how profoundly the economy will evolve by then.

Thanks to accelerating technological change, overall wealth will be far greater, and given the long-term stability of our social safety net regardless of the governing party, it is very likely to remain in place—and at substantially higher levels than today.[152] Remember, though, that technological abundance doesn't automatically benefit everyone equally at once. For example, in 2022, $1.00 could buy more than 50,000 times as much computing power as it could in 2000 (adjusted for overall inflation).[153] By contrast, according to official statistics, $1.00 in 2022 could buy only about 81 percent of the health care it could buy in 2000 (adjusted for overall inflation).[154] And although some medical treatments, like cancer immunotherapies, got qualitatively better in that time, most health-care expenditures, such as hospital stays and X-rays, remained roughly the same. Thus, people who spend a lot of their money on computers, like students and young people, got a lot of benefit from falling computing prices. Meanwhile, those who spend a big part of their income on health care, like the elderly and those with chronic illnesses, may have even been worse off overall.

Therefore, we'll need smart governmental policies to ease the transition and ensure that prosperity is broadly shared. Although it will be technologically and economically possible for everyone to enjoy a standard of living that is high by today's measures, whether we actually provide this support to everyone who needs it will be a political decision. By analogy, although there are occasional famines around the world today, they are not a result of insufficient food production, or because the secret to good agriculture is being kept in a few elite hands. Rather, famines usually happen because of bad governance or civil war. Under these conditions it is much harder for people to compensate for local droughts and other natural disasters, and more challenging for international assistance to be effective.[155] In a similar way, if we are not careful as a society, toxic politics could interfere with rising living standards.

As COVID-19 showed, this is an especially urgent concern in medicine. Although innovation will unlock transformative capabilities to deliver affordable and effective treatments, this does not guarantee outcomes as if by magic. We will need an engaged public and sensible governance to manage a safe, fair, and orderly transition to more advanced health care. One could imagine, for example, a future where lifesaving technologies are widely distrusted. Just as misinformation and conspiracy theories about vaccines thrive online today, people might spread similar rumors about decision support AI, gene therapy, or medical nanotechnology in the coming decades. Given valid concerns about cybersecurity, it's not hard to see how exaggerated fears of secret genetic manipulation or government-controlled nanobots could cause people in 2030 or 2050 to reject crucial treatments. Public understanding of these issues is our best defense against those dynamics costing human lives needlessly.

If we can get these political challenges right, human life will be utterly transformed. Historically we've had to compete to meet the physical needs of life. But as we enter an era of abundance, and the availability of material necessities eventually becomes universal—while many traditional jobs go away—our main struggle will be for purpose and meaning. In effect, we are moving up Maslow's hierarchy of needs.[156] This is already evident in the generation now deciding on careers. I mentor and speak to young people from ages eight to twenty, and their focus is usually on forging a path that is meaningful, whether pursuing creative expression through the arts or helping to overcome the grand challenges—social, psychological, and otherwise—that humanity has struggled with for millennia.

And so, considering the role of jobs in our lives forces us to reconsider our broader search for meaning. People often say that it is death and the brevity of our existence that gives meaning to life. But my view, rather, is that this perspective is an attempt to rationalize the tragedy of death as a good thing. The reality is that the death of a loved one literally robs us of part of ourselves. The neocortical modules that were wired to interact with and enjoy the company of that person now generate loss, emptiness, and pain. Death takes from us all the things that in my view give life meaning—skills, experiences, memories, and relationships. It prevents us from enjoying more of the transcendent moments that define us—producing and appreciating creative works, expressing a loving sentiment, sharing humor.

All of these abilities will be greatly enhanced as we extend our neocortex into the cloud. Imagine trying to explain music, literature, a YouTube video, or a joke to primates that missed out on the neocortical expansion that the large foreheads of hominids made possible two million years ago. This analogy helps us to at least glimpse how, when we digitally augment our neocortex starting sometime in the 2030s, we will be able to create meaningful expressions that we cannot imagine or understand today.

But a key issue remains: What happens between now and then? In a personal dialogue I had with Daniel Kahneman, he concurred with my view that information technology has been growing and will continue to grow exponentially in price-performance and capacity and that this will ultimately encompass physical products such as clothing and food. He also agreed that we are headed toward an era of abundance that will meet our physical needs and that the primary struggle will then be to satisfy higher levels of Maslow's hierarchy. However, he envisioned a protracted period of conflict, and even violence, between now and then. He pointed out that there will invariably be winners and losers as automation continues its impact. A driver losing his job is not going to be satisfied by the promise of humanity moving up the hierarchy of life, because he as an individual may not, in fact, be able to make that transition.

One of the great challenges of adapting to technological changes is that they tend to bring diffuse benefits to a large population, but concentrated harms to a small group. For example, autonomous vehicles will bring massive benefits to society—from saved lives to lower pollution, eased congestion, more free time, and lower transportation costs. All of the United States' population—projected to be almost 400 million by 2050—will share to varying degrees in these improvements.[157] Depending on which assumptions are used, estimates of the total potential benefits are valued at between $642 billion and $7 trillion per year.[158] Suffice it to say, it's an enormous number. Yet the benefits to any single person are unlikely to totally transform their life. And even though we know statistically that tens of thousands of lives will be spared annually, there'll be no way to identify which individual people avoid death each year.[159]

By contrast, the harms from autonomous vehicles will be mainly confined to the several million people who work as drivers and will lose their livelihoods. These people will be specifically identifiable, and the disruption to their lives may be quite serious. When someone is in this position, it's not enough to know that the overall benefits to society outweigh their individual suffering. They'll need policies in place that help reduce their economic pain and ease their transition to something else that provides meaning, dignity, and economic security.

One phenomenon that supports Kahneman's concern is that popular fears are typically far worse than the reality, as I discussed previously. Losing your job when you already feel like everything in society is getting worse is a good formula for alienation. Kahneman and I agree that a lot of the polarization we see in politics today is the product of automation, both actual and anticipated, rather than coming from traditional political issues such as immigration.[160] If you have a high level of anxiety about your own economic stability, you'll probably be hostile toward anything that appears as if it might compound your problems.

Yet history shows that society is better at adapting to even dramatic changes than we expect. We have replaced most of the economy's jobs repeatedly over the past two centuries and have done so without causing much long-term social dislocation, let alone violent revolutions. Mass communication and law enforcement have been effective in preventing or quickly suppressing violence, as the Luddites were suppressed two centuries ago. Yes, individuals can become violent for a variety of reasons, including mental illness. Given American gun culture, there are too many cases where this turns deadly.

However, headline-grabbing tragedies do not change the fact that overall violence has been declining dramatically over the centuries.[161] As discussed in the previous chapter, although violence in different countries may fluctuate from year to year and decade to decade, the long-term trend around the developed world is a dramatic and sustained reduction. As my friend Steven Pinker argued in his 2011 book *The Better Angels of Our Nature: Why Violence Has Declined*, this decline is the result of deep civilizational trends—factors like states based on the rule of law, growing literacy, and economic development.[162] All of these, it's worth noting, are strengthened by the exponential progress of information technologies. So we have solid reason for optimism about the future.

We should recall that new and innovative opportunities accompany the negative effects of job loss through automation, and that is not going to change. Further, the social safety net is substantial and growing, and as I pointed out, is not much dependent on the political climate. It has deep support that transcends the apparent swings of public opinion. These programs are an expression of our natural compassion for fellow citizens, but they are also a political response intended to ameliorate societal disruptions from technological change.

But the social safety net doesn't replace the sense of purpose jobs give, and as Kahneman argued, there are going to be many losers in the labor market. While it's true that the United States has managed several rounds of automation without severe social dislocation, the difference this time will be the breadth, depth, and rapidity of the change. Kahneman believes that people need time to adapt to change and to take advantage of new opportunities, and many will be unable to rapidly retrain for new types of employment or alternative personal business models.

I think Kahneman is right in some respects, but we should keep in mind that there are many areas of technological change where losers don't exist, or at least do not make themselves known. Take, for example, a new cure for a disease. Companies and individuals who profited from treating that disease lose a long-term stream of income. But the benefit to society is so great that the cure is almost universally celebrated—even by most of the people who treated that disease. They know firsthand the suffering that is being alleviated, and they don't dwell on lost business. At any rate, society recognizes that it is better to find ways of softening the economic blow for these people than to block cures for the sake of jobs and profits.

Throughout the entire two-hundred-year history of automation, many have imagined that jobs disappear from an otherwise unchanged world. This phenomenon applies to all aspects of anticipating the future—people envision one change as if nothing else will be different. The reality is that many positive changes attend each type of job loss, and these positive changes will come as quickly as the disruptive ones.

People actually adapt very quickly to change, especially change for the better. In the late 1980s, when the internet was still mainly limited to universities and governments, I predicted that a vast worldwide network for communication and information sharing would eventually be available to everyone, even schoolchildren, by the late 1990s.[163] I also predicted the advent of mobile devices that people would use to harness this network by the early twenty-first century.[164] Such predictions seemed daunting and disruptive (not to mention unlikely) when I made them, but they actually came true, and those technologies were very rapidly adopted and accepted. As just one example, the entire app economy barely existed a decade and a half ago, yet it is now so deeply established that people can hardly remember when it wasn't around.

And the effects aren't just economic. Stanford research found that an estimated 39 percent of American heterosexual couples in 2017 had met online—many of those through mobile apps like Tinder and Hinge.[165] That means that many of the children now in elementary school exist only because of a technology that's just a few years older than they are. When you take a step back to look at these changes in perspective, the speed with which apps have affected society is truly amazing.

People also frequently envision unenhanced humans struggling to compete with machines, but this is a misconception. Imagining a world where humans are largely in competition with AI-powered machines is the wrong way of thinking about the future. To illustrate this, imagine a time traveler with a 2024 smartphone going back to 1924.[166] This person's intelligence would seem truly superhuman to the people of Calvin Coolidge's day. They could do advanced math effortlessly, translate any major language passably well, play chess better than any grandmaster, and command a whole Wikipedia's worth of facts. To the people of 1924, it would seem obvious that the time traveler's capabilities were radically enhanced by the phone. But to us in the 2020s it's easy to lose this perspective. We don't *feel* augmented. In a similar way, we will harness the advances of 2030 and 2040 to seamlessly augment our own capabilities—and this will feel even more natural as our brains interface directly with computers. When it comes to tackling the cognitive challenges of our abundant future, in most respects we won't be competing with AI any more than we currently compete with our smartphones.[167] Indeed, this symbiosis is nothing new: it has been the purpose of technology since stone tools to extend our reach physically and intellectually.

All of that being said, I do think the specter of troublesome social dislocation—including violence—during this transition is a possibility that we should anticipate and work to mitigate. But my expectation is that a violent transition is unlikely, given the powerful long-term trends I've discussed that will promote stability.

The most important reason for optimism about the coming social transitions is that growing material abundance will lower the incentives for violence. When people don't have the necessities of life, or when crime is already high, citizens can feel that they have nothing to lose by violence. But the same technologies that will cause these social disruptions will also be making food, housing, transportation, and health care much cheaper. And crime will likely keep falling through a combination of better education, smarter policing, and a reduction in environmental toxins like lead that damage people's brains. When people feel that they have long, safe lives ahead of them, they have stronger incentives to work out their differences through politics instead of risking everything by resorting to force.

Kahneman and I postulated that our differing outlooks on the nature of the coming transition are probably influenced by our contrasting childhoods. He spent his formative years fleeing the Nazis with his family in France. I was born after World War II in the relative safety of New York City, although I was nonetheless influenced by the Holocaust as a member of the "generation after." Thus, Kahneman experienced firsthand the extraordinary conflict, uprooting, and hatred that arose, arguably, from the deprivation of post–First World War Europe, especially in Germany.

Nonetheless, orchestrating a continued constructive transition will require enlightened political strategies and policy decisions. Because policy and social organization will remain critical factors, there will continue to be an important role for politicians and civic leaders. Yet the opportunity inherent in this technological progress is immense. It is nothing less than overcoming the age-old afflictions of humanity.

CHAPTER 6

THE NEXT THIRTY YEARS IN HEALTH AND WELL-BEING

THE 2020S: COMBINING AI WITH BIOTECHNOLOGY

When you take your car to the shop to get it fixed, the mechanic has a full understanding of its parts and how they work together. Automotive engineering is effectively an exact science. Thus, well-maintained cars can last almost indefinitely, and even the worst wrecks are technically possible to repair. The same is not true of the human body. Despite all the marvelous advances of scientific medicine over the past two hundred years, medicine is not yet an exact science. Doctors still do many things that are known to work without fully understanding *how* they work. Much of medicine is built on messy approximations that are usually mostly right for most patients but probably aren't totally right for *you*.

Turning medicine into an exact science will require transforming it into an information technology—allowing it to benefit from the exponential progress of information technologies. This profound paradigm shift is now well underway, and it involves combining biotechnology with AI and digital simulations. We are already seeing immediate benefits, as I'll describe in this chapter—from drug discovery to disease surveillance and robotic surgery. For example, in 2023 the first drug designed end-to-end by AI entered phase II clinical trials to treat a rare lung disease.[1] But the most fundamental benefit of AI–biotech convergence is even more significant.

When medicine relied solely on painstaking laboratory experimentation and human doctors passing their expertise down to the next generation, innovation made plodding, linear progress. But AI can learn from more data than a human doctor ever could and can amass experience from billions of procedures instead of the thousands a human doctor can perform in a career. And since artificial intelligence benefits from exponential improvements to its underlying hardware, as AI plays an ever greater role in medicine, health care will reap the exponential benefits as well. With these tools we've already begun finding answers to biochemical problems by digitally searching through every possible option and identifying solutions in hours rather than years.[2]

Perhaps the most important class of problems at present is designing treatments for emerging viral threats. This challenge is like finding which key will open a given virus's chemical lock—from a pile of keys that could fill a swimming pool. A human researcher using her own knowledge and cognitive skills might be able to identify a few dozen molecules with potential to treat the disease, but the actual number of possibly relevant molecules is generally in the trillions.[3] When these are sifted through, most will obviously be inappropriate and won't warrant full simulation, but billions of possibilities may warrant a more robust computational examination. At the other extreme, the space of physically possible potential drug molecules has been estimated to contain some one million billion billion billion billion billion possibilities![4] However one frames the exact number, AI now lets scientists sort through that gigantic pile to focus on those keys most likely to fit for a given virus.

Think of the advantages of this kind of exhaustive search. In our current paradigm, once we have a potentially feasible disease-fighting agent, we can organize a few dozen or a few hundred human subjects and then test them in clinical trials over the course of months or years at a cost of tens or hundreds of millions of dollars. Very often this first option is not an ideal treatment: it requires exploration of alternatives, which will also take a few years to test. Not much further progress can be made until those results are available. The US regulatory process involves three main phases of clinical trials, and according to a recent MIT study, only 13.8 percent of candidate drugs make it all the way through to FDA approval.[5] The ultimate result is a process that typically takes a decade to bring a new drug to market, at an average cost estimated between $1.3 billion and $2.6 billion.[6]

In just the past few years, the pace of AI-assisted breakthroughs has increased noticeably. In 2019 researchers at Flinders University, in Australia, created a "turbocharged" flu vaccine by using a biology simulator to discover substances that activate the human immune system.[7] It digitally generated trillions of chemicals, and the researchers, seeking the ideal formulation, used another simulator to determine whether each of them would be useful as an immune-boosting drug against the virus.[8]

In 2020 a team at MIT used AI to develop a powerful antibiotic that kills some of the most dangerous drug-resistant bacteria in existence. Rather than evaluate just a few types of antibiotics, it analyzed 107 million of them in a matter of hours and returned twenty-three potential candidates, highlighting two that appear to be the most effective.[9] According to University of Pittsburgh drug design researcher Jacob Durrant, "The work really is remarkable. This approach highlights the power of computer-aided drug discovery. It would be impossible to physically test over 100 million compounds for antibiotic activity."[10] The MIT researchers have since started applying this method to design effective new antibiotics from scratch.

But by far the most important application of AI to medicine in 2020 was the key role it played in designing safe and effective COVID-19 vaccines in record time. On January 11, 2020, Chinese authorities released the virus's genetic sequence.[11] Moderna scientists got to work with powerful machine-learning tools that analyzed what vaccine would work best against it, and just two days later they had created the sequence for its mRNA vaccine.[12] On February 7 the first clinical batch was produced. After preliminary testing, it was sent to the National Institutes of Health on February 24. And on March 16—just sixty-three days after sequence selection—the first dose went into a trial participant's arm. Before the pandemic, vaccines typically took five to ten years to develop. Achieving this breakthrough so quickly surely saved millions of lives.

But the war isn't over. In 2021, with COVID-19 variants looming, researchers at USC developed an innovative AI tool to speed adaptive development of vaccines that may be needed as the virus continues to mutate.[13] Thanks to simulation, candidate vaccines can be designed in less than a minute and digitally validated within one hour. By the time you read this, even more advanced methods will likely be available.

All the applications I've described are instances of a much more fundamental challenge in biology: predicting how proteins fold. The DNA instructions in our genome produce sequences of amino acids, which fold up into a protein whose three-dimensional features largely control how the protein actually works. Our bodies are mostly made of proteins, so understanding the relationship between their composition and function is key to developing new medicines and curing disease. Unfortunately, humans have had a fairly low accuracy rate at predicting protein folding, as the complexity involved defies any single easy-to-conceptualize rule. Thus, discoveries still depend on luck and laborious effort, and optimal solutions may remain undiscovered. This has long been one of the main obstacles to achieving new pharmaceutical breakthroughs.[14]

This is where the pattern recognition capabilities of AI offer a profound advantage. In 2018 Alphabet's DeepMind created a program called AlphaFold, which competed against the leading protein-folding predictors, including both human scientists and earlier software-driven approaches.[15] DeepMind did not use the usual method of drawing on a catalog of protein shapes to be used as models. Like AlphaGo Zero, it dispensed with established human knowledge. AlphaFold placed a prominent first out of ninety-eight competing programs, having accurately predicted twenty-five out of forty-three proteins, whereas the second-place competitor got only three out of forty-three.[16]

Yet the AI predictions still weren't as accurate as lab experiments, so DeepMind went back to the drawing board and incorporated transformers—the deep-learning technique that powers GPT-3. In 2021 DeepMind publicly released AlphaFold 2, which achieved a truly stunning breakthrough.[17] The AI is now able to achieve nearly experimental-level accuracy for almost any protein it is given. This suddenly expands the number of protein structures available to biologists from over 180,000[18] to hundreds of millions, and it will soon reach the billions.[19] This will greatly accelerate the pace of biomedical discoveries.

At present, AI drug discovery is a human-guided process—scientists have to identify the problem they are trying to solve, formulate the problem in chemical terms, and set the parameters of the simulation. Over the coming decades, though, AI will gain the capacity to search more creatively. For example, it might identify a problem that human clinicians hadn't even noticed (e.g., that a particular subset of people with a certain disease don't respond well to standard treatments) and propose complex and novel therapies.

Meanwhile, AI will scale up to modeling ever larger systems in simulation—from proteins to protein complexes, organelles, cells, tissues, and whole organs. Doing so will enable us to cure diseases whose complexity puts them out of the reach of today's medicine. For example, the past decade has seen the introduction of many promising cancer treatments, including immunotherapies like CAR-T, BiTEs, and immune checkpoint inhibitors.[20] These have saved thousands of lives, but they frequently still fail because cancers learn to resist them. Often this involves tumors altering their local environment in ways we can't fully understand with current techniques.[21] When AI can robustly simulate the tumor and its microenvironment, though, we'll be able to tailor therapies to overcome this resistance.

Likewise, such neurodegenerative diseases as Alzheimer's and Parkinson's involve subtle, complex processes that cause misfolded proteins to build up in the brain and inflict harm.[22] Because it's impossible to study these effects thoroughly in a living brain, research has been extremely slow and difficult. With AI simulations we'll be able to understand their root causes and treat patients effectively long before they become debilitated. Those same brain-simulation tools will also let us achieve breakthroughs for mental health disorders, which are expected to affect more than half the US population at some point in their lives.[23] So far doctors have relied on blunt-approach psychiatric drugs like SSRIs and SNRIs, which temporarily adjust chemical imbalances but often have modest benefits, don't work at all for some patients, and carry long lists of side effects.[24] Once AI gives us a full functional understanding of the human brain—the most complex structure in the known universe!—we'll be able to target many mental health problems at their source.

In addition to the promise of AI for discovering new therapies, we are also moving toward a revolution in the trials we use to validate them. The FDA is now incorporating simulation results in its regulatory approval process.[25] In the coming years this will be especially important in cases similar to the COVID-19 pandemic—where a new viral threat emerges suddenly and millions of lives can be saved through accelerated vaccine development.[26]

But suppose we could digitize the trials process altogether—using AI to assess how a drug would work for tens of thousands of (simulated) patients for a (simulated) period of years, and do all of this in a matter of hours or days. This would enable much richer, faster, and more accurate trial results than the relatively slow, underpowered human trials we use today. A major drawback of human trials is that (depending on the type of drug and the stage of the trial) they involve only about a dozen to a few thousand subjects.[27] This means that in any given group of subjects, few of them—if any—are statistically likely to react to the drug in exactly the way your body would. Many factors can affect how well a pharmaceutical works for you, such as genetics, diet, lifestyle, hormone balance, microbiome, disease subtype, other drugs you're taking, and other diseases you may have. If no one in the clinical trials matches you along all those dimensions, it might be the case that even though a drug is good for the average person, it's bad for you.

Today a trial might result in an average 15 percent improvement in a certain condition for 3,000 people. But simulated trials could reveal hidden details. For example, a certain subset of 250 people from that group (e.g., those with a certain gene) will actually be harmed by the drug, experiencing 50 percent worse conditions, while a different subset of 500 (e.g., those who also have kidney disease) will see a 70 percent improvement. Simulations will be able to find numerous such correlations, yielding highly specific risk-benefit profiles for each individual patient.

The introduction of this technology will be gradual because the computational demands of biological simulations will vary among applications. Drugs consisting mainly of a single molecule are at the easier end of the spectrum and will be first to be simulated. Meanwhile, techniques like CRISPR and therapies intended to affect gene expression involve extremely complex interactions between many kinds of biological molecules and structures and will accordingly take longer to simulate satisfactorily *in silico*. To replace human trials as the primary testing method, AI simulations will need to model not just the direct action of a given therapeutic agent, but how it fits into a whole body's complex systems over an extended period.

It is unclear how much detail will ultimately be required for such simulations. For example, it seems unlikely that skin cells on your thumb are relevant for testing a liver cancer drug. But to validate these tools as safe, we'll likely need to digitize the entire human body at essentially molecular resolution. Only then will researchers be able to robustly determine which factors can be confidently abstracted away for a given application. This is a long-term goal, but it is one of the most profoundly important lifesaving objectives for AI—and we will be making meaningful progress on it by the end of the 2020s.

There will likely be substantial resistance in the medical community to increasing reliance on simulations for drug trials—for a variety of reasons. It is very sensible to be cautious about the risks. Doctors won't want to change approval protocols in a way that could endanger patients, so simulations will need a very solid track record of performing as well as or better than current trial methods. But another factor is liability. Nobody will want to be the person who approved a new and promising treatment on the chance that it turns out to be a disaster. Thus, regulators will need to anticipate these emerging approaches and be proactive to make sure that the incentives are balanced between appropriate caution and lifesaving innovation.

Even before we have robust biosimulation, though, AI is already making an impact in genetic biology. The 98 percent of genes that do not code for proteins were once dismissed as "junk" DNA.[28] We now know that they are critical to gene expression (which genes are actively used and to what extent), but it is very hard to determine these relationships from the noncoding DNA itself. Yet because it can detect very subtle patterns, AI is starting to break this logjam, as it did with the 2019 discovery by New York scientists of links between noncoding DNA and autism.[29] Olga Troyanskaya, the project's lead researcher, said that it is "the first clear demonstration of non-inherited, noncoding mutations causing any complex human disease or disorder."[30]

In the wake of the COVID-19 pandemic, there is also new urgency to the challenge of monitoring infectious diseases. In the past, epidemiologists had to choose from among several imperfect types of data when trying to predict viral outbreaks across the United States. A new AI system called ARGONet integrates disparate kinds of data in real time and weights them based on their predictive power.[31] ARGONet combines electronic medical records, historical data, live Google searches by worried members of the public, and spatial-temporal patterns of how flu spreads from place to place.[32] Lead researcher Mauricio Santillana of Harvard explained, "The system continuously evaluates the predictive power of each independent method, and recalibrates how this information should be used to produce improved flu estimates."[33] Indeed, 2019 research showed that ARGONet outperformed all previous approaches. It bested Google Flu Trends in 75 percent of states studied and was able to predict statewide flu activity a week ahead of the CDC's normal methods.[34] More new AI-driven approaches are now being developed to help stop the next major outbreak.

In addition to scientific applications, AI is gaining the ability to surpass human doctors in clinical medicine. In a 2018 speech I predicted that within a year or two a neural net would be able to analyze radiology images as well as human doctors do. Just two weeks later, Stanford researchers announced CheXNet, which used 100,000 X-ray images to train a 121-layer convolutional neural network to diagnose fourteen different diseases. It outperformed the human doctors to whom it was compared, providing preliminary but encouraging evidence of huge diagnostic potential.[35] Other neural networks have shown similar capabilities. A 2019 study showed that a neural net analyzing natural-language clinical metrics was able to diagnose pediatric diseases better than eight junior physicians exposed to the same data—and outperformed all twenty human doctors in some areas.[36] In 2021 a Johns Hopkins team developed an AI system called DELFI that is able to recognize subtle patterns of DNA fragments in a person's blood to detect 94 percent of lung cancers via a simple lab test—something even expert humans cannot do alone.[37]

Such clinical tools are rapidly making the jump from proof of concept to large-scale deployment. In July 2022, *Nature Medicine* published results of a massive study of more than 590,000 hospital patients who were monitored with an AI-powered system called the Targeted Real-Time Early Warning System (TREWS) to detect sepsis—a life-threatening infection response that kills around 270,000 Americans a year.[38] TREWS gave doctors early warning to begin treatment, lowering sepsis deaths among patients by 18.7 percent—indicating the potential to save tens of thousands of lives annually as adoption widens. Increasingly, such models will incorporate richer forms of information like data from our wearable fitness trackers and will be able to suggest treatment before someone even knows they're sick.

As the 2020s progress, AI-powered tools will reach superhuman performance levels at virtually all diagnostic tasks.[39] Interpreting medical imaging is a task that allows neural networks to use their natural strengths most powerfully. Clinically significant information can be buried in images too subtly for a human to visually detect, yet can be obvious to an AI system. And unlike other forms of diagnosis, which require integrating many disparate and qualitative kinds of information, the pixel patterns in images are totally reducible to quantifiable data—AI's strong suit. This is why medical imaging is one of the first fields to see AI reach such remarkable levels of performance. For the same reason, it will be relatively easy to generalize systems like CheXNet and its cousin CheXpert to other kinds of medical image analysis. Ultimately AI will likely be able to unlock vast untapped potential in medical images—perhaps identifying risk factors hidden in apparently healthy organs, which might allow lifesaving preventive measures long before a problem causes damage.

Surgeries will also benefit from this revolution, as both the amount of quality data about surgeries and available computational resources are rapidly growing.[40] For years robots have been used to assist human doctors, but they are now demonstrating an ability to perform without human participation. In the United States in 2016, the Smart Tissue Autonomous Robot (STAR) achieved better outcomes than human surgeons at an intestinal stitching task in animal testing.[41] In 2017, a Chinese robot completed a full dental implant surgery on its own, a very high-precision procedure.[42] Then, in 2020, Neuralink debuted a surgical robot that automates much of the process of implanting a brain–computer interface, and the company is working toward full autonomy.[43]

The average human surgeon may perform several hundred surgeries per year, amounting to, at most, a few tens of thousands over a full career. In many cases, such as surgeons in specialties requiring longer and more complex procedures, this number may be even smaller. By contrast, the AI powering robotic surgeons will be able to learn from the experience of any surgery that system performs, anywhere in the world. This will cover a much wider range of clinical circumstances than any human could encounter—potentially many millions of surgeries. In addition, the AI will be able to perform billions of simulated surgeries, tinkering with unusual variables that would be impossible or unethical to train on in a clinical setting. For example, simulated surgeries could train robotic surgeons in dealing with rare combinations of diseases, or in pushing the limits of trauma medicine with complex injuries that most surgeons won't see even once in a career. This will make surgery much safer and more effective than it is today.[44]

THE 2030S AND 2040S: DEVELOPING AND PERFECTING NANOTECHNOLOGY

It is remarkable that biology has created a creature as elaborate as a human being, one with both the intellectual dexterity and the physical coordination (e.g., opposable thumbs) to enable technology. However, we are far from optimal, especially with regard to thinking. As Hans Moravec argued back in 1988, when contemplating the implications of technological progress, no matter how much we fine-tune our DNA-based biology, our flesh-and-blood systems will be at a disadvantage relative to our purpose-engineered creations.[45] As writer Peter Weibel put it, Moravec understood that in this regard humans can only be "second-class robots."[46] This means that even if we work at optimizing and perfecting what our biological brains are capable of, they will be billions of times slower and far less capable of what a fully engineered body will be able to achieve.

A combination of AI and the nanotechnology revolution will enable us to redesign and rebuild—molecule by molecule—our bodies and brains and the worlds with which we interact. Human neurons fire around two hundred times per second at most (with one thousand as an absolute theoretical maximum), and in reality most probably sustain averages of less than one fire per second.[47] By contrast, transistors can now cycle over one trillion times per second, and retail computer chips exceed five billion cycles per second.[48] This disparity is so great because the cellular computing in our brains uses a much slower, clunkier architecture than what precision engineering makes possible in digital computing. And as nanotechnology advances, the digital realm will be able to pull even further ahead.

Also, the size of the human brain limits its total processing power to, at most, about 10^{14} operations per second, according to my estimate in *The Singularity Is Near*—which is within an order of magnitude of Hans Moravec's estimate based on a different analysis.[49] The US super-computer Frontier can already top 10^{18} operations per second in an AI-relevant performance benchmark.[50] Because computers can pack transistors more densely and efficiently than the brain's neurons, and because they can both be physically larger than the brain and network together remotely, they will leave unaugmented biological brains in the dust. The future is clear: minds based only on the organic sub-strates of biological brains can't hope to keep up with minds aug-mented by nonbiological precision nanoengineering.

The first reference to nanotechnology was made by the physicist Richard Feynman (1918–1988) in his seminal 1959 lecture "There's Plenty of Room at the Bottom," in which he described the inevitability of creating machines at the scale of individual atoms, as well as the profound implications of doing so.[51] As Feynman said: "The princi-ples of physics, as far as I can see, do not speak against the possibility of maneuvering things atom by atom. . . . It would be, in principle, possible . . . for a physicist to synthesize any chemical substance that the chemist writes down. . . . How? Put the atoms down where the chemist says, and so you make the substance."[52] Feynmann was opti-mistic: "The problems of chemistry and biology can be greatly helped if our ability to see what we are doing, and to do things on an atomic level, is ultimately developed—a development which I think cannot be avoided."

In order for nanotechnology to have an impact on large objects, it needs to have a self-replication system. The idea of how to create a self-replicating module was first formalized by legendary mathemati-cian John von Neumann (1903–1957) in a series of lectures during the late 1940s and in a 1955 *Scientific American* article.[53] But the full range of his ideas was not collected and widely published until 1966, almost a decade after his death. Von Neumann's approach was highly abstract and mathematical, and focused mostly on the logical underpinnings rather than the detailed physical practicalities of building self-replicating machines. In his concept, a self-replicator includes a "universal computer" and a "universal constructor." The computer runs a program that controls the constructor, and the constructor can copy both the whole self-replicator and the program—so the copies can do likewise indefinitely.[54]

In the mid-1980s, engineer K. Eric Drexler founded the modern field of nanotechnology, building upon this concept from von Neumann.[55] Drexler designed an abstract machine that used atoms and molecular fragments found in ordinary substances to provide the materials for his von Neumann–style constructor, which would feature a computer able to direct the placement of atoms.[56] Drexler's "assembler" could essentially make anything in the world, so long as its structure is atomically stable. It is this flexibility and generalizability that distinguishes the molecular mechanosynthesis approach Drexler pioneered from biology-based approaches, which would assemble objects at nanoscales but be much more limited in the designs and materials available.

Drexler outlined a very simple computer using molecular "interlocks" rather than transistor gates. (These were conceptual; he hasn't actually built them yet.)[57] Each interlock would require only six cubic nanometers of space and could switch its state in one ten-billionth of a second—allowing a feasible computing speed around one billion operations per second.[58] Many variations of this computer have been proposed, with increasing refinement. In 2018 an all-mechanical computing system suitable for nanoscale implementations was devised by Ralph Merkle and several collaborators.[59] Their detailed design (again conceptual) provides for about 10^{20} logic gates per liter and would operate at 100 MHz resulting in up to 10^{28} computing operations per second per liter of computer volume (though heat dissipation would require that this volume have high surface area).[60] The amount of power expended by this design would be on the order of one hundred watts.[61] Since there are around eight billion people in the world, emulating the brain computation of *all* human beings together would thus take fewer than 10^{24} operations per second (10^{14} per person times 10^{10} people).[62]

As discussed in chapter 2, my 10^{14} estimate is for a simulation of every neuron. Yet the brain employs massive parallelism. Because the wet biological environment inside our skulls is (at least at the molecular level) a very turbulent place, any single neuron may die or simply fail to fire at the correct instant. If human cognition depended heavily on the performance of any single neuron, it would be very unreliable. But when many neurons work together in parallel, the "noise" gets canceled out and we're able to think just fine.

When building nonbiological computers, though, we can control the internal environment much more precisely. The interior of a computer chip is much more clean and stable than brain tissue, so all that parallelism won't be necessary. This will allow for more efficient computation, so it's plausible that a mind could be simulated with even fewer than 10^{14} operations per second. But because it remains unclear how much parallelism brains have, I use this larger estimate to be conservative. Theoretically, then, a perfectly efficient one-liter nanologic computer would provide the equivalent of about 10,000 times 10 billion human beings (or about 100 trillion human beings) in terms of brain capability. To be clear, I'm not arguing that this is achievable in practice. The point is that nanoscale engineering offers *staggering* amounts of headroom for future progress. Even a tiny fraction of a percent of theoretical maximums would be an utterly revolutionary new paradigm for computing—and allow nanoscale machines to attain useful amounts of computing power.

In the context of self-replicating nanobots, this would enable the massive coordination required to achieve macroscopic results. The control system would be similar to the computer instruction architecture called SIMD (single instruction, multiple data), meaning that a single computational unit would read the program instructions and then transmit them to the trillions of molecular-size assemblers (each with its own simple computer) simultaneously.[63]

Using this "broadcast" architecture would also address a key safety concern. If the self-replication process gets out of control, or in the event of a bug or security breach, the source of the replication instructions could be immediately shut down, preventing any further nanobot activity.[64] As will be discussed further in chapter 7, the worst-case scenario for nanotechnology would be so-called gray goo—self-replicating nanobots that form an uncontrollable chain reaction.[65] In theory, this could consume most of the biomass of the earth and turn it into more nanobots. But the "broadcast" architecture advocated by Ralph Merkle is a strong defense against this. If the instructions must all come from a central source, shutting off the broadcast in an emergency would render the nanobots inactive and physically unable to continue self-replicating.

The actual construction machine taking these instructions would be a simple molecular robot with a single arm, similar to von Neumann's universal constructor, built on a tiny scale.[66] The feasibility of building molecular-scale robot arms, gears, rotors, and motors has already been demonstrated repeatedly.[67]

Physics does not allow a molecular-scale arm moving atoms around to grasp and carry them the way a human hand would. This makes the future of nanotechnology controversial. In 2001, American physicist and chemist Richard Smalley began a public debate with Eric Drexler over whether atomically precise manufacturing using "molecular assemblers" will ever be possible.[68] Both Smalley and Drexler had made essential contributions to the field of nanotechnology, but they differed in their approaches. Drexler argued that "top-down" methods are the ultimate goal of nanotech, allowing fabricator machines to build nanobots from scratch. Smalley argued that the laws of physics make this an impossibility—that "bottom-up" approaches of biology-style self-assembly are the only sensible goal. Smalley's objection was twofold: the "fat fingers problem," in which the manipulator arms necessary for moving a reaction's atoms into place would be too bulky to work effectively at the nanoscale, and the "sticky fingers problem," in which atoms being moved around would adhere to the manipulator arms.

Drexler responded that techniques designed to use a single manipulator arm wouldn't confront the former problem, and biological machines like enzymes and ribosomes already demonstrate that the latter is possible to overcome. As the debate heated up in 2003, I offered my own comments, mainly coming down on Drexler's side of the argument.[69] Looking back almost two decades later, I am pleased to say that recent advances in nanotechnology are making the "top-down" view look more and more plausible—even though it will be at least a decade before the field starts maturing, likely aided by advances in AI. Scientists have already made significant strides in precision control of atoms, and the 2020s will see many more key breakthroughs.

Drexler's design for a nanoscale constructor arm still appears to be the most promising. Instead of an awkward and complex grasping claw, it would have a *single* tip that uses a mechanical and electric function to pick up an atom or small molecule and then release it at a different location.[70] Drexler's 1992 book *Nanosystems* provides a number of different chemistries that could achieve this.[71] One approach is to move carbon atoms around to build objects out of a nano-size diamond substance called diamondoid.[72]

Diamondoids are tiny cages of carbon atoms (as few as ten), arranged as the most basic types of diamond crystal—with hydrogen atoms bonded to the outside of the cage. These may be able to form the building blocks of extremely light, strong nanoengineered structures. When Drexler explored the ideas of nanotechnology and diamondoid manufacturing in his 1986 book *Engines of Creation* and in *Nanosystems*, he inspired science fiction author Neal Stephenson to write the Hugo Award–winning 1995 novel *The Diamond Age*, imagining a future where diamond-based nanotechnology defines civilization, much like bronze defined the Bronze Age and iron the Iron Age.[73] More than a quarter century beyond the novel's release, diamondoid research has made huge advances, and scientists are beginning to see practical applications in laboratory research. Over the next decade, though, AI will allow detailed chemistry simulations that unlock much faster progress.

Many proposed nanotech designs use this approach. We know from the chemical vapor deposition process that artificial diamonds can be created this way.[74] Not only is diamondoid extremely strong, but it can have impurities precisely added via "doping" to alter physical properties like thermal conductivity or to create electronic components like transistors.[75] Research over the past decade has shown promising methods of engineering both electronic and mechanical systems from various arrangements of carbon atoms at the nanoscale.[76] This area of nanotech is now attracting serious attention all over the world, but some of the most intriguing proposals have come from Ralph Merkle and his co-authors.[77] As early as 1997 Merkle devised a "metabolism" for an assembler that could construct hydrocarbons like diamondoids from a "feedstock solution" of butadiyne.[78]

Since *The Singularity Is Near* was published in 2005, there have also been exciting new breakthroughs in graphene (a one-atom-thick hexagonal lattice of carbon), carbon nanotubes (essentially rolled tubes of graphene), and carbon nanothreads (nearly one-dimensional strands of carbon surrounded by hydrogen), all of which will see a huge variety of practical applications over the next two decades.[79]

Many pathways to this type of mechanosynthesis and other nanotechnologies are currently being pursued.[80] Among them are DNA origami[81] and DNA nanorobotics,[82] bio-inspired molecular machines,[83] molecular Lego,[84] single-atom qubits for quantum computing,[85] electron beam–based atom placement,[86] hydrogen depassivation lithography,[87] and scanning tunneling microscope–based manufacturing.[88] There are several stealth projects within these spaces that are making steady progress, and—considering that superhuman engineering AI will be available by the end of the 2020s to solve remaining problems—we are on track for nanotechnology concepts using atom-by-atom placement to be implemented sometime in the 2030s.

In practice, this will entail information being imparted to "dumb" raw materials through an exponential process. A central computer would broadcast commands simultaneously to a small core of starting nanobots, placed amid the feedstock of atoms or basic molecules they need. They would be directed to self-replicate, iteratively creating a cascade of copies of themselves, soon reaching into the trillions of assemblers. Then the computer would give these molecular robots commands for how to build the desired structures.

In the technology's mature form, a molecular assembler might be a tabletop-size unit capable of manufacturing virtually any physical product for which it has the requisite atoms. This will require mastering one of the great challenges of nanoscale manufacturing: generalizability. It is one thing to design an assembler that can make one particular kind of substance (like diamondoid), but it is another to design one capable of mastering very diverse chemistries. Unlocking the latter capabilities will require extremely advanced AI. Thus, we can expect to see relatively chemically homogenous objects (like gemstones, furniture, or clothing) built by assemblers well before those with widely varied chemical compositions and highly complex microstructures (like a cooked meal, a bionic organ, or a computer more powerful than all unenhanced human brains combined).

Once we have advanced nanomanufacturing, the incremental cost of manufacturing any physical object (including molecular assemblers themselves) would be only pennies per pound—essentially just the cost of the atomic precursor materials.[89] Drexler's 2013 estimates for the total cost for a molecular manufacturing process fall around $2 per kilogram, no matter what's being made, whether diamonds or food.[90] And because nanoengineered materials can be much stronger than steel or plastic, most structures could be built from around a tenth the mass. Even when the raw materials used in finished goods are expensive—for example, gold, copper, and rare earth metals in electronics—it will often be possible in the future to substitute components built with cheaper, more abundant elements like carbon.

The true value of products, then, would lie in the information they contain—in essence, all the innovation that has gone into them, from creative ideas to lines of software code that control their manufacture. This has already taken place for goods that can be digitized. Think of e-books. When books were first invented, they had to be copied by hand, so labor was a massive component of their value. With the advent of the printing press, physical materials like paper, binding, and ink took on the dominant share of the price. But with e-books, the costs of energy and computation to copy, store, and transmit a book are effectively zero. What you're paying for is creative assembly of information into something worth reading (and often some ancillary factors, like marketing). One way to see this difference for yourself is by browsing Amazon for blank journals and then browsing for hardcover novels with roughly similar bindings. If you look only at the prices, you won't be immediately and consistently able to tell which are which. On the other hand, e-book novels are usually priced at several dollars, while the idea of paying any money at all for a blank e-book is laughable. This is what it means for a product's value to be entirely information.

The nanotech revolution will bring this transformative shift into the physical world. In 2023 the value of physical products comes from many sources, especially raw materials, manufacturing labor, factory machine time, energy costs, and transportation. But convergent innovations will be dramatically reducing most of those costs in the coming decades. Raw materials will be cheaper to extract or synthesize with automation, robotics will replace expensive human labor, high-priced factory machines will themselves become cheaper, energy prices will fall due to better solar photovoltaics and energy storage (and eventually fusion), and autonomous electric vehicles will drive down transportation costs. As all these components of value become less expensive, the proportional value of the information contained in products will increase. Indeed, we are already going in that direction, as the "information content" of most products is rapidly increasing—and will ultimately get very close to 100 percent of their value.

In many cases, this will make products cheap enough that they can be free to consumers. Again, we can look to the digital economy to see how this has already played out. As discussed in chapter 5, platforms like Google and Facebook spend billions of dollars on their infrastructure, but the average cost per search or per Like is so low that it makes more sense to make them totally free for users—with other revenue sources like ads being used to make money. In a similar way, it's possible to imagine a future where people watch political ads or share personal data in order to get free nano-manufactured products. Governments might also offer such products as incentives for volunteer service, continuing education, or maintaining healthy habits.

This dramatic reduction of physical scarcity will finally allow us to easily provide for the needs of everyone. Note that this is a prediction about technological capabilities, but culture and politics will play a large role in determining how rapidly the economy changes. It will be a challenge to guarantee that these benefits are shared widely and fairly. That said, I am optimistic. The idea that wealthy elites would simply hoard this new abundance is grounded in a misunderstanding. When goods are truly abundant, hoarding them is pointless. Nobody bottles up air for themselves, because it's easy to get and there's enough for everyone. Similarly, when other people use Wikipedia, it doesn't make any of the information less available to you. The next step is simply extending that kind of abundance into the world of material goods.

While nanotechnology will allow the alleviation of many kinds of physical scarcity, economic scarcity is also partly driven by culture—especially when it comes to luxury goods. For example, to the naked eye, artificial diamonds are already indistinguishable from natural diamonds, but they sell for about 30 to 40 percent less.[91] This component of the price has nothing to do with the ornamental beauty of the diamonds, but rather with our cultural conventions that assign more value to diamonds that formed naturally. Likewise, paintings by the old masters aren't really any better at sprucing up a living room than high-quality reproductions, but because people value their status as originals, they may sell for roughly a *million* times more.[92] So the nanotech manufacturing revolution won't eliminate all economic scarcity.

Historic diamonds and Rembrandts will remain scarce. But over the scale of generations, cultural values do change. Who's to say whether the people who are now children will choose different values when they are adults? Or their own children?

APPLYING NANOTECHNOLOGY TO HEALTH AND LONGEVITY

As I discussed in my life extension book *Transcend*,[93] we are now in the later stages of the first generation of life extension, which involves applying the current class of pharmaceutical and nutritional knowledge to overcoming health challenges. This has been an evolving process that constantly applies new ideas, and it is the basis for the regimen I've followed for my own health in recent decades.

In the 2020s we are starting the second phase of life extension, which is the merger of biotechnology with AI. This will involve developing and testing breakthrough treatments in digital biology simulators. Early stages of this have already begun, and with these techniques we will be able to discover very powerful new therapies in days rather than years.

The 2030s will usher in the third phase of life extension, which will be to use nanotechnology to overcome the limitations of our biological organs altogether. As we enter this phase, we'll greatly extend our lives, allowing people to greatly transcend the normal human limit of 120 years.[94]

Only one person, Jeanne Calment—a Frenchwoman who survived to age 122—is documented to have lived longer than 120 years.[95] So why is this such a hard limit to human longevity? One might guess that the reasons people don't make it past this age are statistical—that elderly people face a certain risk of Alzheimer's, stroke, heart attack, or cancer every year, and that after enough years of being exposed to these risks, everyone eventually dies of something. But that's not what's happening. Actuarial data shows that from age 90 to age 110, a person's chances of dying in the following year increase by about 2 percentage points annually.[96] For example, an American man at age 97 has about a 30 percent chance of dying before 98, and if he makes it that far he will have a 32 percent chance of dying before 99. But from age 110 onward, the risk of death rises by about 3.5 percentage points a year.

Doctors have offered an explanation: At around age 110, the bodies of the oldest people start breaking down in ways that are qualitatively different from the aging of younger senior citizens.[97] Supercentenarian (110-plus) aging is not simply a continuation or worsening of the same kinds of statistical risks of late adulthood. While people at that age also have an annual risk from ordinary diseases (although the worsening of these risks may decelerate in the very old), they additionally face new challenges like kidney failure and respiratory failure. These often seem to happen spontaneously—not as a result of lifestyle factors or any disease onset. The body apparently just starts breaking down.

Over the past decade, scientists and investors have started giving much more serious attention to finding out why. One of the leading researchers in this field is biogerontologist Aubrey de Grey, founder of the LEV (Longevity Escape Velocity) foundation.[98] As de Grey explains, aging is like the wear on the engine of an automobile—it is damage that accumulates as a result of the system's normal operation. In the human body's case, that damage largely comes from a combination of cellular metabolism (using energy to stay alive) and cellular reproduction (mechanisms for self-replication). Metabolism creates waste in and around cells and damages structures through oxidation (much like the rusting of a car!).

When we're young, our bodies are able to remove this waste and repair the damage efficiently. But as we get older, most of our cells reproduce over and over, and errors accumulate. Eventually the damage starts piling up faster than the body can fix it.

In a person in their seventies, eighties, or nineties, this damage will likely cause one fatal problem a significant amount of time before it causes several. So if science develops a drug to successfully treat an otherwise fatal cancer in an eighty-year-old, that person might expect to live almost another decade before something else kills him. But eventually everything starts failing at once and it is no longer effective to treat the symptoms of the damage caused by aging. Instead, longevity researchers argue, the only solution is to cure aging itself. The SENS (Strategies for Engineered Negligible Senescence) Research Foundation has proposed a detailed research agenda for how to do this (even though it will certainly take decades to fully accomplish).[99]

In short, we need the ability to repair damage from aging at the level of individual cells and local tissues. There are a number of possibilities being explored for how to achieve this, but I believe the most promising ultimate solution is nanorobots capable of entering the body and carrying out this repair directly. This wouldn't make people immortal. We could still be killed by accidents and mishaps, but the annual risk of death would no longer increase as we got older—so many people could live well past 120 in good health.

And we don't need to wait until these technologies are fully mature in order to benefit. If you can live long enough for anti-aging research to start adding at least one year to your remaining life expectancy annually, that will buy enough time for nanomedicine to cure any remaining facets of aging. This is longevity escape velocity.[100] This is why there is sound logic behind Aubrey de Grey's sensational declaration that the first person to live to 1,000 years has likely already been born. If the nanotechnology of 2050 solves enough issues of aging for 100-year-olds to start living to 150, we'll then have until 2100 to solve whatever new problems may crop up at that age. With AI playing a key role in research by then, progress during that time will be exponential. So even though these projections are admittedly startling, and even sound absurd to our intuitive linear thinking—we have solid reasons to see this as a likely future.

I've had many conversations over the years about life extension, and the idea often meets resistance. People become upset when they hear of an individual whose life has been cut short by a disease, yet when confronted with the possibility of generally extending all human life, they react negatively. "Life is too difficult to contemplate going on indefinitely" is a common response. But people generally do not want to end their lives at any point unless they are in enormous pain—physically, mentally, or spiritually. And if they were to absorb the ongoing improvements of life in all its dimensions, as elaborated on in chapter 4, most such afflictions would be alleviated. That is, extending human life would also mean vastly *improving* it.

To imagine how life extension improves quality of life, it's helpful to think back to a century ago. In 1924 life expectancy in the United States averaged about 58.5 years, so babies born that year were statistically expected to die in 1982.[101] But medicine saw so many improvements during that interval that many of these individuals lived into the 2000s or 2010s. Thanks to this life extension, they got to enjoy retirements during an age with cheap air travel, safer cars, cable television, and the internet. For babies born in 2024, the technological advances during the years that get added to their lives will be exponentially faster than those of the previous century. In addition to these enormous material advantages, they will also enjoy richer culture—all the art, music, literature, television, and video games created by humanity during those extra years. Perhaps most importantly, they will get to enjoy more time with family and friends, loving and being loved. All this, in my view, is what gives life its greatest meaning.

But how will nanotechnology actually make this possible? As I see it, the long-term goal is medical nanorobots. These will be made from diamondoid parts with onboard sensors, manipulators, computers, communicators, and possibly power supplies.[102] It is intuitive to imagine nanobots as tiny metal robotic submarines chugging through the bloodstream, but physics at the nanoscale requires a substantially different approach. At this scale, water is a powerful solvent, and oxidant molecules are highly reactive, so strong materials like diamondoid will be needed.

And whereas macro-scale submarines can smoothly propel themselves through liquids, for nanoscale objects, fluid dynamics are dominated by sticky frictional forces.[103] Imagine trying to swim through peanut butter! So nanobots will need to harness different principles of propulsion. Likewise, nanobots probably won't be able to store enough onboard energy or computing power to accomplish all their tasks independently, so they will need to be designed to draw energy from their surroundings and either obey outside control signals or collaborate with one another to do computation.

To maintain our bodies and otherwise counteract health problems, we will all need a huge number of nanobots, each about the size of a cell. The best available estimates say that the human body is made of several tens of trillions of biological cells.[104] If we augment ourselves with just one nanobot per one hundred cells, this would amount to several hundred billion nanobots. It remains to be seen, though, what ratio is optimal. It might turn out, for example, that advanced nanobots could be effective even at a cell-to-nanobot ratio several orders of magnitude greater.

One of the main effects of aging is degrading organ performance, so a key role of these nanobots will be to repair and augment them. Other than expanding our neocortex, as discussed in chapter 2, this will mainly involve helping our nonsensory organs to efficiently place substances into the blood supply (or lymph system) or remove them.[105] For example, the lungs put in oxygen and take out carbon dioxide.[106] The liver and kidneys take out toxins.[107] The entire digestive tract puts nutrients into our blood supply.[108] Various organs such as the pancreas produce hormones that control metabolism.[109] Changes in hormone levels can result in diseases like diabetes. (There are already devices[110] that can measure blood insulin levels and transfer insulin into the bloodstream, much like a real pancreas.)[111] By monitoring the supply of these vital substances, adjusting their levels as needed, and maintaining organ structures, nanobots can keep a person's body in good health indefinitely. Ultimately, nanobots will be able to replace biological organs altogether, if needed or desired.

But nanobots won't be limited to preserving the body's normal function. They could also be used to adjust concentrations of various substances in our blood to levels more optimal than what would normally occur in the body. Hormones could be tweaked to give us more energy and focus, or speed up the body's natural healing and repair. If optimizing hormones could make our sleep more efficient,[112] that would in effect be "backdoor life extension." If you just go from needing eight hours of sleep a night to seven hours, that adds as much waking existence to the average life as five more years of life span!

Eventually, using nanobots for body maintenance and optimization should prevent major diseases from even arising. Of course, there will likely be a period when nanobots are available but not everyone has used them yet for this purpose. Thus, conditions like cancer will need to be addressed once they have already been diagnosed.

Part of why cancer can be so hard to eliminate is because each cancer cell has the ability to self-replicate, so every single cell has to be removed.[113] Although the immune system is often able to control the very earliest phases of cancerous cell division, once a tumor does get established, it can develop resistance to the body's immune cells. At that point, even if a treatment destroys most of the cancer cells, the survivors can start growing new tumors. A subpopulation known as cancer stem cells are especially likely to be these dangerous survivors.[114]

Although cancer medicine has made amazing strides over the past decade and will make even greater breakthroughs with the help of AI during this decade, we're still using relatively blunt tools to treat it. Chemotherapy often fails to eradicate cancer entirely and causes serious collateral damage to other, noncancerous cells all over the body.[115] Not only does this result in brutal side effects for many cancer patients, but it weakens the immune system and makes them more vulnerable to other health risks. Even advanced immunotherapies and targeted drugs fall well short of complete effectiveness and precision.[116] By contrast, medical nanobots will be able to examine each individual cell and determine whether it is cancerous or not, and then destroy all of the malignant ones. Recall the automobile mechanic analogy from the beginning of this chapter. Once nanobots can selectively repair or destroy individual cells, we will fully master our biology, and medicine will become the exact science it has long aspired to be.

Achieving this will also entail gaining complete control over our genes. In our natural state, cells reproduce by copying the DNA in each nucleus.[117] If there is a problem with the DNA sequence in a group of cells, there is no way to address it without updating it in every individual cell.[118] This is an advantage in unenhanced biological organisms, because random mutations within individual cells are unlikely to cause fatal damage to the whole body. If any mutation in any cell in our bodies were instantly copied to every other cell, we wouldn't be able to survive. But the decentralized robustness of biology is a major challenge to a species (like ours) that can edit the DNA of individual cells fairly well but has not yet mastered the nanotechnology needed to edit DNA effectively throughout the whole body.

If instead each cell's DNA code were controlled by a central server (as many electronic systems are), then we could change the DNA code by simply updating it once from that "central server." To do this, we would augment each cell's nucleus with a nanoengineered counterpart—a system that would receive the DNA code from the central server and then produce a sequence of amino acids from this code.[119] I use "central server" here as a shorthand for a more centralized broadcast architecture, but this probably does not mean every nanobot getting direct instructions from literally one computer. The physical challenges of nanoscale engineering might ultimately dictate that a more localized broadcast system is preferable. But even if there are hundreds or thousands of micro-scale (as opposed to nanoscale) control units placed around our bodies (which would be large enough for more complex communications with an overall control computer), this would be orders of magnitude more centralization than the status quo: independent functioning by tens of trillions of cells.

The other parts of the protein synthesis system, such as the ribosome, could be augmented in the same fashion. In this way we could simply turn off activity from malfunctioning DNA, whether it is responsible for cancer or genetic disorders. The nanocomputer maintaining this process would also implement the biological algorithms that govern epigenetics—how genes are expressed and activated.[120] As of the early 2020s, we still have a lot to learn about gene expression, but AI will allow us to simulate it in enough detail by the time nanotechnology is mature that nanobots will be able to precisely regulate it. With this technology we'll also be able to prevent and reverse the accumulation of DNA transcription errors, which are a major cause of aging.[121]

Nanobots will also be useful for neutralizing urgent threats to the body—destroying bacteria and viruses, halting autoimmune reactions, or drilling through clogged arteries. In fact, recent research by Stanford and Michigan State University has already created a nanoparticle that finds the monocytes and macrophages that cause atherosclerotic plaque and eliminates those cells.[122] Smart nanobots will be vastly more effective. At first such treatments would be initiated by humans, but ultimately they will be carried out autonomously; the nanobots will perform tasks on their own and report their activities (via a controlling AI interface) to humans monitoring them.

As AI gains greater ability to understand human biology, it will be possible to send nanobots to address problems at the cellular level long before they would be detectable by today's doctors. In many cases this will allow prevention of conditions that remain unexplained in 2023. Today, for example, about 25 percent of ischemic strokes are "cryptogenic"—they have no detectable cause.[123] But we know they must happen for some reason. Nanobots patrolling the bloodstream could detect small plaques or structural defects at risk of creating stroke-causing clots, break up forming clots, or raise the alarm if a stroke is silently unfolding.

Just as with hormone optimization, though, nanomaterials will allow us to not just restore normal body function but augment it beyond what our biology alone makes possible. Biological systems are limited in strength and speed because they must be constructed from protein. Although these proteins are three-dimensional, they have to be folded from a one-dimensional string of amino acids.[124] Engineered nanomaterials won't have this limitation. Nanobots built from diamondoid gears and rotors would be thousands of times faster and stronger than biological materials, and designed from scratch to perform optimally.[125]

Thanks to these advantages, even our blood supply may be replaced by nanobots. A design by founding Singularity University nanotechnology cochair Robert A. Freitas called the respirocyte is an artificial red blood cell.[126] According to Freitas's calculations, someone with respirocytes in his bloodstream could hold his breath for about four hours.[127] In addition to artificial blood cells, we'll eventually be able to engineer artificial lungs to oxygenate them more efficiently than the respiratory system that biology has given us. Ultimately, even hearts made from nanomaterials will make people immune to heart attacks and make cardiac arrest due to trauma much rarer.

Nanobots will also allow people to change their cosmetic appearance as never before. It is already possible for people to freely customize their avatars in digital environments like chat rooms and online role-playing games. People often use this as an outlet for expressing creativity and personality. In addition to personal appearance choices and fashion statements, users may embody virtual characters that are of a different age, gender, and even species from themselves. It remains to be seen how this will carry over to real life when nanotechnology gives people the ability to radically customize their physical bodies. Will it be just as common to make radical cosmetic changes in reality as it currently is in games? Or will psychological and cultural forces make people more conservative about these choices?

Yet the most important role of nanotech in our bodies will be augmenting the brain—which will eventually become more than 99.9 percent nonbiological. There are two distinct pathways by which this will happen. One is the gradual introduction of nanobots to the brain tissue itself. These may be used to repair damage or replace neurons that have stopped working. The other is connecting the brain to computers, which will both provide the ability to control machines directly with our thoughts and allow us to integrate digital layers of neocortex in the cloud. As described in more detail in chapter 2, this will go far beyond just better memory or faster thinking.

A deeper virtual neocortex will give us the ability to think thoughts more complex and abstract than we can currently comprehend. As a dimly suggestive example, imagine being able to clearly and intuitively visualize and reason about ten-dimensional shapes. That sort of facility will be possible across many domains of cognition. For comparison, the cerebral cortex (which is mainly made up of the neocortex) has an average of 16 billion neurons in a volume of roughly half a liter.[128] Ralph Merkle's design for a nanoscale mechanical computing system, described earlier in this chapter, could theoretically pack more than 80 quintillion logic gates into the same amount of space. And the speed advantage would be enormous: the electrochemical switching speed of mammalian neuron firing probably averages within an order of magnitude of once per second, as compared with likely around 100 million to one billion cycles per second for nanoengineered computation.[129] Even if only a minuscule fraction of these values are achievable in practice, it is clear that such technology will allow the digital parts of our brain (stored on nonbiological computing substrates) to vastly outnumber and outperform the biological ones.

Recall my estimate that the computation inside the human brain (at the level of neurons) is on the order of 10^{14} per second. As of 2023, $1,000 of computing power could perform up to 130 trillion computations per second.[130] Based on the 2000–2023 trend, by 2053 about $1,000 of computing power (in 2023 dollars) will be enough to perform around 7 million times as many computations per second as the unenhanced human brain.[131] If it turns out, as I suspect, that only a fraction of the brain's neurons are necessary to digitize the conscious mind (e.g., if we don't have to simulate the actions of many cells that govern the body's other organs), this point could be reached several years sooner. And even if it turns out that digitizing our conscious minds requires simulating every protein in every neuron (which I think is unlikely), it might take a few more decades to reach that level of affordability—but it's still something that would happen within the lifetimes of many people living today. In other words, because this future depends on fundamental exponential trends, even if we greatly change our assumptions about how easy it will be to affordably digitize ourselves, that won't dramatically change the date by which this milestone will be reached.

In the 2040s and 2050s, we will rebuild our bodies and brains to go vastly beyond what our biology is capable of, including their backup and survival. As nanotechnology takes off, we will be able to produce an optimized body at will: we'll be able to run much faster and longer, swim and breathe under the ocean like fish, and even give ourselves working wings if we want them. We will think millions of times faster, but most importantly, we will not be dependent on the survival of any of our bodies for our *selves* to survive.

CHAPTER 7

PERIL

"Environmentalists must now grapple squarely with the idea of a world that has enough wealth and enough technological capability, and should not pursue more."[1]

BILL McKIBBEN, ENVIRONMENTALIST AND
AUTHOR ON GLOBAL WARMING

"I just think that their flight from and hatred of technology is self-defeating. The Buddha, the Godhead, resides quite as comfortably in the circuits of a digital computer or the gears of a cycle transmission as he does at the top of a mountain or in the petals of a flower. To think otherwise is to demean the Buddha—which is to demean oneself."[2]

ROBERT M. PIRSIG, ZEN AND THE ART OF MOTORCYCLE MAINTENANCE

PROMISE AND PERIL

So far this book has explored the many ways that the final years until the Singularity will bring rapidly increasing human prosperity. But just as this progress will improve billions of lives, it will also heighten peril for our species. New, destabilizing nuclear weapons, breakthroughs in synthetic biology, and emerging nanotechnologies will all introduce threats we must deal with. And as AI itself reaches and surpasses human capabilities, it will need to be carefully aligned with beneficial purposes and specifically designed to avert accidents and thwart misuse. There is good reason to believe that our civilization will overcome these perils—not because the threats aren't real, but precisely because the stakes are so high. Not only does peril bring out the best of human ingenuity, but the same technological fields that give rise to danger are also creating powerful new tools to protect against it.

NUCLEAR WEAPONS

The very first time that humanity created a technology that could wipe out civilization was at the birth of my generation. I remember how in primary school we had to go under our desks for civil defense drills and hold our arms behind our heads to protect ourselves from a thermonuclear blast. We made it through intact, so this safety measure must have worked.

Humanity currently has roughly 12,700 nuclear warheads, around 9,440 of which are active and could be used in a nuclear war.[3] The United States and Russia each maintain around 1,000 large warheads that could be launched with less than a half hour's notice.[4] A major nuclear exchange could quickly kill several hundred million people from the weapons' direct effects.[5] But this does not include secondary effects that could kill billions.

Because the world's human population is so spread out, even an all-out nuclear exchange couldn't kill everyone via the initial warhead explosions.[6] But nuclear fallout could spread radioactive material over large areas of the globe, and the fires from burning cities would throw gigantic amounts of soot into the atmosphere, causing severe global cooling and mass starvation. Combined with catastrophic disruption to technologies like medicine and sanitation, this would extend the death toll greatly beyond the initial casualties. Future nuclear arsenals might include warheads "salted" with cobalt or other elements that could horribly worsen the lingering radioactivity. In 2008, Anders Sandberg and Nick Bostrom surveyed experts at the Global Catastrophic Risks Conference of Oxford University's Future of Humanity Institute. The experts' median responses estimated a 30 percent chance of at least one million dead in nuclear wars before the year 2100, a 10 percent chance of at least one billion dead, and a 1 percent chance of total extinction.[7]

As of 2023, there are five nations known to have a full "triad" of nuclear weapons (intercontinental ballistic missiles, air-delivery bombs, and submarine-launched ballistic missiles): the United States (5,244 warheads), Russia (5,889), China (410), Pakistan (170), and India (164).[8] Three other nations are known to have a more limited form of delivery system: France (290), the United Kingdom (225), and North Korea (around 30). Israel has not officially acknowledged having nuclear weapons, but is widely believed to have a full triad of around 90 warheads.

The global community has negotiated a number of international treaties[9] that have successfully reduced the total number of active warheads to fewer than 9,500 from a peak of 64,449 in 1986,[10] halted environmentally harmful aboveground testing,[11] and kept outer space nuclear-free.[12] But the current number of active weapons is still sufficient to end our civilization.[13] And even if the annual risk of nuclear war is low, the cumulative risks over decades or a century become extremely serious. As long as high-alert arsenals are maintained in their present form, it is likely only a matter of time before these weapons are used somewhere in the world, whether deliberately—by a government, terrorists, or rogue military officers—or by accident.

Mutually assured destruction (MAD), the most well-known strategy for reducing nuclear risk, was used by both the United States and the Soviet Union for most of the Cold War.[14] It entails sending potential enemies a credible message that, if they use nuclear weapons, they will be met with an overwhelming retaliatory response in kind. This approach is based on game theory. If a country realizes that using even one nuclear weapon will cause its opponent to launch a full-scale retaliation, there is no incentive to use those weapons because doing so would be suicidal. For MAD to work, each side must have the ability to use its nukes against the other without being stopped by defensive countermeasures.[15] The reason for this is that if one nation can stop incoming nuclear warheads, it is no longer suicidal to use its own offensively (though some theorists have proposed that the fallout would still ruin the attacking country, leading to the acronym SAD—self-assured destruction).[16]

Partly because of the risks of disrupting the stable MAD equilibrium, the world's militaries have put fairly limited effort into developing missile defense systems, and as of 2023 no nation has defenses strong enough to be able to confidently weather a large-scale nuclear attack. But in recent years, new delivery technologies have started to upset the balance of power. Russia is working to build underwater drones to carry nuclear weapons, as well as nuclear-powered cruise missiles designed to loiter for an extended period just outside a target country and strike from unpredictable angles.[17] Russia, China, and the United States are all racing to develop hypersonic vehicles capable of evasive maneuvers to thwart defenses as they deliver their warheads.[18] Because these systems are so new, they increase the risk of miscalculation if rival militaries draw different conclusions about their potential effectiveness.

Even with such a compelling deterrent as MAD, the potential remains for catastrophe resulting from miscalculation or misunderstanding.[19] We have grown so accustomed to this situation, however, that it is barely discussed.

Still, there is reason for measured optimism about the trajectory of nuclear risk. MAD has been successful for more than seventy years, and nuclear states' arsenals continue to shrink. The risk of nuclear terrorism or a dirty bomb remains a major concern, but advances in AI are leading to more effective tools for detecting and countering such threats.[20] And while AI cannot eliminate the risk of nuclear war, smarter command-and-control systems can significantly reduce the risk of sensor malfunctions causing inadvertent use of these terrible weapons.[21]

BIOTECHNOLOGY

We now have another technology that can threaten all of humanity. Consider that there are many naturally occurring pathogens that can make us sick but that most people survive. Conversely, there are a small number that are more likely to cause death but that do not spread very easily. Malevolent plagues like the Black Death arose from a combination of fast spread and severe mortality—killing about one third of Europe's population[22] and reducing the world population from around 450 million to about 350 million by the end of the fourteenth century.[23] Yet thanks in part to variations in DNA, some people's immune systems were better at fighting the plague. One benefit of sexual reproduction is that each of us has a different genetic makeup.[24]

But advances in genetic engineering[25] (which can edit viruses by manipulating their genes) could allow the creation—either intentionally or accidentally—of a supervirus that would have both extreme lethality and high transmissibility. Perhaps it would even be a stealth infection that people would catch and spread long before they realized they had contracted it. No one would have preexisting immunity, and the result would be a pandemic capable of ravaging the human population.[26] The 2019–2023 coronavirus pandemic offers us a pale glimpse of what such a catastrophe could be like.

The specter of this possibility was the impetus for the original Asilomar Conference on Recombinant DNA in 1975, fifteen years before the Human Genome Project was initiated.[27] It drew up a set of standards to prevent accidental problems and to guard against intentional ones. These "Asilomar guidelines" have been continually updated, and some of their principles are now baked into legal regulations governing the biotechnology industry.[28]

There have also been efforts to create a rapid response system to counteract a suddenly emerging biological virus, whether released accidentally or intentionally.[29] Before COVID-19, perhaps the most notable effort to improve epidemic reaction times was the US government's June 2015 establishment of the Global Rapid Response Team at the Centers for Disease Control. The GRRT, as it is known, was formed in response to the 2014–2016 Ebola virus outbreak in West Africa. The team is able to rapidly deploy anywhere in the world and provide high-level expertise to assist local authorities in the identification, containment, and treatment of threatening disease outbreaks.

As for deliberately released viruses, the overall federal bioterrorism defense efforts of the United States are coordinated through the National Interagency Confederation for Biological Research (NICBR). One of the most important institutions in this work is the United States Army Medical Research Institute of Infectious Diseases (USAMRIID). I have worked with them (via the Army Science Board) to provide advice on developing better capabilities to quickly respond in the event of such an outbreak.[30]

When such an outbreak occurs, millions of lives depend on how quickly authorities can analyze the virus and form a strategy for containment and treatment. Fortunately, the speed of virus sequencing is following a long-term trend of acceleration. It took thirteen years after its discovery to sequence full-length genome of HIV in 1996, and only thirty-one days to sequence the SARS virus in 2003, and we can now sequence many biological viruses in a single day.[31] A rapid response system would entail capturing a new virus, sequencing it in about a day, and then quickly designing medical countermeasures.

One strategy for treatment is to use RNA interference, which consists of small pieces of RNA that can destroy the messenger RNA expressing a gene (based on the observation that viruses are analogous to disease-causing genes).[32] Another approach is an antigen-based vaccine that targets distinctive protein structures on the surface of a virus.[33] As discussed in the previous chapter, AI-augmented drug discovery can already enable potential vaccines or therapies for a newly emerging viral outbreak to be identified in a matter of days or weeks— hastening the start of the much longer process of clinical trials. Later in the 2020s, though, we will have the technology to accelerate an increasing proportion of the clinical trial pipeline via simulated biology.

In May 2020 I wrote an article for *Wired* arguing that we should leverage artificial intelligence in order to create vaccines—for example, against the SARS-CoV-2 virus that causes COVID-19.[34] As it turned out, that is exactly how successful vaccines like Moderna's were created in record time. The company used a wide range of advanced AI tools to design and optimize mRNA sequences, as well as to speed up the manufacturing and testing process.[35] Thus, within sixty-five days of receiving the virus's genetic sequence, Moderna dosed the first human subject with its vaccine—and received FDA emergency authorization just 277 days after that.[36] This is stunning progress, considering that before COVID-19 the fastest anyone had ever created a vaccine was about four years.[37]

As this book is being written, there is ongoing scientific investigation into the possibility that the COVID-19 virus might have been accidentally released after genetic engineering research in a lab.[38] Because there has been a great deal of misinformation surrounding lab-leak theories, it is important to base our inferences on high-quality scientific sources. Yet the possibility itself underscores a real danger: it could have been far worse. The virus could have been extremely transmissible and at the same time very lethal, so it is not likely that it was created with malicious intentions. But because the technology to create something much deadlier than COVID-19 already exists, AI-driven countermeasures will be critical to mitigating the risk to our civilization.

NANOTECHNOLOGY

Most risks in biotechnology have to do with self-replication. A problem with any one cell is unlikely to be a threat. The same is true for nanotechnology: no matter how destructive an individual nanobot

might be, it has to be able to self-replicate to create a truly global catastrophe. Nanotechnology will make possible a wide range of offensive
weapons, many of which could be extremely destructive. In addition,
once nanotechnology is mature, such weapons could be manufactured
very cheaply, unlike nuclear arsenals today, which require large amounts
of resources to construct. (For a rough sense of how much it would
cost rogue actors to build nuclear weapons, consider the example of
North Korea, a pariah state denied most access to outside assistance.
The South Korean government estimates that the North's nuclear
weapons program cost between $1.1 billion and $3.2 billion in 2016,
the year it successfully developed nuclear-armed missiles.)[39]

By contrast, biological weapons can be very cheap. According to a
1996 NATO report, such weapons could be developed for $100,000
(around $190,000 in 2023 money) by a team of just five biologists in
the space of a few weeks, without any exotic equipment.[40] When it
comes to impact, a 1969 expert panel reported to the United Nations
that biological weapons were about eight hundred times as cost-
effective as nuclear weapons at targeting civilians—and biotech advances in the five decades since have almost surely increased that ratio
significantly.[41] While we cannot say for certain how much mature
nanotechnology will cost in the future, because it will operate on self-
replication principles similar to biology, we should regard the costs of
biological weapons as a first-order approximation. Since nanotech will
be taking advantage of AI-optimized manufacturing processes, the
costs could well be even lower.

Nano-based weapons could include tiny drones that deliver poisons
to targets without being detected, nanobots that enter the body in
water or as an aerosol and tear it apart from within, or systems that
selectively target certain groups of people of any description.[42] As
nanotechnology pioneer Eric Drexler wrote in 1986, "'Plants' with
'leaves' no more efficient than today's solar cells could out-compete
real plants, crowding the biosphere with an inedible foliage. Tough,
omnivorous 'bacteria' could out-compete real bacteria: they could
spread like blowing pollen, replicate swiftly, and reduce the biosphere

to dust in a matter of days. Dangerous replicators could easily be too tough, small, and rapidly spreading to stop—at least if we made no preparation. We have trouble enough controlling viruses and fruit flies."[43]

The most commonly discussed worst-case scenario is the potential creation of "gray goo"—self-replicating machines that consume carbon-based matter and turn it into more self-replicating machines.[44] Such a process could lead to a runaway chain reaction, potentially converting the entire biomass of the earth to such machines.

Let's consider how long it might take for the earth's entire biomass to be destroyed. The available biomass has on the order of 10^{40} carbon atoms.[45] The carbon atoms within a single replicating nanobot might be on the order of 10^7.[46] The nanobot therefore would need to create 10^{33} copies of itself—not all directly, I should stress, but via iterated replication. The nanobots in each "generation" might just create two copies of themselves, or another small number. The astoundingly large numbers come from this process being repeated again and again with the copies and the copies of the copies. So that's about 110 generations of the nanobots (since $2^{110} = 10^{33}$), or 109 if the nanobots in previous generations stay active.[47] Nanotechnology expert Robert Freitas estimates a replication time of about one hundred seconds, so under ideal conditions the gray goo wipeout time for that much carbon would be around three hours.[48]

However, the actual rate of destruction would be much slower, because the world's biomass is not laid out in a continuous block. The limiting factor would be the actual movement of the front of destruction. Nanobots cannot travel very quickly because of their small size, so it would likely take weeks for such a destructive process to circle the globe.

Yet a two-phase attack could circumvent this restriction. Over some period of time an insidious process could convert a tiny portion of carbon atoms throughout the world so that one out of every thousand trillion (10^{15}) becomes part of a "sleeping" force of gray goo nanobots. These nanobots would not be noticeable, as they would be in such a small concentration. However, because they would be *every-where*, they would not have to travel far once the attack started. Then, at some kind of predetermined signal (perhaps relayed from a small number of nanobots self-assembled into antennas long enough to pick up long-range radio waves), the prepositioned nanobots would just re-produce rapidly in place. Each nanobot multiplying itself a thousand trillionfold would require fifty binary replications—less than ninety minutes.[49] The movement speed of a destructive wave front would no longer be a limiting factor.

Sometimes this scenario is imagined as the result of malicious ac-tion by humans—perhaps as a terrorist weapon intended to destroy life on earth. Yet this need not happen out of malice. We can imagine cases where nanobots accidentally enter a runaway self-replication process, perhaps due to an error in their programming. For example, if carelessly designed, nanobots intended to consume only a certain type of matter or operate within a limited area could malfunction and cause a global disaster. So instead of trying to add security features to inherently dangerous systems, we must only build nanobots that are naturally fail-safe.

One strong protection against unintended replication would be to design any self-replicating nanobots with a "broadcast architecture."[50] This means that instead of carrying their own programming, they would rely on a signal (perhaps carried by radio waves) to give them all their instructions. This way, the signal could be turned off or modified in an emergency, stopping the self-replication chain reaction.

Yet even if responsible people design safe nanobots, bad actors could still design dangerous ones. Therefore, we will need a nanotech-nology "immune system" already in place before these scenarios can even become a possibility. This immune system would have to be ca-pable of contending not just with scenarios that cause obvious destruc-tion but with any potentially dangerous stealthy replication, even at very low concentrations.

It is encouraging that the field is already taking safety seriously. Nanotechnology safety guidelines have been in existence for about two decades, having emerged from the 1999 Workshop on Molecular Nanotechnology Research Policy Guidelines (which I attended), and have been revised and updated since.[51] It appears that the main immune system defense against gray goo would be "blue goo"—nanobots that would neutralize their gray counterparts.[52]

According to Freitas's calculations, if dispersed optimally around the world, 88,000 metric tons of "blue goo"–type defensive nanobots could be sufficient to sweep the entire atmosphere in about twenty-four hours.[53] For perspective, this is less than the weight of the water displaced by a large aircraft carrier—huge, but very small in comparison with the mass of the entire planet. Still, these figures assume ideal efficiency and deployment conditions, which would likely not be achieved in practice. As of 2023, with so much nanotechnology development yet to be done, it is very difficult to judge how much the actual blue goo requirements would differ from this theoretical estimate.

One requirement is clear, though: the blue goo would not be created using only abundant natural ingredients (which is what gray goo would be composed of). These nanobots would have to be made of special materials, so that the blue goo could not be converted to gray goo. There are a number of tricky issues to making this method work robustly, as well as theoretical issues to solve to ensure safe, failproof blue goo, but I do believe it will prove a workable approach. Ultimately there is no fundamental reason why harmful nanobots would have an asymmetric advantage over well-designed defensive systems. The key is to ensure that good nanobots are deployed around the world before bad ones so that self-replication chain reactions can be detected and neutralized before they have a chance to get out of control.

My friend Bill Joy's 2000 essay "Why the Future Doesn't Need Us" has an excellent discussion of nanotechnology risks, including the gray goo scenario.[54] Most nanotechnology experts consider a gray goo catastrophe to be unlikely, and so do I. Yet because it would be an extinction-level event, it is very important to keep these risks in mind as nanotechnology develops in the coming decades. I am hopeful that with proper precautions—and AI assistance in designing secure systems—humanity can keep such scenarios in the realm of science fiction.

ARTIFICIAL INTELLIGENCE

With biotechnology risks, we can still be subject to a pandemic such as COVID-19, which caused nearly seven million deaths in the world through 2023,[55] but we are quickly developing the means to rapidly sequence a new virus and develop medicines to avert civilization-threatening catastrophe. With nanotechnology, while gray goo isn't yet a threat, we do already have an overall strategy that should provide defense against even the ultimate two-phase attack. But superintelligent AI entails a fundamentally different kind of peril—in fact, the primary peril. If AI is smarter than its human creators, it could potentially find a way around any precautionary measures that have been put in place. There is no general strategy that can definitively overcome that.

There are three broad categories of peril from superintelligent AI, and with focused research on each, we can at least mitigate the risk. *Misuse* encompasses cases where the AI functions as its human operators intend, but those operators deploy it to deliberately cause harm to others.[56] For example, terrorists might use an AI's biochemistry abilities to design a new virus that causes a deadly pandemic.

Next is *outer misalignment*, which refers to cases where there's a mismatch between the programmers' actual intentions and the goals they teach the AI in hopes of achieving them.[57] This is the classic problem depicted in stories about genies—it's hard to specify exactly what you want to someone who takes your commands literally. Imagine programmers intending to cure cancer, so they instruct the AI to design a virus that kills all cells with a certain oncogenic DNA mutation. The AI does this successfully, but the programmers didn't realize that this mutation is also present in many healthy cells, so the virus kills the patients who receive it.

Finally, *inner misalignment* occurs when the methods the AI learns to achieve its goal produce undesirable behavior, at least in some cases.[58] For example, training an AI to identify genetic changes unique to cancerous cells might reveal a spurious pattern that works on the sample data but not when deployed in the real world. Perhaps cancerous cells in the training data were stored for a longer time prior to analysis than healthy cells, and the AI learns to recognize subtle genetic alterations that result. If the AI designs a cancer-killing virus based on this information, it won't work on live patients. These examples are relatively simple, but as AI models are given increasingly complex tasks, it will become more challenging to detect misalignments.

There is a field of technical research that is actively seeking ways to prevent both kinds of AI misalignment. There are many promising theoretical approaches, though much work remains to be done. "Imitative generalization" involves training AI to imitate how humans draw inferences, so as to make it safer and more reliable when applying its knowledge in unfamiliar situations.[59] "AI safety via debate" uses competing AIs to point out flaws in each other's ideas, allowing humans to judge issues too complex to properly evaluate unassisted.[60] "Iterated amplification" involves using weaker AIs to assist humans in creating well-aligned stronger AIs, and repeating this process to eventually align AIs much stronger than unaided humans could ever align on their own.[61]

And so, while the AI alignment problem will be very hard to solve,[62] we will not have to solve it on our own—with the right techniques, we can use AI itself to dramatically augment our own alignment capabilities. This also applies to designing AI that resists misuse. In the biochemistry example described previously, a safely aligned AI would have to recognize the dangerous request and refuse to comply. But we will also need ethical bulwarks against misuse—strong international norms favoring safe and responsible deployment of AI.

As AI systems have become dramatically more powerful over the past decade, limiting perils from misuse has taken on greater global priority. Over the past several years we have seen a concerted effort to create an ethical prescription for artificial intelligence. In 2017 I attended the Asilomar Conference on Beneficial AI, inspired by the biotechnology guidelines established at its counterpart four decades earlier.[63] Some useful principles were established there, and I have signed on to them. However, it remains easy to see how entities with ideas that are undemocratic and opposed to free expression could still use advanced AI for their own objectives, even if most of the world follows the Asilomar proposals. Notably, the major military powers have not signed these guidelines—and they have historically been among the most powerful forces in promoting advanced technology. For example, the internet came from our Defense Advanced Research Projects Agency.[64]

Still, the Asilomar AI Principles provide a foundation for responsible AI development that has been shaping the field in a positive direction. Six of the document's twenty-three principles promote "human" values or "humanity." For example, principle 10, Value Alignment, states that "[h]ighly autonomous AI systems should be designed so that their goals and behaviors can be assured to align with human values throughout their operation."[65]

Another document, the Lethal Autonomous Weapons Pledge, promotes the same concept: "We the undersigned agree that the decision to take a human life should never be delegated to a machine. There is a moral component to this position, that we should not allow machines to make life-taking decisions for which others—or nobody—will be culpable."[66] While such influential figures as Stephen Hawking, Elon Musk, Martin Rees, and Noam Chomsky have signed the ban on LAWs, top military powers, including the United States, Russia, the United Kingdom, France, and Israel, have rejected it.

Even though the US military does not endorse these guidelines, it has its own "human directive" policy stating that systems targeting humans must be controlled by humans.[67] A 2012 Pentagon directive established that "autonomous and semi-autonomous weapon systems shall be designed to allow commanders and operators to exercise appropriate levels of human judgment over the use of force."[68] In 2016

US Deputy Secretary of Defense Robert Work stated that the American military will "not delegate lethal authority to a machine to make a decision" about the use of such force.[69] Still, he left open the possibility that at some point in the future this policy could be reversed if it were necessary to compete with a rival nation that is "more willing to delegate authority to machines than we are."[70] I was involved in discussions that shaped this policy, just as I was on the policy board to guide implementations toward combating the use of biological hazards.

In early 2023, following an international conference that included dialogue with China, the United States released a "Political Declaration on Responsible Military Use of Artificial Intelligence and Autonomy," urging states to adopt sensible policies that include ensuring ultimate human control over nuclear weapons.[71] Yet the notion of "human control" itself is hazier than it might seem. If humans authorized a future AI system to "stop an incoming nuclear attack," how much discretion should it have over how to do so? Note that an AI general enough to successfully thwart such an attack could also be used for offensive purposes.

And so we need to recognize the fact that AI technologies are inherently dual-use. This is true even of systems already deployed. The very same drone that delivers medication to a hospital that is inaccessible by road during a rainy season could later carry an explosive to that hospital. Keep in mind that military operations have for more than a decade been using drones so precise that they can send a missile through a particular window that is literally on the other side of the earth from its operators.[72]

We also have to think through whether we would really want our side to observe a LAW ban if hostile military forces are not doing so. What if an enemy nation sent an AI-controlled contingent of advanced war machines to threaten your security? Wouldn't you want your side to have an even more intelligent capability to defeat them and keep you safe? This is the primary reason why the "Campaign to Stop Killer Robots" has failed to gain major traction.[73] As of 2023, all major military powers have declined to endorse the campaign, with the notable exception of China, which did so in 2018 but later clarified that it supported a ban on only use, not development[74]—although even this is likely more for strategic and political reasons than moral ones, as autonomous weapons used by the United States and its allies could disadvantage Beijing militarily. My own view is that if we were attacked by such weapons, we would want to have a counterweapon, which would necessarily involve violating this prohibition.

Further, what will "human" even ultimately mean in the context of human control when we introduce a nonbiological addition to our own decision-making starting in the 2030s using brain–computer interfaces? That nonbiological component will only grow exponentially, while our biological intelligence will stay the same. So as we get to the late 2030s, our thinking itself will be largely nonbiological. So where will the human decision-making be when our own thought largely uses nonbiological systems?

Some of the other Asilomar principles also leave open questions. For example, principle 7, Failure Transparency: "If an AI system causes harm, it should be possible to ascertain why." And principle 8, Judicial Transparency: "Any involvement by an autonomous system in judicial decision-making should provide a satisfactory explanation auditable by a competent human authority."

These efforts to render AI decisions more comprehensible are valuable, but the basic problem is that, regardless of any explanation they provide, we simply won't have the capacity to fully understand most of the decisions made by superintelligent AI. If a Go-playing program that is far beyond the best human were to explain its strategic decisions, not even the best player in the world (without the assistance of a cybernetic enhancement) would entirely grasp them.[75] One promising line of research aimed at reducing risks from opaque AI systems is "eliciting latent knowledge."[76] This project is trying to develop techniques that can ensure that if we ask an AI a question, it gives us all the relevant information it knows, instead of just telling us what it thinks we want to hear—which will be a growing risk as machine-learning systems become more powerful.

The principles also laudably promote noncompetitive dynamics around AI development, notably principle 18, AI Arms Race—"An arms race in lethal autonomous weapons should be avoided"—and principle 23, Common Good: "Superintelligence should only be developed in the service of widely shared ethical ideals, and for the benefit of all humanity rather than one state or organization." Yet because superintelligent AI could be a decisive advantage in warfare and bring tremendous economic benefits, military powers will have strong incentives to engage in an arms race for it.[77] Not only does this worsen risks of misuse, but it also increases the chances that safety precautions around AI alignment could be neglected.

Recall the value alignment issue addressed in principle 10. The next principle, Human Value, fleshes out which values are intended: "AI systems should be designed and operated so as to be compatible with ideals of human dignity, rights, freedoms, and cultural diversity."

Yet having this goal does not guarantee achieving it—and this gets to the whole point of the peril from AI. An AI that had a harmful objective would not have difficulty explaining why it makes sense for some wider purpose, even justifying it in terms of values people widely share.

It is very difficult to usefully restrict development of any fundamental AI capability, especially since the basic idea behind general intelligence is so broad. There are encouraging signs as this book goes to press that major governments are taking the challenge seriously—like the Bletchley Declaration following the 2023 AI Safety Summit in the UK—but much will depend on how such initiatives are actually implemented.[78] One optimistic argument, which is based on the principle of the free market, is that each step toward superintelligence is subject to market acceptance. In other words, artificial general intelligence will be created by humans to solve real human problems, and there are strong incentives to optimize it for beneficial purposes. Since AI is emerging from a deeply integrated economic infrastructure, it will reflect our values because in an important sense it will *be* us. We are already a human-machine civilization. Ultimately, the most important approach we can take to keep AI safe is to protect and improve on our human governance and social institutions. The best way to avoid destructive conflict in the future is to continue the advance of our ethical ideals, which has already profoundly reduced violence in recent centuries and decades.[79]

I do think we also need to take seriously the misguided and increasingly strident Luddite voices that advocate broad relinquishment of technological progress to avoid the genuine dangers of genetics, nanotechnology, and robotics (GNR).[80] Delays in overcoming human suffering are still of great consequence—for example, the worsening of famine in Africa resulting from opposition to any food aid that might contain GMOs (genetically modified organisms).[81]

With technologies now beginning to modify our bodies and brains, another type of opposition to progress has emerged in the form of "fundamentalist humanism": opposition to any change in the nature of what it means to be human.[82] This would include modifying our genes and our protein folding, and taking other steps toward radical life extension. This opposition will ultimately fail, however, because the demand for therapies that can overcome the pain, disease, and short life spans inherent in our version 1.0 bodies will ultimately prove irresistible.

When people are presented with the prospect of radical life extension, two objections are quickly raised. The first is the probability of running out of material resources to support an expanding biological population. We frequently hear that we are running out of energy, clean water, housing, land, and the other resources we need to support a growing population, and that this problem will only be exacerbated when the death rate starts to plummet. But as I articulated in chapter 4, as we begin to optimize our use of the earth's resources, we'll find they are thousands of times greater than we require. For example, we have nearly ten thousand times the sunlight we need to theoretically meet all of our current energy needs.[83]

The second objection to radical life extension is that we will become profoundly bored doing the same things over and over again for centuries. But in the 2020s we will have virtual and augmented reality delivered in very compact external devices, and in the 2030s we will have VR and AR connected directly to our nervous systems by nanobots feeding signals to our senses. We will thereby have radical life expansion in addition to radical life extension. We will inhabit vast virtual and augmented realities limited only by our imagination—which itself will be expanded. Even if we lived hundreds of years, we would not exhaust all the knowledge there is to gain, and all the culture there is to consume.

AI is the pivotal technology that will allow us to meet the pressing challenges that confront us, including overcoming disease, poverty, environmental degradation, and all of our human frailties. We have a moral imperative to realize this promise of new technologies while mitigating the peril. But it won't be the first time we've succeeded in doing so. At the beginning of this chapter, I mentioned the civil defense drills I experienced as a boy in preparation for a potential nuclear war. When I was growing up, most people around me assumed that nuclear war was almost inevitable. The fact that our species found the wisdom to refrain from using these terrible weapons shines as an example of how we have it in our power to likewise use biotechnology, nanotechnology, and superintelligent AI responsibly. We are not doomed to failure in controlling these perils.

Overall, we should be cautiously optimistic. While AI is creating new technical threats, it will also radically enhance our ability to deal with those threats. As for abuse, since these methods will enhance our intelligence regardless of our values, they can be used for both promise and peril.[84] We should thus work toward a world where the powers of AI are broadly distributed, so that its effects reflect the values of humanity as a whole.

CHAPTER 8

DIALOGUE WITH CASSANDRA

CASSANDRA: So you anticipate a neural net with sufficient processing power to be able to exceed all capabilities by humans by 2029.

RAY: Correct. They are already doing that with one capability after another.

CASSANDRA: And when they do that, they will be far better than any human in every skill possessed by any human.

RAY: Correct. In one area after another, they will be better than all humans by 2029.

CASSANDRA: And to pass a Turing test, an AI will have to be made less smart.

RAY: Yes, otherwise we would know they would not be an unenhanced human.

CASSANDRA: And you are also expecting that by the early 2030s we will have a means of going inside the brain and connecting to the top levels of the neocortex, both to tell what is going on and to activate connections.

RAY: Right.

CASSANDRA: And thus this superintelligence we are creating will directly be part of our brain, at least through these connections to the cloud.

RAY: Correct.

CASSANDRA: Okay, but these two advances—teaching a neural net everything that all humans can do and beyond and connecting internally to the brain with effective two-way connections—are in very different fields.

RAY: Well, yes.

CASSANDRA: One field involves experimentation with computers, which are largely not regulated. Experiments take days, and one advance can come right after another. Progress goes very quickly. On the other hand, a process like placing an attachment involving a million wires into the brain is something else entirely. It requires all kinds of supervision and regulation. This is putting something not only into the human body but into the brain itself, which is probably the most physically sensitive part of the body. And it's not even clear to regulators that it is necessary. If we can prevent a profound brain disease, for example, that might offer a distinct benefit, but to connect to an external computer would be very difficult.

RAY: But it will still happen, partly driven by the goal of fixing the significant brain disorders you mention.

CASSANDRA: Yes, I agree it's possible, but it is likely to be substantially delayed.

RAY: That's why I have predicted its arrival in the 2030s.

CASSANDRA: But any regulation regarding inserting foreign objects into the brain could postpone its happening for, say, ten years, until the 2040s. That would dramatically change your timeline for the interaction between super-intelligent machines and people. For one thing, the machines would take all of the jobs rather than just becoming an extension of people's intelligence.

RAY: Well, a mind extension directly in our brains would be convenient—you wouldn't lose it that way, like you might your cell phone. But even while such devices are not yet connected directly, they still function as an extension of human intelligence. A kid today can access all of human knowledge with her mobile device. And AI still augments far more workers than it replaces. Although the brain extenders are outside our bodies, we do jobs now that would be impossible without the ones we already have, even though they are not physically attached to our brains.

CASSANDRA: Yes, but you predicted that we would need millions of circuits to connect to the top layer of the neocortex. By contrast, extending our intelligence via external devices requires inputs typed from our keyboards, which is several orders of magnitude slower. It would certainly compromise the interaction a lot. And why would an AI even want to deal with a human with such a slow communication speed? It might as well just do everything itself.

RAY: By the middle of the 2020s we will have a means of interacting with a computer that is thousands of times faster than keyboarding: fully immersive virtual reality with full-screen video and audio. We will see and hear ordinary reality, but it will be interlaced with two-way communication with our computers. That is almost as fast as a connection with the top layers of our neocortex. This will ultimately replace interaction by keyboards.

CASSANDRA: Okay, we will advance in our ability to communicate with our computers, but it is still not the same as actually extending our neocortex.

RAY: But people will still have the work they need to get done in order to get food, shelter, and other needs. Even before internal brain extenders enable us to think more abstract thoughts, external brain extenders with advanced AI will let us perform difficult tasks and solve hard problems.

CASSANDRA: But people need deeper purpose. If AIs can do *everything* that humans can do in every intellectual sphere, and do it far better than the best humans and at far greater speeds, what is there for humans to do that will give us meaning?

RAY: Well, that's why we want to merge with the intelligence we are creating. The AIs will become part of us, and thus it is we who will be doing those things.

CASSANDRA: Right. And that is why I am still concerned at what could be a one-decade delay in getting a brain extender to work, given the extraordinary challenge of putting a device with millions of connections inside our skulls. I can accept the feasibility of all kinds of changes outside our bodies, including VR, but that is not the same as actually extending our neocortex.

RAY: That was the concern that Daniel Kahneman expressed, and he was also concerned about the potential for violence between people who lost their jobs and others.

CASSANDRA: By "others" do you mean the computers, since it will be the computers that are superior to humans in every skill?

RAY: Not the computers, as we will be dependent on the computers for our well-being, but rather humans who may be perceived as using AI to expand their own wealth and power at the expense of displaced workers.

CASSANDRA: Yes, I suspect Kahneman was thinking of an in-between period where some humans retain power, and AI has not yet created enough material abundance to avoid conflict.

RAY: Okay, but conflict can be minimized when people feel they have purpose. And extending our neocortex into the cloud will be critical to humans keeping a sense of purpose. Just as growing more neocortex hundreds of thousands of years ago elevated our primate ancestors from survival instinct to contemplating philosophy, extended humans will have even more capacity for empathy and ethics.

CASSANDRA: I agree, but extending the neocortex into the cloud is a very different kind of progress than better external brain extenders.

RAY: Yes, your point is valid, but I do think we will achieve an extension of the neocortex by the early 2030s. So the in-between period probably won't be very long.

CASSANDRA: But the timeline for connecting to the neocortex is a key concern. It could be a huge problem if that were delayed.

RAY: Well, yes, that is true.

CASSANDRA: Also, if an AI were to emulate you, and we replaced the biological you with the emulation, it might seem to be you, and would appear as you to everyone else, but *you* would effectively be gone.

RAY: Okay, but we are not talking about that. We are not emulating your biological brain. We are *adding* to it. Your biological brain would remain as it was, only now with added intelligence.

CASSANDRA: But the nonbiological intelligence would ultimately be many times more powerful than your biological brain, ultimately thousands to millions of times.

RAY: Yes, but still, nothing is being taken away. There will be a great deal added.

CASSANDRA: You've argued, though, that within a few years our brains will effectively be cloud extensions.

RAY: We're actually already doing that. And whatever philosophical significance you see for our biological brain, we are not taking that away, either.

CASSANDRA: But the biological brain at that point will be pretty in-
significant.

RAY: But it is still there, and it will retain all its fundamental qualities.

CASSANDRA: Well, I see very profound changes in a very short time.

RAY: That we agree on.

PRICE-PERFORMANCE OF COMPUTATION, 1939–2023 CHART SOURCES

MACHINE SELECTION METHODOLOGY

The machines in this chart are selected as major programmable computing machines whose computation price-performance surpassed all previous machines. If multiple machines achieved this in a given calendar year, only the machine with the best price-performance is included, irrespective of release date within that year. Machines not sold or rented commercially are assigned to the year of their first normal operation. Machines that were sold or rented commercially are assigned to their first calendar year of general public availability (as opposed to initial design or prototyping). For consumer-scale machines, only devices in mass production for retail sale were considered for inclusion—individually customized machines or chimeric "homebrew" computers made from disparate retail components would muddy the broader analysis. Machines that were designed but not built, such as Charles Babbage's Analytical Engine, and machines that were built but were not reliably functional, such as Konrad Zuse's Z1, are not included. Likewise, this chart ignores some highly specialized devices like digital signal processors, which are technically capable of performing a given number of digital operations per second but are not widely used as general-purpose CPUs.

PRICE DATA METHODOLOGY

Nominal prices are adjusted to real February 2023 prices according to CPI data from the US Bureau of Labor Statistics (chained CPI-U, 1982–1984 = 100). CPI for each year is presented as an annual average. Thus, although the underlying real-price calculations do not use rounded numbers, they should never be regarded as precise to within a dollar, and in general should be regarded as precise only to within several percentage points. For machines created with non-dollar currencies, the vagaries of exchange rates add an additional several percentage points of uncertainty.

Where multiple retail prices for a given machine are attested in a certain year, the lowest open-market prices are preferred here, in order to reflect the best price-performance available in that year. One point of moderate incommensurability is that prior to the mid-1990s, almost all computing was done through discrete computers with limited upgradability. As such, unit prices inevitably included components other than the processor itself, such as the hard drive and display. By contrast, it is now common for chips to be sold separately, and it is feasible for retail users to connect several or many CPUs/GPUs together for more performance-intensive tasks. As a result, assessing chip prices is now a better guide to overall computing price-performance than including components that are not integral to computation—even though this modestly exaggerates the price-performance improvement during the 1990s.

Pricing for Google Cloud TPU v4-4096 is very loosely approximated on the basis of 4,000 rental hours, in order to be roughly commensurable with the rest of the dataset, which since the 1950s has exclusively consisted of equipment available for purchase. This dramatically underestimates price-performance for small machine-learning projects, where brief access to enormous amounts of computing power is useful but the capital costs of purchasing that computing power would be totally prohibitive. This is a major and underappreciated effect of the cloud computing revolution.

Prices cited here are construction price, retail purchase price, or rental price, as applicable. Other separate costs, such as delivery and installation, electricity, maintenance, operator labor, taxes, and depreciation, are excluded. That is because those costs are highly variable among users and cannot be effectively averaged for a given machine. Based on available evidence, though, these factors would not greatly change the overall analysis—and, to the extent that they would, would likely serve to drive down price-performance of older, more logistics-intensive machines and therefore increase the apparent rate of progress across the graph (see page 165, "Price-Performance of Computation, 1939–2023"). Thus, omitting these costs is the more analytically conservative choice.

PERFORMANCE DATA METHODOLOGY

"Computations per second" is a synthetic metric derived from splicing together several datasets across eighty-four years. Because the computational capabilities of these machines have not only improved quantitatively but also qualitatively changed over time, it is impossible to devise a rigorously commensurable metric for comparing their performance across the entire period in question. Put another way, even given unlimited time, the 1939 Z2 computer could not do everything a 2023 Tensor Processing Unit does in a fraction of a second—they are simply not commensurable.

As such, any attempts to convert all performance statistics in this dataset into a fully commensurable metric would be misleading. For example, while Anders Sandberg and Nick Bostrom (2008) estimated the equivalence between millions of instructions per second (MIPS) and millions of floating-point operations per second (MFLOPS), they do not scale linearly, making this an inappropriate method to use in a dataset with a performance range as large as this one. Artificially converting recent computers' FLOPS ratings to IPS would exaggerate the actual performance capabilities of newer machines, while rating old computers in FLOPS would misleadingly underrate them.

Likewise, the information-theoretic approaches favored by Hans Moravec (1988) and William Nordhaus (2001)—while useful—do not capture the qualitative evolution of computing performance and applications. For example, Nordhaus's MSOPS (million standard operations per second) metric prescribes a fixed ratio of additions and multiplications that is not realistically applicable when comparing 1960s computers calculating rocket ballistics with modern GPUs and TPUs using low-precision computations for machine learning.

For this reason, the methodology used here favors using the metrics that machines were originally assessed by. This means that the more lenient instructions-per-second paradigm is favored from 1939 (where it refers to basic additions in Konrad Zuse's Z2 computer) until the introduction of the Pentium 4 in 2001, when the floating-point-operations-per-second paradigm became dominant in measuring modern computing performance. This reflects the fact that the applications of computing power have changed over time, and for some uses, high floating-point performance is at a premium over integer performance or other metrics. Specialization has also increased in some areas. For example, GPUs and specialized AI/deep-learning chips excel at their roles with high FLOPS ratings, but measuring their ability to perform general CPU tasks in an attempt at commensurability with older general computing chips would be misleading.

This chart favors best-achieved performance statistics or, for earlier machines, derives best performance from performance at operations comparable to addition. While this will be above what these machines achieved on average in practice in daily operation, this is a more broadly commensurable form of assessment than average-performance statistics, which depend on many factors outside raw computing speed that vary irregularly between machines.

ADDITIONAL RESOURCES

Anders Sandberg and Nick Bostrom, *Whole Brain Emulation: A Roadmap*, technical report 2008-3, Future of Humanity Institute, Oxford University (2008), https://www.fhi.ox.ac.uk/brain-emulation-roadmap-report.pdf.

William D. Nordhaus, "The Progress of Computing," discussion paper 1324, Cowles Foundation (September 2001), https://ssrn.com/abstract=285168.

Hans Moravec, "MIPS Equivalents," Field Robotics Center, Carnegie Mellon Robotics Institute, accessed December 2, 2021, https://web.archive.org/web/2021060905 2024/https://frc.ri.cmu.edu/~hpm/book97/ch3/processor.list.

Hans Moravec, *Mind Children: The Future of Robot and Human Intelligence* (Cambridge, MA: Harvard University Press, 1988).

LISTED MACHINES, DATA, AND SOURCES

CPI DATA SOURCES

"Consumer Price Index, 1913–," Federal Reserve Bank of Minneapolis, accessed April 20, 2023, https://www.minneapolisfed.org/about-us/monetary-policy/inflation -calculator/consumer-price-index-1913-; US Bureau of Labor Statistics, "Consumer Price Index for All Urban Consumers: All Items in U.S. City Average (CPIAUCSL)," retrieved from FRED, Federal Reserve Bank of St. Louis, updated April 12, 2023, https://fred.stlouisfed.org/series/CPIAUCSL.

1939 Z2

Real price: $50,489.31
Computations per second: 0.33
Computations/second/dollar: 0.0000065

Price source: Jane Smiley, *The Man Who Invented the Computer: The Biography of John Atanasoff, Digital Pioneer* (New York: Doubleday, 2010), loc. 638, Kindle (v3.1_r1); "Purchasing Power Comparisons of Historical Monetary Amounts," Deutsche Bundesbank, accessed December 20, 2021, https://www.bundesbank.de/en/statistics /economic-activity-and-prices/producer-and-consumer-prices/purchasing-power -comparisons-of-historical-monetary-amounts-795290#tar-5; "Purchasing Power Equivalents of Historical Amounts in German Currencies," Deutsche Bundesbank, 2021, https://www.bundesbank.de/resource/blob/622372/154f0fc435da99ee935666983 a5146a2/mL/purchaising-power-equivalents-data.pdf; Lawrence H. Officer, "Exchange Rates," in *Historical Statistics of the United States, Millennial Edition*, ed. Susan B. Carter et al. (Cambridge, UK: Cambridge University Press, 2002), reproduced in Harold Marcuse, "Historical Dollar-to-Marks Currency Conversion Page," University of California, Santa Barbara, updated October 7, 2018, https://marcuse.faculty .history.ucsb.edu/projects/currency.htm; "Euro to US Dollar Spot Exchange Rates for 2020," Exchange Rates UK, accessed December 20, 2021, https://www.exchange

rates.org.uk/EUR-USD-spot-exchange-rates-history-2020.html. In purchasing power, 7,000 reichsmarks was equivalent to about 30,100 euros in 2020. This averages $40,124 in early 2023 US dollars. This has the advantage of avoiding the commensurability problem posed by purchasing power differences between Nazi Germany and the United States. But it has the disadvantage of focusing on price levels, which were largely set by the totalitarian government, with rationing and black market transactions limiting the relevance of nominal prices. By exchange rates, 7,000 reichsmarks in 1939 averaged $2,800, which is the equivalent of $60,853 in early 2023. This has the advantage of avoiding distortions from Germany's totalitarian war economy, but the disadvantage of introducing uncertainty due to differing purchasing power between the two currencies. Because the advantages and disadvantages of both figures are complementary, and because there is no clear principle by which to judge which is the most representative of relevant ground truth, this chart uses the average of the two: $50,489.

Performance source: Horst Zuse, "Z2," Horst-Zuse.Homepage.t-online.de, accessed December 20, 2021, http://www.horst-zuse.homepage.t-online.de/z2.html.

1941 Z3

Real price: $136,849.13
Computations per second: 1.25
Computations/second/dollar: 0.0000091

Price source: Jack Copeland and Giovanni Sommaruga, "The Stored-Program Universal Computer: Did Zuse Anticipate Turing and von Neumann?," in *Turing's Revolution: The Impact of His Ideas About Computability*, ed. Giovanni Sommaruga and Thomas Strahm (Cham, Switzerland: Springer International Publishing, 2016; corrected 2021 publication), 53, https://www.google.com/books/edition/Turing_s_Revolution/M8ZyCwAAQBAJ; "Purchasing Power Comparisons of Historical Monetary Amounts," Deutsche Bundesbank, accessed December 20, 2021, https://www.bundesbank.de/en/statistics/economic-activity-and-prices/producer-and-consumer-prices/purchasing-power-comparisons-of-historical-monetary-amounts-795290#tar-5; "Purchasing Power Equivalents of Historical Amounts in German Currencies," Deutsche Bundesbank, 2021, https://www.bundesbank.de/resource/blob/622372/154f0fc435da99ee935666983a5146a2/mL/purchaising-power-equivalents-data.pdf; Lawrence H. Officer, "Exchange Rates," in *Historical Statistics of the United States, Millennial Edition*, ed. Susan B. Carter et al. (Cambridge, UK: Cambridge University Press, 2002), reproduced in Harold Marcuse, "Historical Dollar-to-Marks Currency Conversion Page," University of California, Santa Barbara, updated October 7, 2018, https://marcuse.faculty.history.ucsb.edu/projects/currency.htm; "Euro to US Dollar

Spot Exchange Rates for 2020," Exchange Rates UK, accessed December 20, 2021, https://www.exchangerates.org.uk/EUR-USD-spot-exchange-rates-history-2020 .html; "Consumer Price Index, 1913–," Federal Reserve Bank of Minneapolis, accessed October 11, 2021, https://www.minneapolisfed.org/about-us/monetary-policy /inflation-calculator/consumer-price-index-1913-. In purchasing power, 20,000 reichsmarks was equivalent to about 82,000 euros in 2020. This averages $109,290 in early 2023 US dollars. This has the advantage of avoiding the commensurability problem posed by purchasing power differences between Nazi Germany and the United States. But it has the disadvantage of focusing on price levels, which were largely set by the totalitarian government, with rationing and black market transactions limiting the relevance of nominal prices. By exchange rates, 20,000 reichsmarks in 1941 averaged $8,000, which is the equivalent of $164,408 in early 2023. This has the advantage of avoiding distortions from Germany's totalitarian war economy, but the disadvantage of introducing uncertainty due to differing purchasing power between the two currencies. Because the advantages and disadvantages of both figures are complementary, and because there is no clear principle by which to judge which is the most representative of relevant ground truth, this chart uses the average of the two: $136,849.

Performance source: Horst Zuse, "Z3," Horst-Zuse.Homepage.t-online.de, accessed December 20, 2021, http://www.horst-zuse.homepage.t-online.de/z3-detail.html.

1943 COLOSSUS MARK 1

Real price: $33,811,510.61
Computations per second: 5,000
Computations/second/dollar: 0.00015

Price source: Chris Smith, "Cracking the Enigma Code: How Turing's Bombe Turned the Tide of WWII," BT, November 2, 2017, http://web.archive.org/web/2018032103 5325/http://home.bt.com/tech-gadgets/cracking-the-enigma-code-how-turings -bombe-turned-the-tide-of-wwii-11363990654704; Jack Copeland (computing history expert), email to author, January 12, 2018; "Inflation Calculator," Bank of England, January 20, 2021, https://www.bankofengland.co.uk/monetary-policy/inflation /inflation-calculator; "Historical Rates for the GBP/USD Currency Conversion on 01 July 2020 (01/07/2020)," Pound Sterling Live, accessed November 11, 2021, https:// www.poundsterlinglive.com/best-exchange-rates/british-pound-to-us-dollar -exchange-rate-on-2020-07-01. No unit cost figures for Colossus are directly available, as it was not built for commercial purposes. We know that the earlier Bombe machines were constructed at a cost of around £100,000 each. Although no precise declassified figures for the construction of Colossus are available, computing history expert Jack Copeland suggests as a very rough approximation that its costs were about five times as much as a single Bombe. This corresponds to £23,314,516 in 2020 British pounds, or $33,811,510 in early 2023 US dollars. Keep in mind that due to the uncertainty of the underlying estimates, only the first two significant digits should be regarded as meaningful.

Performance source: B. Jack Copeland, ed., *Colossus: The Secrets of Bletchley Park's Codebreaking Computers* (Oxford, UK: Oxford University Press, 2010), 282.

1946 ENIAC

Real price: $11,601,846.15
Computations per second: 5,000
Computations/second/dollar: 0.00043

Price source: Martin H. Weik, *A Survey of Domestic Electronic Digital Computing Systems*, report no. 971 (Aberdeen Proving Ground, MD: Ballistic Research Laboratories, December 1955), 42, https://books.google.com/books?id=-BPSAAAAMAAJ.

Performance source: Brendan I. Koerner, "How the World's First Computer Was Rescued from the Scrap Heap," *Wired*, November 25, 2014, https://www.wired.com/2014/11/eniac-unearthed.

1949 BINAC

Real price: $3,523,451.43
Computations per second: 3,500
Computations/second/dollar: 0.00099

Price source: William R. Nester, *American Industrial Policy: Free or Managed Markets?* (New York: St. Martin's, 1997), 106, https://books.google.com/books?id=hCi_DAAAQBAJ.

Performance source: Eckert-Mauchly Computer Corp., *The BINAC* (Philadelphia: Eckert-Mauchly Computer Corp., 1949), 2, http://s3data.computerhistory.org/brochures/eckertmauchly.binac.1949.102646200.pdf.

1953 UNIVAC 1103

Real price: $10,356,138.62
Computations per second: 50,000
Computations/second/dollar: 0.0048

Price source: Martin H. Weik, *A Third Survey of Domestic Electronic Digital Computing Systems*, report no. 1115 (Aberdeen, MD: Ballistic Research Laboratories, March 1961), 913, http://web.archive.org/web/20160403031739/http://www.textfiles.com/bit savers/pdf/brl/compSurvey_Mar1961/brlReport1115_0900.pdf; https://bitsavers.org/pdf /brl/compSurvey_Mar1961/brlReport1115_0000.pdf.

Performance source: Martin H. Weik, *A Third Survey of Domestic Electronic Digital Computing Systems*, report no. 1115 (Aberdeen, MD: Ballistic Research Laboratories, March 1961), 906, http://web.archive.org/web/20160403031739/http://www.textfiles .com/bitsavers/pdf/brl/compSurvey_Mar1961/brlReport1115_0900.pdf.

1959 DEC PDP-1

Real price: $1,239,649.32
Computations per second: 100,000
Computations/second/dollar: 0.081

Price source: "PDP 1 Price List," Digital Equipment Corporation, February 1, 1963, https://www.computerhistory.org/pdp-1/_media/pdf/DEC.pdp_1.1963.102652408 .pdf.

Performance source: Digital Equipment Corporation, *PDP-1 Handbook* (Maynard, MA: Digital Equipment Corporation, 1963), 10, http://s3data.computerhistory.org /pdp-1/DEC.pdp_1.1963.102636240.pdf.

1962 DEC PDP-4

Real price: $647,099.67
Computations per second: 62,500
Computations/second/dollar: 0.097

Price source: Digital Equipment Corporation, *Nineteen Fifty-Seven to the Present* (Maynard, MA: Digital Equipment Corporation, 1978), 3, http://s3data.computerhistory .org/pdp-1/dec.digital_1957_to_the_present_(1978).1957-1978.102630349.pdf.

Performance source: Digital Equipment Corporation, *PDP-4 Manual* (Maynard, MA: Digital Equipment Corporation, 1962), 18, 57, http://gordonbell.azurewebsites.net /digital/pdp%204%20manual%201962.pdf.

1965 DEC PDP-8

Real price: $172,370.29
Computations per second: 312,500
Computations/second/dollar: 1.81

Price source: Tony Hey and Gyuri Pápay, *The Computing Universe: A Journey Through a Revolution* (New York: Cambridge University Press, 2015), 165, https://books.goo gle.com/books?id=q4FIBQAAQBAJ.

Performance source: Digital Equipment Corporation, *PDP-8* (Maynard, MA: Digital Equipment Corporation, 1965), 10, http://archive.computerhistory.org/resources/ac cess/text/2009/11/102683307.05.01.acc.pdf.

1969 DATA GENERAL NOVA

Real price: $65,754.33
Computations per second: 169,492
Computations/second/dollar: 2.58

Price source: "Timeline of Computer History—Data General Corporation Introduces the Nova Minicomputer," Computer History Museum, accessed November 10, 2021, https://www.computerhistory.org/timeline/1968.

Performance source: NOVA brochure, Data General Corporation, 1968, 12, http:// s3data.computerhistory.org/brochures/dgc.nova.1968.102646102.pdf.

1973 INTELLEC 8

Real price: $16,291.71
Computations per second: 80,000
Computations/second/dollar: 4.91

Price source: "Intellec 8," Centre for Computing History, accessed November 10, 2021, http://www.computinghistory.org.uk/det/3366/intellec-8.

Performance source: Intel, *Intellec 8 Reference Manual*, rev. 1 (Santa Clara, CA: Intel, 1974), xxxxiii, https://ia802603.us.archive.org/14/items/bitsavers_intelMCS8InceMan ualRev1Jun74_14022374/Intel_Intellec_8_Reference_Manual_Rev_1_Jun74.pdf.

1975 ALTAIR 8800

Real price: $3,481.85
Computations per second: 500,000
Computations/second/dollar: 144

Price source: "MITS Altair 8800: Price List," CTI Data Systems, July 1, 1975, http://vtda.org/docs/computing/DataSystems/MITS_Altair8800_PriceList01Jul75.pdf.

Performance source: MITS, *Altair 8800 Operator's Manual* (Albuquerque, NM: MITS, 1975), 21, 90, http://www.classiccmp.org/dunfield/altair/d/88opman.pdf.

1984 APPLE MACINTOSH

Real price: $7,243.62
Computations per second: 1,600,000
Computations/second/dollar: 221

Price source: Regis McKenna Public Relations, "Apple Introduces Macintosh Advanced Personal Computer," press release, January 24, 1984, https://web.stanford.edu/dept/SUL/sites/mac/primary/docs/pr1.html.

Performance source: Motorola, *Motorola Semiconductor Master Selection Guide*, rev. 10 (Chicago: Motorola, 1996), 2.2-2, http://www.bitsavers.org/components/motorola/_catalogs/1996_Motorola_Master_Selection_Guide.pdf.

1986 COMPAQ DESKPRO 386 (16 MHZ)

Real price: $17,886.96
Computations per second: 4,000,000
Computations/second/dollar: 224

Price source: Peter H. Lewis, "Compaq's Gamble on an Advanced Chip Pays Off," *New York Times*, September 20, 1987, https://www.nytimes.com/1987/09/20/business/the-executive-computer-compaq-s-gamble-on-an-advanced-chip-pays-off.html.

Performance source: Peter H. Lewis, "Compaq's Gamble on an Advanced Chip Pays Off," *New York Times*, September 20, 1987, https://www.nytimes.com/1987/09/20/business/the-executive-computer-compaq-s-gamble-on-an-advanced-chip-pays-off.html.

1987 PC'S LIMITED 386 (16 MHZ)

Real price: $11,946.43
Computations per second: 4,000,000
Computations/second/dollar: 335

Price source: Peter H. Lewis, "Compaq's Gamble on an Advanced Chip Pays Off," *New York Times*, September 20, 1987, https://www.nytimes.com/1987/09/20/business/the-executive-computer-compaq-s-gamble-on-an-advanced-chip-pays-off.html.

Performance source: Peter H. Lewis, "Compaq's Gamble on an Advanced Chip Pays Off," *New York Times*, September 20, 1987, https://www.nytimes.com/1987/09/20/business/the-executive-computer-compaq-s-gamble-on-an-advanced-chip-pays-off.html.

1988 COMPAQ DESKPRO 386/25

Real price: $20,396.30
Computations per second: 8,500,000
Computations/second/dollar: 417

Price source: "Compaq Deskpro 386/25 Type 38," Centre for Computing History, accessed November 10, 2021, http://www.computinghistory.org.uk/det/16967/Compaq-Deskpro-386-25-Type-38.

Performance source: Jeffrey A. Dubin, *Empirical Studies in Applied Economics* (New York: Springer Science+Business Media, 2012), 72–73, https://www.google.com/books/edition/Empirical_Studies_in_Applied_Economics/41_lBwAAQBAJ.

1990 MT 486DX

Real price: $11,537.40
Computations per second: 20,000,000
Computations/second/dollar: 1,733

Price source: Bruce Brown, "Micro Telesis Inc. MT 486DX," *PC Magazine* 9, no. 15 (September 11, 1990), 140, https://books.google.co.uk/books?id=NsgmyHnvDmUC.

Performance source: Owen Linderholm, "Intel Cuts Cost, Capabilities of 9486; Will Offer Companion Math Chip," *Byte*, June 1991, 26, https://worldradiohistory.com /hd2/IDX-Consumer/Archive-Byte-IDX/IDX/90s/Byte-1991-06-IDX-32.pdf.

1992 GATEWAY 486DX2/66

Real price: $6,439.31
Computations per second: 54,000,000
Computations/second/dollar: 8,386

Price source: Jim Seymour, "The 486 Buyers' Guide," *PC Magazine* 12, no. 21 (December 7, 1993), 226, https://books.google.com/books?id=7k7q-wS0t00C.

Performance source: Mike Feibus, "P6 and Beyond," *PC Magazine* 12, no. 12 (June 29, 1993), 164, https://books.google.co.uk/books?id=gCfzPMoPJWgC&pg=PA164.

1994 PENTIUM (75 MHZ)

Real price: $4,477.91
Computations per second: 87,100,000
Computations/second/dollar: 19,451

Price source: Bob Francis, "75-MHz Pentiums Deskbound," *Info World* 16, no. 44 (October 31, 1994), 5, https://books.google.com/books?id=cTgEAAAAMBAJ&pg=PA5.

Performance source: Roy Longbottom, "Dhrystone Benchmark Results on PCs," Roy Longbottom's PC Benchmark Collection, February 2017, http://www.roylongbottom .org.uk/dhrystone%20results.htm.

1996 PENTIUM PRO (166 MHZ)

Real price: $3,233.73
Computations per second: 242,000,000
Computations/second/dollar: 74,836

Price source: Michael Slater, "Intel Boosts Pentium Pro to 200 MHz," *Microprocessor Report* 9, no. 15 (November 13, 1995), 2, https://www.cl.cam.ac.uk/~pb22/test.pdf.

Performance source: Roy Longbottom, "Dhrystone Benchmark Results on PCs," Roy Longbottom's PC Benchmark Collection, February 2017, http://www.roylongbottom.org.uk/dhrystone%20results.htm.

1997 MOBILE PENTIUM MMX (133 MHZ)

Real price: $533.76
Computations per second: 184,092,000
Computations/second/dollar: 344,898

Price source: "Intel Mobile Pentium MMX 133 MHz Specifications," CPU-World, accessed November 10, 2021, https://web.archive.org/web/20140912204405/http://www.cpu-world.com/CPUs/Pentium/Intel-Mobile%20Pentium%20MMX%20133%20-%20FV80503133.html.

Performance source: "Intel Mobile Pentium MMX 133 MHz vs Pentium MMX 200 MHz," CPU-World, accessed November 11, 2021, http://www.cpu-world.com/Compare/347/Intel_Mobile_Pentium_MMX_133_MHz_(FV80503133)_vs_Intel_Pentium_MMX_200_MHz_(FV80503200).html; Roy Longbottom, "Dhrystone Benchmark Results on PCs," Roy Longbottom's PC Benchmark Collection, February 2017, http://www.roylongbottom.org.uk/dhrystone%20results.htm. Per CPU-World's testing, Mobile Pentium MMX 133 MHz achieved 69.9 percent of the performance (Dhrystone 2.1 VAX MIPS) of the Pentium MMX 200 MHz. For the latter, this was 276 MIPS in Roy Longbottom's testing, corresponding to an estimated 192,924,000 instructions per second for the former.

1998 PENTIUM II (450 MHZ)

Real price: $1,238.05
Computations per second: 713,000,000
Computations/second/dollar: 575,905

Price source: "Intel Pentium II 450 MHz Specifications," CPU-World, accessed November 10, 2021, https://web.archive.org/web/20150428111439/http://www.cpu-world.com:80/CPUs/Pentium-II/Intel-Pentium%20II%20450%20-%2080523PY450512PE%20(B80523P450512E).html.

Performance source: Roy Longbottom, "Dhrystone Benchmark Results on PCs," Roy Longbottom's PC Benchmark Collection, February 2017, http://www.roylongbottom .org.uk/dhrystone%20results.htm.

1999 PENTIUM III (450 MHZ)

Real price: $898.06
Computations per second: 722,000,000
Computations/second/dollar: 803,952

Price source: "Intel Pentium III 450 MHz Specifications," CPU-World, accessed November 10, 2021, https://web.archive.org/web/20140831044834/http://www.cpu -world.com/CPUs/Pentium-III/Intel-Pentium%20III%20450%20-%2080525 PY450512%20(BX80525U450512%20-%20BX80525U450512E).html.

Performance source: Roy Longbottom, "Dhrystone Benchmark Results on PCs," Roy Longbottom's PC Benchmark Collection, February 2017, http://www.roylongbottom .org.uk/dhrystone%20results.htm.

2000 PENTIUM III (1.0 GHZ)

Real price: $1,734.21
Computations per second: 1,595,000,000
Computations/second/dollar: 919,725

Price source: "Intel Pentium III 1BGHz (Socket 370) Specifications," CPU-World, accessed November 10, 2021, https://web.archive.org/web/20160529005115/http:// www.cpu-world.com/CPUs/Pentium-III/Intel-Pentium%20III%201000%20 -%20RB80526PZ001256%20(BX80526C1000256).html.

Performance source: Roy Longbottom, "Dhrystone Benchmark Results on PCs," Roy Longbottom's PC Benchmark Collection, February 2017, http://www.roylongbottom .org.uk/dhrystone%20results.htm.

2001 PENTIUM 4 (1700 MHZ)

Real price: $599.55
Computations per second: 1,843,000,000
Computations/second/dollar: 3,073,978

Price source: "Intel Pentium 4 1.7 GHz Specifications," CPU-World, accessed November 10, 2021, https://web.archive.org/web/20150429131339/http://www.cpu-world.com/CPUs/Pentium_4/Intel-Pentium%204%201.7%20GHz%20-%20RN80528PC029G0K%20(BX80528JK170G).html.

Performance source: Roy Longbottom, "Dhrystone Benchmark Results on PCs," Roy Longbottom's PC Benchmark Collection, February 2017, http://www.roylongbottom.org.uk/dhrystone%20results.htm.

2002 XEON (2.4 GHZ)

Real price: $392.36
Computations per second: 2,480,000,000
Computations/second/dollar: 6,323,014

Price source: "Intel Xeon 2.4 GHz Specifications," CPU-World, accessed November 10, 2021, https://web.archive.org/web/20150502024039/http://www.cpu-world.com:80/CPUs/Xeon/Intel-Xeon%202.4%20GHz%20-%20RK80532KE056512%20(BX80532KE2400D%20-%20BX80532KE2400DU).html.

Performance source: Jack J. Dongarra, "Performance of Various Computers Using Standard Linear Equations Software," technical report CS-89-85, University of Tennessee, Knoxville, February 5, 2013, 7–29, http://www.icl.utk.edu/files/publications/2013/icl-utk-625-2013.pdf. The Dongarra (2013) data is used here instead of the Longbottom (2017) data because by very roughly 2002, MFLOPS ratings had become the dominant performance standard and that data is more consistent and commensurable with subsequent machines' ratings. Most data drawn from Dongarra uses the "TPP Best Effort" metric, which is most commensurable with performance data from early computers. Because TPP Best Effort data is not available for this CPU, it is approximated here using the average ratio of TPP Best Effort MFLOPS to "LINPACK Benchmark" MFLOPS in the dataset. For the fifteen other single-core, non-EM64T Xeon-powered computers tested by Dongarra, TPP Best Effort value is on average 2.559 times that of the LINPACK Benchmark value. Additionally, the data is averaged across results for this CPU from two different OS/compiler combinations.

2004 PENTIUM 4 (3.0 GHZ)

Real price: $348.12
Computations per second: 3,181,000,000
Computations/second/dollar: 9,137,738

Price source: "Intel Pentium 4 3 GHz Specifications," CPU-World, accessed November 10, 2021, https://web.archive.org/web/20171005171131/http://www.cpu-world.com /CPUs/Pentium_4/Intel-Pentium%204%203.0%20GHz%20-%20RK80546 PG0801M%20(BX80546PG3000E).html.

Performance source: Jack J. Dongarra, "Performance of Various Computers Using Standard Linear Equations Software," technical report CS-89-85, University of Tennessee, Knoxville, February 5, 2013, 10, http://www.icl.utk.edu/files/publications /2013/icl-utk-625-2013.pdf. The Dongarra (2013) data is used here instead of the Longbottom (2017) data because starting with the Pentium 4, MFLOPS ratings became the dominant performance standard and that data is more consistent and commensurable with subsequent machines' ratings.

2005   PENTIUM 4 662 (3.6 GHZ)

Real price: $619.36
Computations per second: 7,200,000,000
Computations/second/dollar: 11,624,919

Price source: "Intel Pentium 4 662 Specifications," CPU-World, accessed November 10, 2021, https://web.archive.org/web/20150710050435/http://www.cpu-world.com:80 /CPUs/Pentium_4/Intel-Pentium%204%20662%203.6%20GHz%20-%20HH 80547PG1042MH.html.

Performance source: "Export Compliance Metrics for Intel Microprocessors Intel Pentium Processors," Intel, April 1, 2018, 4, http://web.archive.org/web/201806 01044504/https://www.intel.com/content/dam/support/us/en/documents/processors /APP-for-Intel-Pentium-Processors.pdf.

2006   CORE 2 DUO E6300

Real price: $273.82
Computations per second: 14,880,000,000
Computations/second/dollar: 54,342,788

Price source: "Intel Core 2 Duo E6300 Specifications," CPU-World, accessed November 10, 2021, https://web.archive.org/web/20160605085626/http://www.cpu-world .com/CPUs/Core_2/Intel-Core%202%20Duo%20E6300%20HH80557PH0362M %20(BX80557E6300).html.

Performance source: "Export Compliance Metrics for Intel Microprocessors Intel Pentium Processors," Intel, April 1, 2018, 12, http://web.archive.org/web/2018060 1044310/https://www.intel.com/content/dam/support/us/en/documents/processors /APP-for-Intel-Core-Processors.pdf.

2007 PENTIUM DUAL-CORE E2180

Real price: $122.23
Computations per second: 16,000,000,000
Computations/second/dollar: 130,899,970

Price source: "Intel Pentium E2180 Specifications," CPU-World, accessed November 10, 2021, https://web.archive.org/web/20170610094616/http://www.cpu-world.com /CPUs/Pentium_Dual-Core/Intel-Pentium%20Dual-Core%20E2180%20HH 80557PG0411M%20(BX80557E2180%20-%20BXC80557E2180).html.

Performance source: "Export Compliance Metrics for Intel Microprocessors Intel Pentium Processors," Intel, April 1, 2018, 7, http://web.archive.org/web/201806010 44504/https://www.intel.com/content/dam/support/us/en/documents/processors /APP-for-Intel-Pentium-Processors.pdf.

2008 GTX 285

Real price: $502.98
Computations per second: 708,500,000,000
Computations/second/dollar: 1,408,604,222

Price source: "NVIDIA GeForce GTX 285," TechPowerUp, accessed November 10, 2021, https://www.techpowerup.com/gpu-specs/geforce-gtx-285.c238.

Performance source: "NVIDIA GeForce GTX 285," TechPowerUp, accessed November 10, 2021, https://www.techpowerup.com/gpu-specs/geforce-gtx-285.c238.

2010 GTX 580

Real price: $690.15
Computations per second: 1,581,000,000,000
Computations/second/dollar: 2,290,796,652

Price source: "NVIDIA GeForce GTX 580," TechPowerUp, accessed November 10, 2021, https://www.techpowerup.com/gpu-specs/geforce-gtx-580.c270.

Performance source: "NVIDIA GeForce GTX 580," TechPowerUp, accessed November 10, 2021, https://www.techpowerup.com/gpu-specs/geforce-gtx-580.c270.

2012 GTX 680

Real price: $655.59
Computations per second: 3,250,000,000,000
Computations/second/dollar: 4,957,403,270

Price source: "NVIDIA GeForce GTX 680," TechPowerUp, accessed November 10, 2021, https://www.techpowerup.com/gpu-specs/geforce-gtx-680.c342.

Performance source: "NVIDIA GeForce GTX 680," TechPowerUp, accessed November 10, 2021, https://www.techpowerup.com/gpu-specs/geforce-gtx-680.c342.

2015 TITAN X (MAXWELL 2.0)

Real price: $1,271.50
Computations per second: 6,691,000,000,000
Computations/second/dollar: 5,262,273,757

Price source: "NVIDIA GeForce GTX TITAN X," TechPowerUp, accessed November 10, 2021, https://www.techpowerup.com/gpu-specs/geforce-gtx-titan-x.c2632.

Performance source: "NVIDIA GeForce GTX TITAN X," TechPowerUp, accessed November 10, 2021, https://www.techpowerup.com/gpu-specs/geforce-gtx-titan-x.c2632.

2016 TITAN X (PASCAL)

Real price: $1,506.98
Computations per second: 10,974,000,000,000
Computations/second/dollar: 7,282,098,756

Price source: "NVIDIA TITAN X Pascal," TechPowerUp, accessed November 10, 2021, https://www.techpowerup.com/gpu-specs/titan-x-pascal.c2863.

Performance source: "NVIDIA TITAN X Pascal," TechPowerUp, accessed November 10, 2021, https://www.techpowerup.com/gpu-specs/titan-x-pascal.c2863.

2017 AMD RADEON RX 580

Real price: $281.83
Computations per second: 6,100,000,000,000
Computations/second/dollar: 21,643,984,475

Price source: "AMD Radeon RX 580," TechPowerUp, accessed November 10, 2021, https://www.techpowerup.com/gpu-specs/radeon-rx-580.c2938.

Performance source: "AMD Radeon RX 580," TechPowerUp, accessed November 10, 2021, https://www.techpowerup.com/gpu-specs/radeon-rx-580.c2938.

2021 GOOGLE CLOUD TPU V4-4096

Real price: $22,796,129.30
Computations per second: 1,100,000,000,000,000,000
Computations/second/dollar: 48,253,805,968

Price source: Wherever possible, this chart uses open-market equipment purchase costs as prices, which best reflects overall civilization-level progress in computing price-performance. Yet Google Cloud TPUs are not sold externally and are made available only on a time-rental basis. Counting the hourly rental costs as the price would reflect a staggeringly high price-performance, and while it would be accurate for some very small projects (e.g., short machine-learning tasks for which buying hardware wouldn't be sensible), it would not reflect most actual use cases. Therefore, as a very rough approximation we can use 4,000 hours of working time as a functional equivalent to purchased hardware—based on plausibly representative usage and common product replacement cycles. (While it would be hard to confidently use such a speculative estimate for a same-year hardware comparison, the long time span and logarithmic scale of this graph result in an overall trend that is relatively insensitive to substantially different methodological assumptions for any given data point.) In practice, cloud rental contracts are subject to negotiation and may vary significantly based on the needs of a given customer and project. But as a plausibly representative figure, Google's v4-4096 TPU might rent for $5,120 an hour, corresponding to $20.48 million for the amount of computation time that a hardware owner might have put on a purchased processor. Price estimates are unofficial at time of writing, and were extrapolated from public information and conversations with a range of industry professionals.

Google is likely to have released more extensive pricing information by press time, but this too is imprecise, as factors specific to each project may play a substantial role in pricing. See Google Cloud, "Cloud TPU," Google, accessed December 10, 2021, https://cloud.google.com/tpu; Google project manager, telephone conversations with author, December 2021.

Performance source: Tao Wang and Aarush Selvan, "Google Demonstrates Leading Performance in Latest MLPerf Benchmarks," Google Cloud, June 30, 2021, https://cloud.google.com/blog/products/ai-machine-learning/google-wins-mlperf-benchmarks-with-tpu-v4; Samuel K. Moore, "Here's How Google's TPU v4 AI Chip Stacked Up in Training Tests," *IEEE Spectrum*, May 19, 2021, https://spectrum.ieee.org/heres-how-googles-tpu-v4-ai-chip-stacked-up-in-training-tests.

2023 GOOGLE CLOUD TPU V5E

Real price: $3,016.46
Computations per second: 393,000,000,000,000
Computations/second/dollar: 130,285,276,114

Price source: Google Cloud estimates that the TPU v5e achieves price-performance 2.7 times that of the TPU v4, measured on the MLPerf™ v3.1 Inference Closed benchmark, which is the gold standard for running large language models. To maximize commensurability with the TPU v4-4096 estimate, which approximates plausible high-volume contract pricing, TPU v5e pricing is estimated here per chip based on the known price-performance improvement. If we instead use only the publicly available TPU v5e price of $1.20 per chip-hour, price-performance would be approximately 82 billion computations per second per constant dollar—but because discounts are common for large cloud rental contracts, this would fall well short of reality. See Amin Vahdat and Mark Lohmeyer, "Helping You Deliver High-Performance, Cost-Efficient AI Inference at Scale with GPUs and TPUs," Google Cloud, September 11, 2023, https://cloud.google.com/blog/products/compute/performance-per-dollar-of-gpus-and-tpus-for-ai-inference.

Performance source: INT8 performance per chip. See Google Cloud, "System Architecture," Google Cloud, accessed November 13, 2023, https://cloud.google.com/tpu/docs/system-architecture-tpu-vm#tpu_v5e.

NOTES

INTRODUCTION

1. See the appendix for the sources used for all the historical cost-of-computation calculations in this book.
2. William D. Nordhaus, "Two Centuries of Productivity Growth in Computing," *Journal of Economic History* 67, no. 1 (March 2007): 128–59, https://doi.org/10.1017/S002205070 7000058.

CHAPTER 1: WHERE ARE WE IN THE SIX STAGES?

1. Alan M. Turing, "Computing Machinery and Intelligence," *Mind* 59, no. 236 (October 1, 1950): 435, https://doi.org/10.1093/mind/LIX.236.433.

CHAPTER 2: REINVENTING INTELLIGENCE

1. Alan M. Turing, "Computing Machinery and Intelligence," *Mind* 59, no. 236 (October 1, 1950): 435, https://doi.org/10.1093/mind/LIX.236.433.
2. Alex Shashkevich, "Stanford Researcher Examines Earliest Concepts of Artificial Intelligence, Robots in Ancient Myths," *Stanford News*, February 28, 2019, https://news.stanford .edu/2019/02/28/ancient-myths-reveal-early-fantasies-artificial-life.
3. John McCarthy et al., "A Proposal for the Dartmouth Summer Research Project on Artificial Intelligence," conference proposal, August 31, 1955, http://www-formal.stanford.edu /jmc/history/dartmouth/dartmouth.html.
4. McCarthy et al., "Proposal for the Dartmouth Summer Research Project."
5. Martin Childs, "John McCarthy: Computer Scientist Known as the Father of AI," *The Independent*, November 1, 2011, https://www.independent.co.uk/news/obituaries/john-mc carthy-computer-scientist-known-as-the-father-of-ai-6255307.html; Nello Christianini, "The Road to Artificial Intelligence: A Case of Data Over Theory," *New Scientist*, October 26, 2016, https://institutions.newscientist.com/article/mg23230971-200-the-irresistible -rise-of-artificial-intelligence.
6. James Vincent, "Tencent Says There Are Only 300,000 AI Engineers Worldwide, but Millions Are Needed," *The Verge*, December 5, 2017, https://www.theverge.com/2017/12/5 /16737224/global-ai-talent-shortfall-tencent-report.
7. Jean-Francois Gagne, Grace Kiser, and Yoan Mantha, *Global AI Talent Report 2019*, Element AI, April 2019, https://jfgagne.ai/talent-2019.
8. Daniel Zhang et al., *The AI Index 2022 Annual Report*, AI Index Steering Committee, Stanford Institute for Human-Centered AI, Stanford University, March 2022, 36, https://aiin dex.stanford.edu/wp-content/uploads/2022/03/2022-AI-Index-Report_Master.pdf;

Nestor Maslej et al., *The AI Index 2023 Annual Report*, AI Index Steering Committee, Stanford Institute for Human-Centered AI, Stanford University, April 2023, 24, https://aiindex.stanford.edu/wp-content/uploads/2023/04/HAI_AI-Index-Report_2023.pdf.

9. There was a 26.7 percent decrease in corporate investment from 2021 to 2022, but this is likely attributable to cyclical macroeconomic trends rather than a change in the long-term trajectory of corporate commitment to AI. See Maslej et al., *AI Index 2023 Annual Report*, 171, 184.

10. Ray Kurzweil, *The Age of Spiritual Machines: When Computers Exceed Human Intelligence* (New York: Penguin, 2000; first published by Viking, 1999), 313; Dale Jacquette, "Who's Afraid of the Turing Test?," *Behavior and Philosophy* 20/21 (1993): 72, https://www.jstor.org/stable/27759284.

11. Katja Grace et al., "Viewpoint: When Will AI Exceed Human Performance? Evidence from AI Experts," *Journal of Artificial Intelligence Research* 62 (July 2018): 729–54, https://doi.org/10.1613/jair.1.11222.

12. For more on the reasoning behind my prediction and how it compares with a wide range of opinions by AI experts, see Ray Kurzweil, "A Wager on the Turing Test: Why I Think I Will Win," KurzweilAI.net, April 9, 2002, https://www.kurzweilai.net/a-wager-on-the-turing-test-why-i-think-i-will-win; Vincent C. Müller and Nick Bostrom, "Future Progress in Artificial Intelligence: A Survey of Expert Opinion," in *Fundamental Issues of Artificial Intelligence*, ed. Vincent C. Müller (Cham, Switzerland: Springer, 2016), 553–71, https://philpapers.org/archive/MLLFPI.pdf; Anthony Aguirre, "Date Weakly General AI Is Publicly Known," Metaculus, accessed April 26, 2023, https://www.metaculus.com/questions/3479/date-weakly-general-ai-system-is-devised.

13. Aguirre, "Date Weakly General AI Is Publicly Known."

14. Raffi Khatchadourian, "The Doomsday Invention," *New Yorker*, November 23, 2015, https://www.newyorker.com/magazine/2015/11/23/doomsday-invention-artificial-intelligence-nick-bostrom.

15. A. Newell, J. C. Shaw, and H. A. Simon, "Report on a General Problem-Solving Program," RAND P-1584, RAND Corporation, February 9, 1959, http://bitsavers.informatik.uni-stuttgart.de/pdf/rand/ipl/P-1584_Report_On_A_General_Problem-Solving_Program_Feb59.pdf. See the appendix for the sources used for all the cost-of-computation calculations in this book.

16. Digital Equipment Corporation, *PDP-1 Handbook* (Maynard, MA: Digital Equipment Corporation, 1963), 10, http://s3data.computerhistory.org/pdp-1/DEC.pdp_1.1963.102636240.pdf.

17. Amin Vahdat and Mark Lohmeyer, "Enabling Next-Generation AI Workloads: Announcing TPU v5p and AI Hypercomputer," Google Cloud, December 6, 2023, https://cloud.google.com/blog/products/ai-machine-learning/introducing-cloud-tpu-v5p-and-ai-hypercomputer.

18. See the appendix for the sources used for all the cost-of-computation calculations in this book.

19. V. L. Yu et al., "Antimicrobial Selection by a Computer: A Blinded Evaluation by Infectious Diseases Experts," *Journal of the American Medical Association* 242, no. 12 (September 21, 1979): 1279–82, https://jamanetwork.com/journals/jama/article-abstract/366606.

20. Bruce G. Buchanan and Edward Hance Shortliffe, eds., *Rule-Based Expert Systems: The MYCIN Experiments of the Stanford Heuristic Programming Project* (Reading, MA: Addison-Wesley, 1984); Edward Edelson, "Programmed to Think," *MOSAIC* 11, no. 5 (September/October 1980): 22, https://books.google.co.uk/books?id=PU79ZK2tXeAC.

21. T. Grandon Gill, "Early Expert Systems: Where Are They Now?," *MIS Quarterly* 19, no. 1 (March 1995): 51–81, https://www.jstor.org/stable/249711.

22. For a short and nontechnical explainer on why machine learning reduces the complexity ceiling problem, see Deepanker Saxena, "Machine Learning vs. Rules Based Systems," Socure, August 6, 2018, https://www.socure.com/blog/machine-learning-vs-rule-based-systems.

23. Cade Metz, "One Genius' Lonely Crusade to Teach a Computer Common Sense," *Wired*, March 24, 2016, https://www.wired.com/2016/03/doug-lenat-artificial-intelligence-com mon-sense-engine; "Frequently Asked Questions," Cycorp, accessed November 20, 2021, https://cyc.com/faq.

24. For further information on the black box problem and AI transparency, see Will Knight, "The Dark Secret at the Heart of AI," *MIT Technology Review*, April 11, 2017, https://www .technologyreview.com/s/604087/the-dark-secret-at-the-heart-of-ai; "AI Detectives Are Cracking Open the Black Box of Deep Learning," *Science Magazine*, YouTube video, July 6, 2017, https://www.youtube.com/watch?v=gB_-LabED68; Paul Voosen, "How AI Detectives Are Cracking Open the Black Box of Deep Learning," *Science*, July 6, 2017, https://doi.org /10.1126/science.aan7059; Harry Shum, "Explaining AI," a16z, YouTube video, January 16, 2020, https://www.youtube.com/watch?v=rI_L95qnVkM; Future of Life Institute, "Neel Nanda on What Is Going On Inside Neural Networks," YouTube video, February 9, 2023, https://www.youtube.com/watch?v=mUhO6st6M_0.

25. For an excellent overview of mechanistic interpretability by researcher Neel Nanda, see Future of Life Institute, "Neel Nanda on What Is Going On Inside Neural Networks."

26. For more on techniques for machine learning with imperfect training data, see Xander Steenbrugge, "An Introduction to Reinforcement Learning," Arxiv Insights, YouTube video, April 2, 2018, https://www.youtube.com/watch?v=JgvyzIkgxF0; Alan Joseph Bekker and Jacob Goldberger, "Training Deep Neural-Networks Based on Unreliable Labels," *2016 IEEE International Conference on Acoustics, Speech and Signal Processing* (Shanghai, 2016), 2682–86, https://doi.org/10.1109/ICASSP.2016.7472164; Nagarajan Natarajan et al., "Learning with Noisy Labels," *Advances in Neural Information Processing Systems* 26 (2013), https://papers.nips.cc/paper/5073-learning-with-noisy-labels; David Rolnick et al., "Deep Learning Is Robust to Massive Label Noise," arXiv:1705.10694v3 [cs.LG], February 26, 2018, https://arxiv.org/pdf/1705.10694.pdf.

27. For further information on the Perceptron, its limitations, and a more detailed explanation of how certain neural networks can overcome them, see Marvin L. Minsky and Seymour A. Papert, *Perceptrons: An Introduction to Computational Geometry* (Cambridge, MA: MIT Press, 1990; reissue of 1988 expanded edition); Melanie Lefkowitz, "Professor's Percep-tron Paved the Way for AI—60 Years Too Soon," *Cornell Chronicle*, September 25, 2019, https://news.cornell.edu/stories/2019/09/professors-perceptron-paved-way-ai-60-years -too-soon; John Durkin, "Tools and Applications," in *Expert Systems: The Technology of Knowledge Management and Decision Making for the 21st Century*, ed. Cornelius T. Leondes (San Diego: Academic Press, 2002), 45, https://books.google.co.uk/books?id=5kSam KhS560C; "Marvin Minsky: The Problem with Perceptrons (121/151)," Web of Stories— Life Stories of Remarkable People, YouTube video, October 17, 2016, https://www.you tube.com/watch?v=QW_srPO-LrI; Heinz Mühlenbein, "Limitations of Multi-Layer Per-ceptron Networks: Steps Towards Genetic Neural Networks," *Parallel Computing* 14, no. 3 (August 1990): 249–60, https://doi.org/10.1016/0167-8191(90)90079-O; Aniruddha Kara-jgi, "How Neural Networks Solve the XOR Problem," *Towards Data Science*, November 4, 2020, https://towardsdatascience.com/how-neural-networks-solve-the-xor-problem-5976 3136bdd7.

28. See the appendix for the sources used for all the cost-of-computation calculations in this book.

29. Tim Fryer, "Da Vinci Drawings Brought to Life," *Engineering & Technology* 14, no. 5 (May 21, 2019): 18, https://eandt.theiet.org/content/articles/2019/05/da-vinci-drawings-brought -to-life.

30. For a more detailed timeline of life on earth and a further look at the underlying science, see Michael Marshall, "Timeline: The Evolution of Life," *New Scientist*, July 14, 2009, https://www.newscientist.com/article/dn17453; Dyani Lewis, "Where Did We Come From? A Primer on Early Human Evolution," *Cosmos*, June 9, 2016, https://cosmos magazine.com/palaeontology/where-did-we-come-from-a-primer-on-early-human -evolution; John Hawks, "How Has the Human Brain Evolved?," *Scientific American*, July 1,

2013, https://www.scientificamerican.com/article/how-has-human-brain-evolved; Laura Freberg, *Discovering Behavioral Neuroscience: An Introduction to Biological Psychology*, 4th ed. (Boston: Cengage Learning, 2018), 62–63, https://books.google.co.uk/books?id=HhBED wAAQBAJ; Jon H. Kaas, "Evolution of the Neocortex," *Current Biology* 16, no. 21 (2006): R910–R914, https://www.cell.com/current-biology/pdf/S0960-9822(06)02290-1.pdf; R. Glenn Northcutt, "Evolution of Centralized Systems: Two Schools of Evolutionary Thought," *Proceedings of the National Academy of Sciences* 109, suppl. 1 (June 22, 2012): 10626–33, https://doi.org/10.1073/pnas.1201889109.

31. Marshall, "Timeline: The Evolution of Life"; Holly C. Betts et al., "Integrated Genomic and Fossil Evidence Illuminates Life's Early Evolution and Eukaryote Origin," *Nature Ecology & Evolution* 2 (August 20, 2018): 1556–62, https://doi.org/10.1038/s41559-018-0644-x; Elizabeth Pennisi, "Life May Have Originated on Earth 4 Billion Years Ago, Study of Controversial Fossils Suggests," *Science*, December 18, 2017, https://www.sciencemag.org/news/2017/12/life-may-have-originated-earth-4-billion-years-ago-study-controversial-fossils-suggests.

32. Ethan Siegel, "Ask Ethan: How Do We Know the Universe Is 13.8 Billion Years Old?," *Big Think*, October 22, 2021, https://bigthink.com/starts-with-a-bang/universe-13-8-billion-years; Mike Wall, "The Big Bang: What Really Happened at Our Universe's Birth?," Space.com, October 21, 2011, https://www.space.com/13347-big-bang-origins-universe-birth.html; Nola Taylor Reed, "How Old Is Earth?," Space.com, February 7, 2019, https://www.space.com/24854-how-old-is-earth.html.

33. Marshall, "Timeline: The Evolution of Life."

34. Marshall, "Timeline: The Evolution of Life."

35. Freberg, *Discovering Behavioral Neuroscience*, 62–63; Kaas, "Evolution of the Neocortex"; R. Northcutt, "Evolution of Centralized Nervous Systems"; Frank Hirth, "On the Origin and Evolution of the Tripartite Brain," *Brain, Behavior and Evolution* 76, no. 1 (October 2010): 3–10, https://doi.org/10.1159/000320218.

36. Kaas, "Evolution of the Neocortex."

37. For two engaging explainers on how natural selection works, see Hank Green, "Natural Selection: Crash Course Biology #14," CrashCourse, YouTube video, April 30, 2012, https://www.youtube.com/watch?v=aTftyFboC_M; Primer, "Simulating Natural Selection," YouTube video, November 14, 2018, https://www.youtube.com/watch?v=0ZGbIK d0XrM.

38. Suzana Herculano-Houzel, "Coordinated Scaling of Cortical and Cerebellar Numbers of Neurons," *Frontiers in Neuroanatomy* 4, no. 12 (March 10, 2010), https://doi.org/10.3389/fnana.2010.00012.

39. For some helpful explainers on how this works, see Ainslie Johnstone, "The Amazing Phenomenon of Muscle Memory," *Medium*, Oxford University, December 14, 2017, https://medium.com/oxford-university/the-amazing-phenomenon-of-muscle-memory-fb1cc4c4726; Sara Chodosh, "Muscle Memory Is Real, But It's Probably Not What You Think," *Popular Science*, January 25, 2019, https://www.popsci.com/what-is-muscle-memory; Merim Bilalić, *The Neuroscience of Expertise* (Cambridge, UK: Cambridge University Press, 2017), 171–72, https://books.google.co.uk/books?id=QILTDQAAQBAJ; The Brain from Top to Bottom, "The Motor Cortex," McGill University, accessed November 20, 2021, https://thebrain.mcgill.ca/flash/i/i_06/i_06_cr/i_06_cr_mou/i_06_cr_mou.html.

40. For more technical lessons on basis functions as relevant to machine learning, see "Lecture 17: Basis Functions," Open Data Science Initiative, YouTube video, November 28, 2011, https://youtu.be/OOpfU3CvUkM?t=151; Yaser Abu-Mostafa, "Lecture 16: Radial Basis Functions," Caltech, YouTube video, May 29, 2012, https://www.youtube.com/watch?v=O8CfrnOPtLc.

41. Mayo Clinic, "Ataxia," Mayo Clinic, accessed November 20, 2021, https://www.mayoclinic.org/diseases-conditions/ataxia/symptoms-causes/syc-20355652; Helen Thomson, "Woman of 24 Found to Have No Cerebellum in Her Brain," *New Scientist*, September 10, 2014, https://institutions.newscientist.com/article/mg22329861-900-woman-of-24-found

-to-have-no-cerebellum-in-her-brain; R. N. Lemon and S. A. Edgley, "Life Without a Cerebellum," *Brain* 133, no. 3 (March 18, 2010): 652–54, https://doi.org/10.1093/brain/awq030.

42. For more on how athletic training takes advantage of the shift to unconscious competence, see Bo Hanson, "Conscious Competence Learning Matrix," Athlete Assessments, accessed November 22, 2021, https://athleteassessments.com/conscious-competence-learning-matrix.

43. Suzana Herculano-Houzel, "The Human Brain in Numbers: A Linearly Scaled-Up Primate Brain," *Frontiers in Human Neuroscience* 3, no. 31 (November 9, 2009), https://doi.org/10.3389/neuro.09.031.2009.

44. Herculano-Houzel, "Human Brain in Numbers"; Richard Apps, "Cerebellar Modules and Their Role as Operational Cerebellar Processing Units," *Cerebellum* 17, no. 5 (June 6, 2018): 654–82, https://doi.org/10.1007/s12311-018-0952-3; Jan Voogd, "What We Do Not Know About Cerebellar Systems Neuroscience," *Frontiers in Systems Neuroscience* 8, no. 227 (December 18, 2014), https://doi.org/10.3389/fnsys.2014.00227; Rhoshel K. Lenroot and Jay N. Giedd, "The Changing Impact of Genes and Environment on Brain Development During Childhood and Adolescence: Initial Findings from a Neuroimaging Study of Pediatric Twins," *Development and Psychopathology* 20, no. 4 (Fall 2008): 1161–75, https://doi.org/10.1017/S0954579408000552; Salvador Martinez et al., "Cellular and Molecular Basis of Cerebellar Development," *Frontiers in Neuroanatomy* 7, no. 18 (June 26, 2013), https://doi.org/10.3389/fnana.2013.00018.

45. Fumiaki Sugahara et al., "Evidence from Cyclostomes for Complex Regionalization of the Ancestral Vertebrate Brain," *Nature* 531, no. 7592 (February 15, 2016): 97–100, https://doi.org/10.1038/nature16518; Leonard F. Koziol, "Consensus Paper: The Cerebellum's Role in Movement and Cognition," *Cerebellum* 13, no. 1 (February 2014): 151–77, https://doi.org/10.1007/s12311-013-0511-x; Robert A. Barton and Chris Venditti, "Rapid Evolution of the Cerebellum in Humans and Other Great Apes," *Current Biology* 24, no. 20 (October 20, 2014): 2440–44, https://doi.org/10.1016/j.cub.2014.08.056.

46. For more detailed information on such hardwired animal behaviors, see Jesse N. Weber, Brant K. Peterson, and Hopi E. Hoekstra, "Discrete Genetic Modules Are Responsible for Complex Burrow Evolution in Peromyscus Mice," *Nature* 493, no. 7432 (January 17, 2013): 402–5, http://dx.doi.org/10.1038/nature11816; Nicole L. Bedford and Hopi E. Hoekstra, "Peromyscus Mice as a Model for Studying Natural Variation," *eLife* 4: e06813 (June 17, 2015), https://doi.org/10.7554/eLife.06813; Do-Hyoung Kim et al., "Rescheduling Behavioral Subunits of a Fixed Action Pattern by Genetic Manipulation of Peptidergic Signaling," *PLoS Genetics* 11, no. 9: e1005513 (September 24, 2015), https://doi.org/10.1371/journal.pgen.1005513.

47. For an interesting talk explaining evolutionary computation, see Keith Downing, "Evolutionary Computation: Keith Downing at TEDxTrondheim," TEDx Talks, YouTube video, November 4, 2013, https://www.youtube.com/watch?v=D3zUmfDd79s.

48. For more on the development and function of the neocortex, see Kaas, "Evolution of the Neocortex"; Jeff Hawkins and Sandra Blakeslee, *On Intelligence: How a New Understanding of the Brain Will Lead to the Creation of Truly Intelligent Machines* (New York: Macmillan, 2007), 97–101, https://books.google.co.uk/books?id=Qg2dmntfxmQC; Clay Reid, "Lecture 3: The Structure of the Neocortex," Allen Institute, YouTube video, September 6, 2012, https://www.youtube.com/watch?v=RhdcYNmW0zY; Joan Stiles et al., *Neural Plasticity and Cognitive Development: Insights from Children with Perinatal Brain Injury* (New York: Oxford University Press, 2012), 41–45, https://books.google.co.uk/books?id=QiNpAgAAQBAJ.

49. Brian K. Hall and Benedikt Hallgrimsson, *Strickberger's Evolution*, 4th ed. (Sudbury, MA: Jones & Bartlett Learning, 2011), 533; Kaas, "Evolution of the Neocortex"; Jon H. Kaas, "The Evolution of Brains from Early Mammals to Humans," *Wiley Interdisciplinary Reviews Cognitive Science* 4, no. 1 (November 8, 2012): 33–45, https://doi.org/10.1002/wcs.1206.

50. For more detail on the Cretaceous–Paleogene extinction event, also known as the K-T extinction, see Michael Greshko and National Geographic Staff, "What Are Mass

Extinctions, and What Causes Them?," *National Geographic*, September 26, 2019, https://www.nationalgeographic.com/science/prehistoric-world/mass-extinction; Victoria Jaggard, "Why Did the Dinosaurs Go Extinct?," *National Geographic*, July 31, 2019, https://www.nationalgeographic.com/science/prehistoric-world/dinosaur-extinction; Emily Singer, "How Dinosaurs Shrank and Became Birds," *Quanta*, June 2, 2015, https://www.quantamagazine.org/how-birds-evolved-from-dinosaurs-20150602.

51. Yasuhiro Itoh, Alexandros Poulopoulos, and Jeffrey D. Macklis, "Unfolding the Folding Problem of the Cerebral Cortex: Movin' and Groovin'," *Developmental Cell* 41, no. 4 (May 22, 2017): 332–34, https://www.sciencedirect.com/science/article/pii/S1534580717303933; Jeff Hawkins, "What Intelligent Machines Need to Learn from the Neocortex," *IEEE Spectrum*, June 2, 2017, https://spectrum.ieee.org/computing/software/what-intelligent-machines-need-to-learn-from-the-neocortex.

52. Jean-Didier Vincent and Pierre-Marie Lledo, *The Custom-Made Brain: Cerebral Plasticity, Regeneration, and Enhancement*, trans. Laurence Garey (New York: Columbia University Press, 2014), 152.

53. For both a nontechnical video and a more academic lecture giving further detail on the neocortex and its minicolumns, see Brains Explained, "The Neocortex," YouTube video, September 16, 2017, https://www.youtube.com/watch?v=x2mYTaJPVnc; Clay Reid, "Lecture 3: The Structure of the Neocortex."

54. V. B. Mountcastle, "The Columnar Organization of the Neocortex," *Brain* 120, no. 4 (April 1997): 701–22, https://doi.org/10.1093/brain/120.4.701; Olaf Sporns, Giulio Tononi, and Rolf Kötter, "The Human Connectome: A Structural Description of the Human Brain," *PLoS Computational Biology* 1, no. 4: e42 (September 30, 2005), https://doi.org/10.1371/journal.pcbi.0010042; David J. Heeger, "Theory of Cortical Function," *Proceedings of the National Academy of Sciences* 114, no. 8 (February 6, 2017): 1773–82, https://doi.org/10.1073/pnas.1619788114.

55. This is somewhat lower than the 300 million I estimated from older research in *How to Create a Mind* but, given that the underlying data is loosely approximated, it is still in the same general range. There may also be some significant variation from one person to another.

56. Jeff Hawkins, Subutai Ahmad, and Yuwei Cui, "A Theory of How Columns in the Neocortex Enable Learning the Structure of the World," *Frontiers in Neural Circuits* 11, no. 81 (October 25, 2017), https://doi.org/10.3389/fncir.2017.00081; Jeff Hawkins, *A Thousand Brains: A New Theory of Intelligence* (New York: Basic Books, 2021).

57. Mountcastle, "Columnar Organization of the Neocortex"; Sporns, Tononi, and Kötter, "The Human Connectome"; Heeger, "Theory of Cortical Function."

58. Malcolm W. Browne, "Who Needs Jokes? Brain Has a Ticklish Spot," *New York Times*, March 10, 1998, https://www.nytimes.com/1998/03/10/science/who-needs-jokes-brain-has-a-ticklish-spot.html; Itzhak Fried et al., "Electric Current Stimulates Laughter," *Scientific Correspondence* 391, no. 650 (February 12, 1998), https://doi.org/10.1038/35536.

59. For an understandable explainer on pain sensing in the brain, see Kristin Muench, "Pain in the Brain," NeuWrite West, November 10, 2015, www.neuwritewest.org/blog/pain-in-the-brain.

60. Browne, "Who Needs Jokes?"

61. Robert Wright, "Scientists Find Brain's Irony-Detection Center!" *Atlantic*, August 5, 2012, https://www.theatlantic.com/health/archive/2012/08/scientists-find-brains-irony-detection-center/260728.

62. "Bigger Brains: Complex Brains for a Complex World," Smithsonian Institution, January 16, 2019, http://humanorigins.si.edu/human-characteristics/brains; David Robson, "A Brief History of the Brain," *New Scientist*, September 21, 2011, https://www.newscientist.com/article/mg21128311-800.

63. Stephanie Musgrave et al., "Tool Transfers Are a Form of Teaching Among Chimpanzees," *Scientific Reports* 6, article 34783 (October 11, 2016), https://doi.org/10.1038/srep34783.

64. Hanoch Ben-Yami, "Can Animals Acquire Language?," *Scientific American*, March 1, 2017, https://blogs.scientificamerican.com/guest-blog/can-animals-acquire-language; Klaus Zu-

berbühler, "Syntax and Compositionality in Animal Communication," *Philosophical Transactions of the Royal Society B* 375, article 20190062 (November 18, 2019), https://doi.org/10.1098/rstb.2019.0062.

65. For an accessible explainer on the evolutionary origin and usefulness of our opposable thumbs, see "Where Do Our Opposable Thumbs Come From?," HHMI BioInteractive, YouTube video, April 24, 2014, https://www.youtube.com/watch?v=lDSkmb4UTlo.

66. Ryan V. Raut et al., "Hierarchical Dynamics as a Macroscopic Organizing Principle of the Human Brain," *Proceedings of the National Academy of Sciences* 117, no. 35 (August 12, 2020): 20890–97, https://doi.org/10.1073/pnas.1201889109.

67. Herculano-Houzel, "Human Brain in Numbers"; Sporns, Tononi, and Kötter, "The Human Connectome"; Ji Yeoun Lee, "Normal and Disordered Formation of the Cerebral Cortex: Normal Embryology, Related Molecules, Types of Migration, Migration Disorders," *Journal of Korean Neurosurgical Society* 62, no. 3 (May 1, 2019): 265–71, https://doi.org/10.3340/jkns.2019.0098; Christopher Johansson and Anders Lansner, "Towards Cortex Sized Artificial Neural Systems," *Neural Networks* 20, no. 1 (January 2007): 48–61, https://doi.org/10.1016/j.neunet.2006.05.029.

68. For a deeper dive on the neocortex and science's evolving understanding of the structural underpinnings of higher cognition, see Matthew Barry Jensen, "Cerebral Cortex," Khan Academy, accessed November 20, 2021, https://www.khanacademy.org/science/health-and-medicine/human-anatomy-and-physiology/nervous-system-introduction/v/cerebral-cortex; Hawkins, Ahmad, and Cui, "Theory of How Columns in the Neocortex Enable Learning"; Jeff Hawkins et al., "A Framework for Intelligence and Cortical Function Based on Grid Cells in the Neocortex," *Frontiers in Neural Circuits* 12, no. 121 (January 11, 2019), https://doi.org/10.3389/fncir.2018.00121; Baoguo Shi et al., "Different Brain Structures Associated with Artistic and Scientific Creativity: A Voxel-Based Morphometry Study," *Scientific Reports* 7, no. 42911 (February 21, 2017), https://doi.org/10.1038/srep42911; Barbara L. Finlay and Kexin Huang, "Developmental Duration as an Organizer of the Evolving Mammalian Brain: Scaling, Adaptations, and Exceptions," *Evolution and Development* 22, nos. 1–2 (December 3, 2019), https://doi.org/10.1111/ede.12329.

69. For one simple overview and one more technical lecture on the associative nature of memory, see Shelly Fan, "How the Brain Makes Memories: Scientists Tap Memory's Neural Code," SingularityHub, July 10, 2015, https://singularityhub.com/2015/07/10/how-the-brain-makes-memories-scientists-tap-memorys-neural-code; Christos Papadimitriou, "Formation and Association of Symbolic Memories in the Brain," Simons Institute, YouTube video, March 31, 2017, https://www.youtube.com/watch?v=IZtYKApSTto.

70. For more on the shift from creationism to evolution by natural selection, see Phillip Sloan, "Evolutionary Thought Before Darwin," in *Stanford Encyclopedia of Philosophy*, ed. Edward N. Zalta (Winter 2019), https://plato.stanford.edu/entries/evolution-before-darwin; Christoph Marty, "Darwin on a Godless Creation: 'It's Like Confessing to a Murder,'" *Scientific American*, February 12, 2009, https://www.scientificamerican.com/article/charles-darwin-confessions.

71. For more on Charles Lyell's work and its influence on Darwin, see Richard A. Fortey, "Charles Lyell and Deep Time," *Geoscience* 21, no. 9 (October 2011), https://www.geolsoc.org.uk/Geoscientist/Archive/October-2011/Charles-Lyell-and-deep-time; Gary Stix, "Darwin's Living Legacy," *Scientific American* 300, no. 1 (January 2009): 38–43, https://www.jstor.org/stable/26001418; Charles Darwin, *On the Origin of Species*, 6th ed. (London: John Murray, 1859; Project Gutenberg, 2013), https://www.gutenberg.org/files/2009/2009-h/2009-h.htm.

72. Walter F. Cannon, "The Uniformitarian-Catastrophist Debate," *Isis* 51, no. 1 (March 1960): 38–55, https://www.jstor.org/stable/227604; Jim Morrison, "The Blasphemous Geologist Who Rocked Our Understanding of Earth's Age," *Smithsonian*, August 29, 2016, https://www.smithsonianmag.com/history/father-modern-geology-youve-never-heard-180960203.

73. Charles Darwin and James T. Costa, *The Annotated Origin: A Facsimile of the First Edition of On the Origin of Species* (Cambridge, MA, and London: Belknap Press of Harvard Uni-

versity Press, 2009), 95, https://www.google.com/books/edition/The_Annotated_i_Origin
_i/C0E03ilhSz4C.

74. Gordon Moore, "Cramming More Components onto Integrated Circuits," *Electronics* 38, no. 8 (April 19, 1965), https://archive.computerhistory.org/resources/access/text/2017/03/102770822-05-01-acc.pdf; Computer History Museum, "1965: 'Moore's Law' Predicts the Future of Integrated Circuits," Computer History Museum, accessed October 12, 2021, https://www.computerhistory.org/siliconengine/moores-law-predicts-the-future-of-integrated-circuits; Fernando J. Corbató et al., *The Compatible Time-Sharing System: A Programmer's Guide* (Cambridge, MA: MIT Press, 1990), http://www.bitsavers.org/pdf/mit/ctss/CTSS_ProgrammersGuide.pdf.

75. No one can say for certain what the next computing paradigm will be, but for some recent promising research, see Jeff Hecht, "Nanomaterials Pave the Way for the Next Computing Generation," *Nature* 608, S2–S3 (2022), https://www.nature.com/articles/d41586-022-02147-3; Peng Lin et al., "Three-Dimensional Memristor Circuits as Complex Neural Networks," *Nature Electronics* 3, no. 4 (April 13, 2020): 225–32, https://doi.org/10.1038/s41928-020-0397-9; Zhihong Chen, "Gate-All-Around Nanosheet Transistors Go 2D," *Nature Electronics* 5, no. 12 (December 12, 2022): 830–31, https://doi.org/10.1038/s41928-022-00899-4.

76. In 1888, the Hollerith tabulating machine represented the first practically deployed device to perform large-scale computation. The exponential price-performance trend has remained remarkably steady right up through the present. See Emile Cheysson, *The Electric Tabulating Machine*, trans. Arthur W. Fergusson (New York: C. C. Shelley, 1892), 2, https://books.google.com/books?id=rJgsAAAAYAAJ; Robert Sobel, *Thomas Watson, Sr.: IBM and the Computer Revolution* (Washington, DC: BeardBooks, 2000; originally published as *I.B.M., Colossus in Transition* by Times Books in 1981), 17, https://www.google.com/books/edition/Thomas_Watson_Sr/H8EFNMBGpY4C; US Bureau of Labor Statistics, "Consumer Price Index for All Urban Consumers: All Items in U.S. City Average (CPIAUCSL)," retrieved from FRED, Federal Reserve Bank of St. Louis, updated April 12, 2023, https://fred.stlouisfed.org/series/CPIAUCSL; Marguerite Zientara, "Herman Hollerith: Punched Cards Come of Age," *Computerworld* 15, no. 36 (September 7, 1981): 35, https://books.google.com/books?id=tk74jLc6HggC&pg=PA35&lpg=PA35; Frank da Cruz, "Hollerith 1890 Census Tabulator," Columbia University Computing History, April 17, 2021, http://www.columbia.edu/cu/computinghistory/census-tabulator.html.

77. Nick Bostrom, "Nick Bostrom The Intelligence Explosion Hypothesis eDay 2012," eAcast55, YouTube video, August 9, 2015, https://www.youtube.com/watch?v=VFE-96XA92w.

78. DeepMind, "AlphaGo," DeepMind, accessed November 20, 2021, https://deepmind.com/research/case-studies/alphago-the-story-so-far.

79. DeepMind, "AlphaGo."

80. For an engaging account of the significance of the Deep Blue–Kasparov match, see Mark Robert Anderson, "Twenty Years On from Deep Blue vs. Kasparov: How a Chess Match Started the Big Data Revolution," *The Conversation*, May 11, 2017, https://theconversation.com/twenty-years-on-from-deep-blue-vs-kasparov-how-a-chess-match-started-the-big-data-revolution-76882.

81. DeepMind, "AlphaGo Zero: Starting from Scratch," DeepMind, October 18, 2017, https://deepmind.com/blog/article/alphago-zero-starting-scratch; DeepMind, "AlphaGo"; Tom Simonite, "This More Powerful Version of AlphaGo Learns on Its Own," *Wired*, October 18, 2017, https://www.wired.com/story/this-more-powerful-version-of-alphago-learns-on-its-own; David Silver et al., "Mastering the Game of Go with Deep Neural Networks and Tree Search," *Nature* 529, no. 7587 (January 27, 2016): 484–89, https://doi.org/10.1038/nature16961.

82. Carl Engelking, "The AI That Dominated Humans in Go Is Already Obsolete," *Discover*, October 18, 2017, https://www.discovermagazine.com/technology/the-ai-that-dominated-humans-in-go-is-already-obsolete; DeepMind, "AlphaGo China," DeepMind, accessed November 20, 2021, https://deepmind.com/alphago-china; DeepMind, "AlphaGo Zero: Starting from Scratch."

83. DeepMind, "AlphaGo Zero: Starting from Scratch."

84. David Silver et al., "AlphaZero: Shedding New Light on Chess, Shogi, and Go," Deep-Mind, December 6, 2018, https://deepmind.com/blog/article/alphazero-shedding-new-light-grand-games-chess-shogi-and-go.

85. Julian Schrittwiese et al., "MuZero: Mastering Go, Chess, Shogi and Atari Without Rules," DeepMind, December 23, 2020, https://deepmind.com/blog/article/muzero-mastering-go-chess-shogi-and-atari-without-rules.

86. AlphaStar Team, "AlphaStar: Mastering the Real-Time Strategy Game StarCraft II," DeepMind, January 24, 2019, https://deepmind.com/blog/article/alphastar-mastering-real-time-strategy-game-starcraft-ii; Noam Brown and Tuomas Sandholm, "Superhuman AI for Heads-Up No-Limit Poker: Libratus Beats Top Professionals," *Science* 359, no. 6374 (January 26, 2018): 418–24, https://doi.org/10.1126/science.aao1733; Cade Metz, "Inside Libratus, the Poker AI That Out-Bluffed the Best Humans," *Wired*, February 1, 2017, https://www.wired.com/2017/02/libratus.

87. For more on the dynamics of *Diplomacy*, see "Diplomacy: Running the Game #40, Politics #3," Matthew Colville, YouTube video, July 15, 2017, https://www.youtube.com/watch?v=HWt0AQWjhPg; Ben Harsh, "Harsh Rules: Let's Learn to Play Diplomacy," Harsh Rules, YouTube video, August 9, 2018, https://www.youtube.com/watch?v=S-sSWsBdbNI; Blake Eskin, "World Domination: The Game," *Washington Post*, November 14, 2004, https://www.washingtonpost.com/archive/lifestyle/magazine/2004/11/14/world-domination-the-game/b65c9d9f-71c7-4846-961f-6dcdd1891e01; David Hill, "The Board Game of the Alpha Nerds," *Grantland*, June 18, 2014, https://grantland.com/features/diplomacy-the-board-game-of-the-alpha-nerds.

88. Matthew Hutson, "AI Learns the Art of Diplomacy," *Science*, November 22, 2022, https://www.science.org/content/article/ai-learns-art-diplomacy-game; Yoram Bachrach and János Kramár, "AI for the Board Game Diplomacy," DeepMind, December 6, 2022, https://www.deepmind.com/blog/ai-for-the-board-game-diplomacy.

89. Ira Boudway and Joshua Brustein, "Waymo's Long-Term Commitment to Safety Drivers in Autonomous Cars," Bloomberg, January 13, 2020, https://www.bloomberg.com/news/articles/2020-01-13/waymo-s-long-term-commitment-to-safety-drivers-in-autonomous-cars.

90. Aaron Pressman, "Google's Waymo Reaches 20 Million Miles of Autonomous Driving," *Fortune*, January 7, 2020, https://fortune.com/2020/01/07/googles-waymo-reaches-20-million-miles-of-autonomous-driving.

91. Darrell Etherington, "Waymo Has Now Driven 10 Billion Autonomous Miles in Simulation," *TechCrunch*, July 10, 2019, https://techcrunch.com/2019/07/10/waymo-has-now-driven-10-billion-autonomous-miles-in-simulation.

92. Gabriel Goh et al., "Multimodal Neurons in Artificial Neural Networks," *Distill*, March 4, 2021, https://distill.pub/2021/multimodal-neurons.

93. For more on the USE, see Yinfei Yang and Amin Ahmad, "Multilingual Universal Sentence Encoder for Semantic Retrieval," *Google Research*, July 12, 2019, https://ai.googleblog.com/2019/07/multilingual-universal-sentence-encoder.html; Yinfei Yang and Chris Tar, "Advances in Semantic Textual Similarity," *Google Research*, May 17, 2018, https://ai.googleblog.com/2018/05/advances-in-semantic-textual-similarity.html; Daniel Cer et al., "Universal Sentence Encoder," arXiv:1803.11175v2 [cs.CL], April 12, 2018, https://arxiv.org/abs/1803.11175.

94. Rachel Syme, "Gmail Smart Replies and the Ever-Growing Pressure to E-Mail Like a Machine," *New Yorker*, November 28, 2018, https://www.newyorker.com/tech/annals-of-technology/gmail-smart-replies-and-the-ever-growing-pressure-to-e-mail-like-a-machine.

95. For a more detailed explainer on how transformers work, and the original technical paper, see Giuliano Giacaglia, "How Transformers Work," *Towards Data Science*, March 10, 2019, https://towardsdatascience.com/transformers-141e32e69591; Ashish Vaswani et al., "Attention Is All You Need," arXiv:1706.03762v5 [cs.CL], December 6, 2017, https://arxiv.org/pdf/1706.03762.pdf.

96. Irene Solaiman et al., "GPT-2: 1.5B Release," OpenAI, November 5, 2019, https://openai.com/blog/gpt-2-1-5b-release.

97. Tom B. Brown et al., "Language Models Are Few-Shot Learners," arXiv:2005.14165 [cs.CL], July 22, 2020, https://arxiv.org/abs/2005.14165.

98. Jack Ray et al., "Language Modelling at Scale: Gopher, Ethical Considerations, and Retrieval," DeepMind, December 8, 2021, https://www.deepmind.com/blog/language-modelling-at-scale-gopher-ethical-considerations-and-retrieval.

99. Pandu Nayak, "Understanding Searches Better Than Ever Before," Google, October 25, 2019, https://blog.google/products/search/search-language-understanding-bert; William Fedus et al., "Switch Transformers: Scaling to Trillion Parameter Models with Simple and Efficient Sparsity," arXiv:2101.03961 [cs.LG], January 11, 2021, https://arxiv.org/abs/2101.03961.

100. For more in-depth information on GPT-3, see Greg Brockman et al., "OpenAI API," OpenAI, June 11, 2020, https://openai.com/blog/openai-api; Brown et al., "Language Models Are Few-Shot Learners"; Kelsey Piper, "GPT-3, Explained: This New Language AI Is Uncanny, Funny—and a Big Deal," *Vox*, August 13, 2020, https://www.vox.com/future-perfect/21355768/gpt-3-ai-openai-turing-test-language; "GPT-3 Demo: New AI Algorithm Changes How We Interact with Technology," Disruption Theory, YouTube video, August 28, 2020, https://www.youtube.com/watch?v=8V20HkoiNtc.

101. David Cole, "The Chinese Room Argument," in *The Stanford Encyclopedia of Philosophy*, ed. Edward N. Zalta (Winter 2020), https://plato.stanford.edu/archives/win2020/entries/chinese-room; Amanda Askell (@amandaaskell), "GPT-3's completion of the Chinese room argument from Searle's 'Minds, Brains, and Programs' (original text is in bold)," Twitter, July 17, 2020, https://twitter.com/AmandaAskell/status/1284186919606251521; David J. Chalmers, *The Conscious Mind: In Search of a Fundamental Theory* (New York: Oxford University Press, 1996), 327.

102. Cade Metz, "Meet GPT-3. It Has Learned to Code (and Blog and Argue)," *New York Times*, November 24, 2020, https://www.nytimes.com/2020/11/24/science/artificial-intelligence-ai-gpt3.html.

103. For more information on LaMDA and a demo of it holding a conversation pretending to be the (dwarf) planet Pluto and then a paper plane, see Eli Collins and Zoubin Ghahramani, "LaMDA: Our Breakthrough Conversation Technology," Google, May 18, 2021, https://blog.google/technology/ai/lamda; "Watch Google's AI LaMDA Program Talk to Itself at Length (Full Conversation)," CNET Highlights, YouTube video, May 18, 2021, https://www.youtube.com/watch?v=aUSSfo5nCdM.

104. Jeff Dean, "Google Research: Themes from 2021 and Beyond," *Google Research*, January 11, 2022, https://ai.googleblog.com/2022/01/google-research-themes-from-2021-and.html.

105. For examples of DALL-E's remarkably creative images, see Aditya Ramesh et al., "Dall-E: Creating Images from Text," OpenAI, January 5, 2021, https://openai.com/research/dall-e.

106. "Dall-E 2," OpenAI, accessed June 30, 2022, https://openai.com/dall-e-2.

107. Chitwan Saharia et al., "Imagen," Google Research, Brain Team, Google, accessed June 30, 2022, https://imagen.research.google.

108. Saharia et al., "Imagen."

109. Scott Reed et al., "A Generalist Agent," DeepMind, May 12, 2022, https://www.deepmind.com/publications/a-generalist-agent.

110. Wojciech Zaremba et al., "OpenAI Codex," OpenAI, August 10, 2021, https://openai.com/blog/openai-codex.

111. AlphaCode Team, "Competitive Programming with AlphaCode," DeepMind, December 8, 2022, https://www.deepmind.com/blog/competitive-programming-with-alphacode; Yujia Li et al., "Competition-Level Code Generation with AlphaCode," arXiv:2203.07814v1 [cs.PL], February 8, 2022, https://arxiv.org/abs/2203.07814.pdf.

112. Aakanksha Chowdhery, "PaLM: Scaling Language Modeling with Pathways," arXiv:2204.02311v3 [cs.CL], April 19, 2022, https://arxiv.org/pdf/2204.02311.pdf; Sharan Narang et al., "Pathways Language Model (PaLM): Scaling to 540 Billion Parameters for Breakthrough Performance," *Google AI Blog*, April 4, 2022, https://ai.googleblog.com/2022/04/pathways-language-model-palm-scaling-to.html.

113. Chowdhery, "PaLM: Scaling Language Modeling with Pathways."

114. Chowdhery, "PaLM: Scaling Language Modeling with Pathways."

115. Chowdhery, "PaLM: Scaling Language Modeling with Pathways."

116. OpenAI, "Introducing ChatGPT," OpenAI, November 30, 2022, https://openai.com/blog /chatgpt#OpenAI.

117. Krystal Hu, "ChatGPT Sets Record for Fastest-Growing User Base—Analyst Note," Reuters, February 2, 2023, https://www.reuters.com/technology/chatgpt-sets-record-fastest -growing-user-base-analyst-note-2023-02-01.

118. Kalley Huang, "Alarmed by A.I. Chatbots, Universities Start Revamping How They Teach," *New York Times*, January 16, 2023, https://www.nytimes.com/2023/01/16/technol ogy/chatgpt-artificial-intelligence-universities.html; Emma Bowman, "A College Student Created an App That Can Tell Whether AI Wrote an Essay," NPR, January 9, 2023, https://www.npr.org/2023/01/09/1147549845/gptzero-ai-chatgpt-edward-tian-plagiarism; Patrick Wood and Mary Louise Kelly, "'Everybody Is Cheating': Why This Teacher Has Adopted an Open ChatGPT Policy," NPR, January 26, 2023, https://www.npr.org/2023 /01/26/1151499213/chatgpt-ai-education-cheating-classroom-wharton-school; Matt O'Brien and Jocelyn Gecker, "Cheaters Beware: ChatGPT Maker Releases AI Detection Tool," Associated Press, January 31, 2023, https://apnews.com/article/technology-education-col leges-and-universities-france-a0ab654549de387316404a7be019116b; Geoffrey A. Fowler, "We Tested a New ChatGPT-Detector for Teachers. It Flagged an Innocent Student," *Washington Post*, April 3, 2023, https://www.washingtonpost.com/technology/2023/04/01 /chatgpt-cheating-detection-turnitin.

119. OpenAI, "GPT-4," OpenAI, March 14, 2023, https://openai.com/research/gpt-4; OpenAI, "GPT-4 Technical Report," arXiv:2303.08774v3 [cs.CL], March 27, 2023, https://arxiv .org/pdf/2303.08774.pdf; OpenAI, "GPT-4 System Card," OpenAI, March 23, 2023, https://cdn.openai.com/papers/gpt-4-system-card.pdf.

120. OpenAI, "Introducing GPT-4," YouTube video, March 15, 2023, https://www.youtube .com/watch?v=--khbXchTeE.

121. Daniel Feldman (@d_feldman), "On the left is GPT-3.5. On the right is GPT-4. If you think the answer on the left indicates that GPT-3.5 does not have a world-model. . . . Then you have to agree that the answer on the right indicates GPT-4 does," Twitter, March 17, 2023, https://twitter.com/d_feldman/status/1636955260680847361.

122. Danny Driess and Pete Florence, "PaLM-E: An Embodied Multimodal Language Model," Google Research, March 10, 2023, https://ai.googleblog.com/2023/03/palm-e-embodied -multimodal-language.html; Danny Driess et al., "PaLM-E: An Embodied Multimodal Language Model," arXiv:2303.03378v1 [cs.LG], March 6, 2023, https://arxiv.org/pdf /2303.03378.pdf.

123. Sundar Pichai and Demis Hassabis, "Introducing Gemini: Our Largest and Most Capable AI Model," Google, December 6, 2023, https://blog.google/technology/ai/google-gemini -ai; Sundar Pichai, "An Important Next Step on Our AI Journey," Google, February 6, 2023, https://blog.google/technology/ai/bard-google-ai-search-updates; Sarah Fielding, "Goo gle Bard Is Switching to a More 'Capable' Language Model, CEO Confirms," *Engadget*, March 31, 2023, https://www.engadget.com/google-bard-is-switching-to-a-more-capable -language-model-ceo-confirms-133028933.html; Yusuf Mehdi, "Confirmed: The New Bing Runs on OpenAI's GPT-4," Microsoft Bing Blogs, March 14, 2023, https://blogs .bing.com/search/march_2023/Confirmed-the-new-Bing-runs-on-OpenAI%E2%80 %99s-GPT-4; Tom Warren, "Hands-on with the New Bing: Microsoft's Step Beyond ChatGPT," *The Verge*, February 8, 2023, https://www.theverge.com/2023/2/8/23590873 /microsoft-new-bing-chatgpt-ai-hands-on.

124. Johanna Voolich Wright, "A New Era for AI and Google Workspace," Google, March 14, 2023, https://workspace.google.com/blog/product-announcements/generative-ai; Jared Spa taro, "Introducing Microsoft 365 Copilot—Your Copilot for Work," *Official Microsoft Blog*, March 16, 2023, https://blogs.microsoft.com/blog/2023/03/16/introducing-microsoft-365 -copilot-your-copilot-for-work.

125. Markus Anderljung et al., "Compute Funds and Pre-Trained Models," Centre for the Gov ernance of AI, April 11, 2022, https://www.governance.ai/post/compute-funds-and-pre

-trained-models; Jaime Sevilla et al., "Compute Trends Across Three Eras of Machine Learning," arXiv:2202.05924v2 [cs.LG], March 9, 2022, https://arxiv.org/pdf/2202.05924 .pdf; Dario Amodei and Danny Hernandez, "AI and Compute," OpenAI, May 16, 2018, https://openai.com/blog/ai-and-compute.

126. Jacob Stern, "GPT-4 Has the Memory of a Goldfish," *Atlantic*, March 17, 2023, https:// www.theatlantic.com/technology/archive/2023/03/gpt-4-has-memory-context-window /673426.

127. Extrapolating forward the long-term trend, which has shown a doubling time of just over 1.34 years since 1983. See the appendix for the sources used for all the cost-of-computation calculations in this book.

128. At time of writing, progress on the MLPerf benchmark for AI training has been nearly five times as fast as would have been enabled by increasing transistor density alone. The balance is a combination of algorithmic improvements to software and architectural improvements that make chips more efficient. See Samuel K. Moore, "AI Training Is Outpacing Moore's Law," *IEEE Spectrum*, December 2, 2021, https://spectrum.ieee.org/ai-training -mlperf.

129. As I write this, prices for the GPT-3.5 API are down to $1.00 per 500,000 tokens, or roughly 370,000 words. Prices will likely be even lower by the time you read this. See Ben Dickson, "OpenAI Is Reducing the Price of the GPT-3 API—Here's Why It Matters," *VentureBeat*, August 25, 2022, https://venturebeat.com/ai/openai-is-reducing-the-price -of-the-gpt-3-api-heres-why-it-matters; OpenAI, "Introducing ChatGPT and Whisper APIs," OpenAI, March 1, 2023, https://openai.com/blog/introducing-chatgpt-and-whisper-apis; OpenAI, "What Are Tokens and How to Count Them?," OpenAI, accessed April 30, 2023, https://help.openai.com/en/articles/4936856-what-are-tokens-and-how-to-count-them.

130. Stephen Nellis, "Nvidia Shows New Research on Using AI to Improve Chip Designs," Reuters, March 27, 2023, https://www.reuters.com/technology/nvidia-shows-new-research -using-ai-improve-chip-designs-2023-03-28.

131. Blaise Aguera y Arcas, "Do Large Language Models Understand Us?," *Medium*, December 16, 2021, https://medium.com/@blaisea/do-large-language-models-understand-us-6f88 1d6d8e75.

132. With better algorithms, the amount of training compute needed to achieve a given level of performance decreases. A growing body of research suggests that for many applications, algorithmic progress is roughly as important as hardware progress. According to a 2022 study, better algorithms halved compute requirements for a given level of performance every nine months from 2012–2021. See Ege Erdil and Tamay Besiroglu, "Algorithmic Progress in Computer Vision," arXiv:2212.05153v4 [cs.CV] August 24, 2023, https://arxiv .org/pdf/2212.05153.pdf; Katja Grace, *Algorithmic Progress in Six Domains*, Machine Intelligence Research Institute technical report 2013-3, December 9, 2013, https://intelligence .org/files/AlgorithmicProgress.pdf.

133. Anderljung et al., "Compute Funds and Pre-Trained Models."

134. Sevilla et al., "Compute Trends Across Three Eras of Machine Learning"; see the appendix for the sources used for all the cost-of-computation calculations in this book.

135. Anderljung et al., "Compute Funds and Pre-Trained Models"; Sevilla et al., "Compute Trends Across Three Eras of Machine Learning"; Amodei and Hernandez, "AI and Compute."

136. "Dallas Fed Energy Survey," Federal Reserve Bank of Dallas, March 27, 2019, https:// www.dallasfed.org/research/surveys/des/2019/1901.aspx#tab-questions.

137. For a clear and accessible overview of big data, see Rebecca Tickle, "What Is Big Data?," Computerphile, YouTube video, May 15, 2019, https://www.youtube.com/watch?v=H4bf _uuMC-g.

138. For a much deeper dive on the potential nature of an intelligence explosion, see Nick Bostrom, "The Intelligence Explosion Hypothesis—eDay 2012," EMERCE, YouTube video, November 19, 2012, https://www.youtube.com/watch?v=g3FMpn321zs; Luke Muehlhauser and Anna Salamon, "Intelligence Explosion: Evidence and Import," in *Singularity Hypotheses: A Scientific and Philosophical Assessment*, ed. Amnon Eden et al. (Berlin:

Springer, 2013), https://intelligence.org/files/IE-EI.pdf; Eliezer Yudkowsky, "Recursive Self-Improvement," LessWrong.com, December 1, 2008, https://www.lesswrong.com /posts/JBadX7rwdcRFzGuju/recursive-self-improvement; Eliezer Yudkowsky, "Hard Take-off," LessWrong.com, December 2, 2008, https://www.lesswrong.com/posts/tjH8XPxAn r6JRbh7k/hard-takeoff; Eliezer Yudkowsky, *Intelligence Explosion Microeconomics*, Machine Intelligence Research Institute technical report 2013-1, September 13, 2013, https://intelli gence.org/files/IEM.pdf; I. J. Good, "Speculations Concerning the First Ultraintelligent Machine," *Advances in Computers* 6 (1966): 31–88, https://doi.org/10.1016/S0065-2458(08) 60418-0; Ephrat Livni, "The Mirror Test for Animal Self-Awareness Reflects the Limits of Human Cognition," *Quartz*, December 19, 2018, https://qz.com/1501318/the-mirror-test -for-animals-reflects-the-limits-of-human-cognition; Darold A. Treffert, "The Savant Syndrome: An Extraordinary Condition. A Synopsis: Past, Present, Future," *Philosophical Transactions of the Royal Society B: Biological Sciences* 364, no. 1522 (May 27, 2009): 1351–57, https://doi.org/10.1098/rstb.2008.0326.

139. Robin Hanson and Eliezer Yudkowsky, *The Hanson-Yudkowsky AI-Foom Debate*, Machine Intelligence Research Institute, 2013, https://intelligence.org/files/AIFoomDebate.pdf.

140. Hanson and Yudkowsky, *Hanson-Yudkowsky AI-Foom Debate*.

141. Jon Brodkin, "1.1 Quintillion Operations per Second: US Has World's Fastest Supercomputer," *Ars Technica*, May 31, 2022, https://arstechnica.com/information-technology/2022 /05/1-1-quintillion-operations-per-second-us-has-worlds-fastest-supercomputer; "November 2022," Top500.org, accessed November 14, 2023, https://www.top500.org/lists/top500 /2022/11.

142. For an outstanding report by Joseph Carlsmith at Open Philanthropy that deeply explores multiple perspectives on the topic and an AI Impacts summary of many estimates with varying methodologies, see Joseph Carlsmith, *How Much Computational Power Does It Take to Match the Human Brain?*, Open Philanthropy, September 11, 2020, https://www.open philanthropy.org/brain-computation-report; "Brain Performance in FLOPS," AI Impacts, July 26, 2015, https://aiimpacts.org/brain-performance-in-flops.

143. Herculano-Houzel, "Human Brain in Numbers"; David A. Drachman, "Do We Have Brain to Spare?," *Neurology* 64, no. 12 (June 27, 2005), https://doi.org/10.1212 /01.WNL.0000166914.38327.BB; Ernest L. Abel, *Behavioral Teratogenesis and Behavioral Mutagenesis: A Primer in Abnormal Development* (New York: Plenum Press, 1989), 113, https://books.google.co.uk/books?id=gV0rBgAAQBAJ.

144. "Neuron Firing Rates in Humans," AI Impacts, April 14, 2015, https://aiimpacts.org/rate -of-neuron-firing; Peter Steinmetz et al., "Firing Behavior and Network Activity of Single Neurons in Human Epileptic Hypothalamic Hamartoma," *Frontiers in Neurology* 2, no. 210 (December 27, 2013), https://doi.org/10.3389/fneur.2013.00210.

145. "Neuron Firing Rates in Humans," AI Impacts.

146. Ray Kurzweil, *The Singularity Is Near* (New York: Viking, 2005), 125; Hans Moravec, *Mind Children: The Future of Robot and Human Intelligence* (Cambridge, MA; Harvard University Press, 1988), 59, https://books.google.co.uk/books?id=56mb7XuSx3QC.

147. Preeti Raghavan, "Stroke Recovery Timeline," Johns Hopkins Medicine, accessed April 27, 2023, https://www.hopkinsmedicine.org/health/conditions-and-diseases/stroke/stroke -recovery-timeline; Apoorva Mandavilli, "The Brain That Wasn't Supposed to Heal," *Atlantic*, April 7, 2016, https://www.theatlantic.com/health/archive/2016/04/brain-injuries/477300.

148. As of early 2023, $1,000 per hour of time on Google Cloud's TPU v5e systems translates to around 328 quadrillion operations per second, which is on the order of 10^{17}. The wide availability of rented cloud computing has effectively brought costs way down for many users, but note that this is not fully commensurable with past computation cost figures, which derive mainly from equipment purchase costs. Rented time cannot be directly compared to time in use on purchased hardware, but a plausible rough comparison (neglecting a lot of minutiae, like IT personnel wages, electricity, and depreciation) is to the cost of 4,000 hours of working time. By that metric, the TPU v5e averages more than a sustained 130 trillion operations per second (on the order of 10^{14}) per $1,000. See the appendix for the sources used for all the cost-of-computation calculations in this book.

149. Extrapolating forward the long-term trend, which has shown a doubling time of just over 1.34 years since 1983. See the appendix for the sources used for all the cost-of-computation calculations in this book.

150. Anders Sandberg and Nick Bostrom, *Whole Brain Emulation: A Roadmap*, technical report 2008-3, Future of Humanity Institute, Oxford University (2008), 80–81, https://www.fhi .ox.ac.uk/brain-emulation-roadmap-report.pdf.

151. Sandberg and Bostrom, *Whole Brain Emulation*.

152. Mitch Kapor and Ray Kurzweil, "A Wager on the Turing Test: The Rules," KurzweilAI .net, April 9, 2002, http://www.kurzweilai.net/a-wager-on-the-turing-test-the-rules.

153. Edward Moore Geist, "It's Already Too Late to Stop the AI Arms Race—We Must Manage It Instead," *Bulletin of the Atomic Scientists* 72, no. 5 (August 15, 2016): 318–21, https://doi .org/10.1080/00963402.2016.1216672.

154. For a representative example by leading science writer John Horgan in *The New York Times*, see John Horgan, "Smarter than Us? Who's Us?," *New York Times*, May 4, 1997, https:// www.nytimes.com/1997/05/04/opinion/smarter-than-us-who-s-us.html.

155. See, for example, Hubert L. Dreyfus, "Why We Do Not Have to Worry About Speaking the Language of the Computer," *Information Technology & People* 11, no. 4 (December 1998): 281–89, https://personal.lse.ac.uk/whitley/allpubs/heideggerspecialissue/heidegger01 .pdf; Selmer Bringsjord, "Chess Is Too Easy," *MIT Technology Review*, March 1, 1998, https://www.technologyreview.com/1998/03/01/237087/chess-is-too-easy.

156. Notably, one of the doctoral students who designed Proverb, the first AI to master crossword puzzles better than most human solvers, was Noam Shazeer. He went on to work at Google, where he was a lead author of "Attention Is All You Need," the paper that invented the transformer architecture for large language models that has powered the latest AI revolution. See Duke University, "Duke Researchers Pit Computer Against Human Crossword Puzzle Players," *ScienceDaily*, April 20, 1999, https://www.sciencedaily.com/releases/1999 /04/990420064821.htm; Vaswani et al., "Attention Is All You Need."

157. For a representative video clip from the matches and analyses of Watson and the competition, see OReilly, "Jeopardy! IBM Challenge Day 3 (HD) Ken Jennings vs. WATSON vs. Brad Rutter (02-16-11)," Vimeo video, June 19, 2017, https://vimeo.com/222234104; Sam Gustin, "Behind IBM's Plan to Beat Humans at Their Own Game," *Wired*, February 14, 2011, https://www.wired.com/2011/02/watson-jeopardy; John Markoff, "Computer Wins on 'Jeopardy!': Trivial, It's Not," *New York Times*, February 16, 2011, https://www.nytimes .com/2011/02/17/science/17jeopardy-watson.html.

158. "Show #6088—Wednesday, February 16, 2011," J! Archive, accessed April 30, 2023, https://j-archive.com/showgame.php?game_id=3577.

159. "Show #6088—Wednesday, February 16, 2011," J! Archive.

160. Jeffrey Grubb, "Google Duplex: A.I. Assistant Calls Local Businesses to Make Appointments," Jeff Grubb's Game Mess, YouTube video, May 8, 2018, https://www.youtube.com /watch?v=D5VN56jQMWM; Georgina Torbet, "Google Duplex Begins International Rollout with a New Zealand pilot," *Engadget*, October 22, 2019, https://www.engadget .com/2019-10-22-google-duplex-pilot-new-zealand.html; IBM, "Man vs. Machine: Highlights from the Debate Between IBM's Project Debater and Harish Natarajan," BusinessWorldTV, YouTube video, February 13, 2019, https://www.youtube.com/watch?v=nJXc FtY9cWY.

161. For more information on the problems LLMs have with hallucination, see Tom Simonite, "AI Has a Hallucination Problem That's Proving Tough to Fix," *Wired*, March 9, 2018, https://www.wired.com/story/ai-has-a-hallucination-problem-thats-proving-tough-to-fix; Craig S. Smith, "Hallucinations Could Blunt ChatGPT's Success," *IEEE Spectrum*, March 13, 2023, https://spectrum.ieee.org/ai-hallucination; Cade Metz, "What Makes A.I. Chatbots Go Wrong?," *New York Times*, March 29, 2023 (updated April 4, 2023), https://www .nytimes.com/2023/03/29/technology/ai-chatbots-hallucinations.html; Ziwei Ji et al., "Survey of Hallucination in Natural Language Generation," *ACM Computing Surveys* 55, no. 12, article 248 (March 3, 2023): 1–38, https://doi.org/10.1145/3571730.

162. Jonathan Cohen, "Right on Track: NVIDIA Open-Source Software Helps Developers Add Guardrails to AI Chatbots," NVIDIA, April 25, 2023, https://blogs.nvidia.com/blog/2023 /04/25/ai-chatbot-guardrails-nemo.

163. Turing, "Computing Machinery and Intelligence."

164. Turing, "Computing Machinery and Intelligence."

165. For example, in 2014 a chatbot called Eugene Goostman gained entirely unearned headlines as having passed the Turing test by imitating a thirteen-year-old Ukrainian boy who spoke poor English. See Doug Aamoth "Interview with Eugene Goostman, the Fake Kid Who Passed the Turing Test," *Time*, June 9, 2014, https://time.com/2847900/eugene -goostman-turing-test.

166. For more information from our Google team on how Talk to Books worked, and for my TED podcast interview with Chris Anderson, see Google AI, "Talk to Books," *Experiments with Google*, September 2018, https://experiments.withgoogle.com/talk-to-books; Chris Anderson, "Ray Kurzweil on What the Future Holds Next," in *The Ted Interview* podcast, December 2018, https://www.ted.com/talks/the_ted_interview_ray_kurzweil_on_what_the _future_holds_next/transcript.

167. For an explainer on fMRI technology, see Mark Stokes, "What Does fMRI Measure?," *Scitable*, May 16, 2015, https://www.nature.com/scitable/blog/brain-metrics/what_does _fmri_measure.

168. Sriranga Kashyap et al., "Resolving Laminar Activation in Human V1 Using Ultra-High Spatial Resolution fMRI at 7T," *Scientific Reports* 8, article 17-63 (November 20, 2018), https://doi .org/10.1038/s41598-018-35333-3; Jozien Goense, Yvette Bohraus, and Nikos K. Logothetis, "fMRI at High Spatial Resolution: Implications for BOLD-Models," *Frontiers in Computational Neuroscience* 10, no. 66 (June 28, 2016), https://doi.org/10.3389/fncom.2016.00066.

169. There are techniques that can achieve 100-millisecond temporal resolution, but they come at the cost of badly degraded spatial resolution, around 5–6 millimeters. See Benjamin Zahneisen et al., "Three-Dimensional MR-Encephalography: Fast Volumetric Brain Imaging Using Rosette Trajectories," *Magnetic Resonance in Medicine* 65, no. 5 (May 2011): 1260–68, https://doi.org/10.1002/mrm.22711; David A. Feinberg et al., "Multiplexed Echo Planar Imaging for Sub-Second Whole Brain FMRI and Fast Diffusion Imaging.," *PloS ONE* 5, no. 12: e15710 (December 20, 2010), https://doi.org/10.1371/journal.pone.0015710.

170. Alexandra List et al., "Pattern Classification of EEG Signals Reveals Perceptual and Attentional States," *PLoS ONE* 12, no. 4: e0176349 (April 26, 2017), https://doi.org/10.1371/journal .pone.0176349; Boris Burle et al., "Spatial and Temporal Resolutions of EEG: Is It Really Black and White? A Scalp Current Density View," *International Journal of Psychophysiology* 97, no. 3 (September 2015): 210–20, https://doi.org/10.1016/j.ijpsycho.2015.05.004.

171. Yahya Aghakhani et al., "Co-Localization Between the BOLD Response and Epileptiform Discharges Recorded by Simultaneous Intracranial EEG-fMRI at 3 T," *NeuroImage: Clinical* 7 (2015): 755–63, https://doi.org/10.1016/j.nicl.2015.03.002; Brigitte Stemmer and Frank A. Rodden, "Functional Brain Imaging of Language Processes," in *International Encyclopedia of the Social & Behavioral Sciences*, ed. James D. Wright, 2nd ed. (Amsterdam: Elsevier Science, 2015), 476–513, https://doi.org/10.1016/B978-0-08-097086-8.54009-4; Burle et al., "Spatial and Temporal Resolutions of EEG); Claudio Babiloni et al., "Fundamentals of Electroencefalography, Magnetoencefalography, and Functional Magnetic Resonance Imaging," in *Brain Machine Interfaces for Space Applications: Enhancing Astronaut Capabilities*, ed. Luca Rossini, Dario Izzo, and Leopold Summerer (New York: Academic Press, 2009), 73, https://books.google.co.uk/books?id=l5Q1bul_ZbEC.

172. For video of the BrainGate system in action, see BrainGate Collaboration, "Thought Control of Robotic Arms Using the BrainGate System," NIHNINDS, YouTube video, May 16, 2012, https://www.youtube.com/watch?v=QRt8QCx3BCo.

173. Tech at Meta, "Imagining a New Interface: Hands-Free Communication Without Saying a Word," *Facebook Reality Labs*, March 30, 2020, https://tech.fb.com/imagining-a-new -interface-hands-free-communication-without-saying-a-word; Tech at Meta, "BCI Milestone: New Research from UCSF with Support from Facebook Shows the Potential of

Brain-Computer Interfaces for Restoring Speech Communication," *Facebook Reality Labs*, July 14, 2021, https://tech.fb.com/ar-vr/2021/07/bci-milestone-new-research-from-ucsf-with -support-from-facebook-shows-the-potential-of-brain-computer-interfaces-for-restoring -speech-communication; Joseph G. Makin et al., "Machine Translation of Cortical Activity to Text with an Encoder–Decoder Framework," *Nature Neuroscience* 23, no. 4 (March 30, 2020): 575–82, https://doi.org/10.1038/s41593-020-0608-8.

174. Antonio Regalado, "Facebook Is Ditching Plans to Make an Interface that Reads the Brain," *MIT Technology Review*, July 14, 2021, https://www.technologyreview.com/2021/07 /14/1028447/facebook-brain-reading-interface-stops-funding.

175. For a long but very accessible conceptual explainer of Neuralink's goals for brain–computer interfaces and the working paper explaining this technology in more technical detail, see Tim Urban, "Neuralink and the Brain's Magical Future (G-Rated Version)," *Wait But Why*, April 20, 2017, https://waitbutwhy.com/2017/04/neuralink-cleanversion.html; Elon Musk and Neuralink, "An Integrated Brain-Machine Interface Platform with Thousands of Channels," Neuralink working paper, July 17, 2019, bioRxiv 703801, https://doi.org /10.1101/703801.

176. John Markoff, "Elon Musk's Neuralink Wants 'Sewing Machine-Like' Robots to Wire Brains to the Internet," *New York Times*, July 16, 2019, https://www.nytimes.com/2019/07 /16/technology/neuralink-elon-musk.html.

177. To watch the monkey in action, see "Monkey MindPong," Neuralink, YouTube video, April 8, 2021, https://www.youtube.com/watch?v=rsCul1sp4hQ.

178. Kelsey Ables, "Musk's Neuralink Implants Brain Chip in its First Human Subject," *Washington Post*, January 30, 2024, https://www.washingtonpost.com/business/2024/01/30/neu ralink-musk-first-human-brain-chip; Neuralink, "Neuralink Clinical Trial," Neuralink, accessed February 6, 2024, https://neuralink.com/pdfs/PRIME-Study-Brochure.pdf; Rachael Levy and Hyunjoo Jin, "Musk Expects Brain Chip Start-up Neuralink to Implant 'First Case' This Year," Reuters, June 20, 2023, https://www.reuters.com/technology/musk -expects-brain-chip-start-up-neuralink-implant-first-case-this-year-2023-06-16; Rachael Levy and Marisa Taylor, "U.S. Regulators Rejected Elon Musk's Bid to Test Brain Chips in Humans, Citing Safety Risks," Reuters, March 2, 2023, https://www.reuters.com/inves tigates/special-report/neuralink-musk-fda; Mary Beth Griggs, "Elon Musk Claims Neuralink Is About 'Six Months' Away from First Human Trial," *The Verge*, November 30, 2022, https://www.theverge.com/2022/11/30/23487307/neuralink-elon-musk-show-and-tell-2022.

179. Andrew Tarantola, "DARPA Is Helping Six Groups Create Neural Interfaces for Our Brains," *Engadget*, July 10, 2017, https://www.engadget.com/2017-07-10-darpa-taps-five -organizations-to-develop-neural-interface-tech.html.

180. "Brown to Receive up to $19M to Engineer Next-Generation Brain-Computer Interface," Brown University, July 10, 2017, https://www.brown.edu/news/2017-07-10/neurograins; Jihun Lee et al., "Wireless Ensembles of Sub-mm Microimplants Communicating as a Network near 1 GHz in a Neural Application," bioRxiv 2020.09.11.293829 (preprint), September 13, 2020, https://www.biorxiv.org/content/10.1101/2020.09.11.293829v1.

181. For more on the hierarchical structure of the neocortex, see Stewart Shipp, "Structure and Function of the Cerebral Cortex," *Current Biology* 17, no. 12 (June 19, 2007): R443–R449, https://www.cell.com/current-biology/pdf/S0960-9822(07)01148-7.pdf; Claus C. Hilgetag and Alexandros Goulas, "'Hierarchy' in the Organization of Brain Networks," *Philosophical Transactions of the Royal Society B*, February 24, 2020, https://doi.org/10.1098/rstb .2019.0319; Jeff Hawkins et al., "A Theory of How Columns in the Neocortex Enable Learning the Structure of the World," *Frontiers in Neural Circuits*, October 25, 2017, https:// doi.org/10.3389/fncir.2017.00081.

182. By "directly" I mean here that the biological and digital neurons would be in direct functional communication in a logical and computational sense. But the physical signals would likely be relayed to and from the cloud by a smaller number of transmission units implanted in and/or worn on the body.

183. For a short explainer on the gap between animal and human language ability, see Ben-Yami, "Can Animals Acquire Language?"

CHAPTER 3: WHO AM I?

1. Samuel Butler, *Erewhon: Or, Over the Range*, 2nd ed. (London: Trübner & Co, 1872), 190, http://www.gutenberg.org/files/1906/1906-h/1906-h.htm.

2. Butler, *Erewhon*, vi.

3. For more on what science and philosophy tell us about animal consciousness, see Colin Allen and Michael Trestman, "Animal Consciousness," in *Stanford Encyclopedia of Philosophy*, ed. Edward N. Zalta (Winter 2017), https://plato.stanford.edu/entries/consciousness -animal; "Just How Smart Are Dolphins?," BBC Earth, YouTube video, October 19, 2014, https://www.youtube.com/watch?v=6M92OA-_5-Y; John Green, "Non-Human Animals," CrashCourse, YouTube video, January 16, 2017, https://www.youtube.com/watch?v=y3 -BX-jN_Ac; Joe Rogan and Roger Penrose, "Are Animals Conscious Like We Are?," JRE Clips, YouTube video, December 18, 2018, https://www.youtube.com/watch?v=TlzY _KvGSZ4.

4. For more on rodent consciousness, see Alla Katnelson, "What the Rat Brain Tells Us About Yours," *Nautilus*, April 13, 2017, http://nautil.us/issue/47/consciousness/what-the -rat-brain-tells-us-about-yours; Jessica Hamzelou, "Zoned-Out Rats May Give Clue to Consciousness," *New Scientist*, October 5, 2011, https://www.newscientist.com/article /mg21128333-700; Cyriel Pennartz et al., "Indicators and Criteria of Consciousness in Animals and Intelligent Machines: An Inside-Out Approach," *Frontiers in Systems Neuroscience* 13, no. 25 (July 16, 2019), https://www.frontiersin.org/articles/10.3389/fnsys.2019 .00025/full; "Scientists Manipulate Consciousness in Rats," National Institutes of Health, December 18, 2015, https://www.nih.gov/news-events/news-releases/scientists-manipulate -consciousness-rats.

5. Douglas Fox, "Consciousness in a Cockroach," *Discover*, January 10, 2007, http://discover magazine.com/2007/jan/cockroach-consciousness-neuron-similarity; Suzana Herculano-Houzel, "The Human Brain in Numbers: A Linearly Scaled-Up Primate Brain," *Frontiers in Human Neuroscience* 3, no. 31 (August 5, 2009), https://www.ncbi.nlm.nih.gov/pmc/arti cles/PMC2776484.

6. Colin Barras, "Smart Amoebas Reveal Origins of Primitive Intelligence," *New Scientist*, October 29, 2008, https://www.newscientist.com/article/dn15068; Yuriv V. Pershin et al., "Memristive Model of Amoeba's Learning," *Physical Review E: Statistical Physics, Plasmas, Fluids, and Related Interdisciplinary Topics* 80, 021926 (July 27, 2009), https://arxiv.org/pdf /0810.4179.pdf.

7. For more on relatively advanced consciousness in animals, see Philip Low et al., "The Cambridge Declaration on Consciousness," Francis Crick Memorial Conference 2012: Consciousness in Animals, University of Cambridge, July 7, 2012, http://fcmconference .org/img/CambridgeDeclarationOnConsciousness.pdf; Virginia Morell, "Monkeys Master a Key Sign of Self-Awareness: Recognizing Their Reflections," *Science*, February 13, 2017, https://www.sciencemag.org/news/2017/02/monkeys-master-key-sign-self-awareness -recognizing-their-reflections; Melanie Boly et al., "Consciousness in Humans and Nonhuman Animals: Recent Advances and Future Directions," *Frontiers in Psychology* 4, no. 625 (October 31, 2013), https://www.ncbi.nlm.nih.gov/pmc/articles/PMC3814086; Marc Bekoff, "Animals Are Conscious and Should be Treated as Such," *New Scientist*, September 19, 2012, https://www.newscientist.com/article/mg21528836-200; Elizabeth Pennisi, "Are Our Primate Cousins 'Conscious'?," *Science* 284, no. 5423 (June 25, 1999): 2073–76.

8. Low et al., "Cambridge Declaration on Consciousness."

9. Danielle S. Bassett and Michael S. Gazzaniga, "Understanding Complexity in the Human Brain," *Trends in Cognitive Science* 15, no. 5 (May 2011): 200–209, https://www.ncbi.nlm .nih.gov/pmc/articles/PMC3170818; Xerxes D. Arsiwalla and Paul Verschure, "Measuring the Complexity of Consciousness," *Frontiers in Neuroscience* 12, no. 424 (June 27, 2018), https://doi.org/10.3389/fnins.2018.00424.

10. For an accessible, engaging video on qualia and a deeper and more technical encyclopedic treatment, see Michael Stevens, "Is Your Red the Same as My Red?," Vsauce, YouTube video, February 17, 2013, https://www.youtube.com/watch?v=evQsOFQju08; Michael

Tye, "Qualia," in *Stanford Encyclopedia of Philosophy*, ed. Edward N. Zalta (Summer 2018), https://plato.stanford.edu/entries/qualia.

11. For more on David Chalmers's concept of zombies and John Searle's related "Chinese room" thought experiment showing why subjective consciousness cannot be proven from behavior, see John Green, "Where Does Your Mind Reside?," CrashCourse, YouTube video, August 1, 2016, https://www.youtube.com/watch?v=3SJROTXnmus; John Green, "Artificial Intelligence & Personhood," CrashCourse, YouTube video, August 8, 2016, https://www.youtube.com/watch?v=39EdqUbj92U; Marcus Du Sautoy, "The Chinese Room Experiment: The Hunt for AI," BBC Studios, YouTube video, September 17, 2015, https://www.youtube.com/watch?v=D0MD4sRHj1M; Robert Kirk, "Zombies," in *Stanford Encyclopedia of Philosophy*, ed. Edward N. Zalta (Spring 2019), https://plato.stanford.edu/entries/zombies; David Cole, "The Chinese Room Argument," in *Stanford Encyclopedia of Philosophy*, ed. Edward N. Zalta (Spring 2019), https://plato.stanford.edu/entries/chinese-room.

12. For Chalmers's more detailed views on the hard and easy problems of consciousness, see David Chalmers, "Hard Problem of Consciousness," Serious Science, YouTube video, July 5, 2016, https://www.youtube.com/watch?v=C5DfnIjZPGw; David Chalmers, "The Meta-Problem of Consciousness," Talks at Google, YouTube video, April 2, 2019, https://www.youtube.com/watch?v=OsYUWtLQBS0; David Chalmers, "Facing Up to the Problem of Consciousness," *Journal of Consciousness Studies* 2, no. 3 (1995): 200–219, http://consc.net/papers/facing.html.

13. For Chalmers's in-depth views on zombies, panprotopsychism, and philosophical zombies, see David J. Chalmers, "Panpsychism and Panprotopsychism," in *Panpsychism*, ed. Godehard Bruntrup and Ludwig Jaskolla (New York: Oxford University Press, 2016), http://consc.net/papers/panpsychism.pdf; David Chalmers, "Panpsychism and Explaining Consciousness," Oppositum, TED Talk, YouTube video, January 18, 2016, https://www.youtube.com/watch?v=SiYfN7-gaLk; David Chalmers, "How Does Panpsychism Fit in Between Dualism and Materialism?," Loyola Productions Munich, YouTube video, November 8, 2011, https://www.youtube.com/watch?v=OSmfhc_8gew.

14. Nobel Prize–winning physicist Sir Roger Penrose has developed, along with physician Stuart Hameroff, a provocative theory called orchestrated objective reduction (Orch OR), which attempts to explain consciousness as arising from quantum processes within molecules inside neurons called microtubules. Yet this theory has so far failed to gain much acceptance within the scientific community. Physicists such as Max Tegmark have observed that quantum effects would likely decohere too quickly in the brain to influence macroscale structures and control behavior. While Orch OR is a fascinating possibility, I have not seen any compelling evidence that quantum physics is needed to explain functional consciousness in the brain. See Steve Paulson, "Roger Penrose on Why Consciousness Does Not Compute," *Nautilus*, April 27, 2017, https://nautil.us/roger-penrose-on-why-consciousness-does-not-compute-236591; Max Tegmark, "The Importance of Quantum Decoherence in Brain Processes," *Physical Review E: Statistical Physics, Plasmas, Fluids, and Related Interdisciplinary Topics* 61 (May 2000): 4194–206, https://arxiv.org/pdf/quant-ph/9907009.pdf; Stuart Hameroff, "How Quantum Brain Biology Can Rescue Conscious Free Will," *Frontiers in Integrative Neuroscience* 6, article 93 (October 12, 2012), https://doi.org/10.3389/fnint.2012.00093.

15. For a deeper dive on free will and competing schools of thought about it in philosophy and politics, see John Green, "Determinism vs. Free Will," Crash Course, YouTube video, August 15, 2016, https://www.youtube.com/watch?v=vCGtkDzELAI; M. S., "Free Will and Politics," *Economist*, January 12, 2012, https://www.economist.com/democracy-in-america/2012/01/12/free-will-and-politics; Timothy O'Connor and Christopher Franklin, "Free Will," in *Stanford Encyclopedia of Philosophy*, ed. Edward N. Zalta (Summer 2019), https://plato.stanford.edu/entries/freewill; Randolph Clarke and Justin Capes, "Incompatibilist (Nondeterministic) Theories of Free Will," in *Stanford Encyclopedia of Philosophy*, ed. Edward N. Zalta (Spring 2017), https://plato.stanford.edu/entries/incompatibilism-theories; Michael McKenna and Justin D. Coates, "Compatibilism," in *Stanford Encyclopedia of*

Philosophy, ed. Edward N. Zalta (Winter 2018), https://plato.stanford.edu/entries/compatibilism.

16. For a range of interesting nontechnical views on free will and predetermination, see Shaun Nichols, "Free Will Versus the Programmed Brain," *Scientific American*, August 19, 2008, https://www.scientificamerican.com/article/free-will-vs-programmed-brain; Bill Nye, "Hey Bill Nye, Do Humans Have Free Will?," *Big Think*, YouTube video, January 19, 2016, https://www.youtube.com/watch?v=ITdMa2bCaVc; Michio Kaku, "Why Physics Ends the Free Will Debate," *Big Think*, YouTube video, May 20, 2011, https://www.youtube.com/watch?v=Jint5kjoy6I; Stephen Cave, "There's No Such Thing as Free Will," *Atlantic*, June 2016, https://www.theatlantic.com/magazine/archive/2016/06/theres-no-such-thing-as-free-will/480750.

17. Simon Blackburn, *Think: A Compelling Introduction to Philosophy* (Oxford, UK: Oxford University Press, 1999), 85, https://books.google.co.uk/books?id=yEEITQSyxAMC.

18. For deeper explanations and demonstrations of cellular automata, see Daniel Shiffman, "Cellular Automata," chap. 7 in *The Nature of Code* (Magic Book Project, 2012), https://natureofcode.com/book/chapter-7-cellular-automata; Devin Acker, "Elementary Cellular Automaton," Github.io, accessed March 10, 2023, http://devinacker.github.io/celldemo; Francesco Berto and Jacopo Tagliabue, "Cellular Automata," in *Stanford Encyclopedia of Philosophy*, ed. Edward N. Zalta (Fall 2017), https://plato.stanford.edu/archives/fall2017/entries/cellular-automata.

19. "John Conway's Game of Life," Bitstorm.org, accessed March 10, 2023, https://bitstorm.org/gameoflife; "Life in Life," Phillip Bradbury, YouTube video, May 13, 2012, https://www.youtube.com/watch?v=xP5-iIeKXE8; Amanda Ghassaei, "OTCA Metapixel—Conway's Game of Life," Trybotics, accessed March 10, 2023, https://trybotics.com/project/OTCA-Metapixel-Conways-Game-of-Life-98534.

20. Stephen Wolfram, *A New Kind of Science* (Champaign, IL: Wolfram Media, 2002), 23–41, 58–70.

21. Wolfram, *New Kind of Science*, 56, wolframscience.com/nks.

22. Wolfram, *New Kind of Science*, 56, wolframscience.com/nks.

23. Wolfram, *New Kind of Science*, 23–27, 31, wolframscience.com/nks.

24. Wolfram, *New Kind of Science*, 23–27, 31, wolframscience.com/nks.

25. For Wolfram's foundational paper on cellular automata and complexity, see Stephen Wolfram, "Cellular Automata as Models of Complexity," *Nature* 311, no. 5985 (October 1984), https://www.stephenwolfram.com/publications/academic/cellular-automata-models-complexity.pdf.

26. For a quick explainer on emergence, see Emily Driscoll and Lottie Kingslake, "What Is Emergence?," *Quanta Magazine*, YouTube video, December 20, 2018, https://www.youtube.com/watch?v=TlysTnxF_6c.

27. Wolfram, *New Kind of Science*, 31.

28. To explore more information about the Wolfram Physics Project, including visualizations and interactive tools, visit www.wolframphysics.org. For a highly detailed technical explainer, see Stephen Wolfram, "A Class of Models with the Potential to Represent Fundamental Physics," *Complex Systems* 29, no. 2 (April 2020): 107–536, https://doi.org/10.25088/ComplexSystems.29.2.107.

29. Wolfram's form of foresight-proof determinism is also compatible with the apparent randomness of quantum events—they would be following deterministic rules, but not in a way that could reveal any usable information about their future behavior. It's worth noting that while the properties of quantum physics can be harnessed to generate numbers that are truly random as far as anyone within the universe is concerned, these don't have emergent properties like class 4 automata do. In other words, if you know you're going to get a random, unweighted distribution of 1s and 0s, nothing more complex will emerge out of that. There might be long strings like 11111111, for example, but each new digit is unrelated to the last. Thus, you can immediately say that the chances of the billionth digit being a 1 are 50 percent, without needing to know anything about the ones that came before. This is consistent with the ordinary scientific paradigm, where laws themselves fully explain the

behavior of the phenomena they describe. See Xiongfeng Ma et al., "Quantum Random Number Generation," *npj Quantum Information* 2, article 16021 (June 28, 2016), https://www.nature.com/articles/npjqi201621.

30. For more explanation of the Blue Brain Project, which discovered the eleven-dimensional brain structures and is working to map and digitally reconstruct animal (and eventually human) brains, see "Blue Brain Team Discovers a Multi-Dimensional Universe in Brain Networks," *Frontiers Science News*, June 12, 2017, https://blog.frontiersin.org/2017/06/12/blue-brain-team-discovers-a-multi-dimensional-universe-in-brain-networks; Michael W. Reimann et al., "Cliques of Neurons Bound into Cavities Provide a Missing Link Between Structure and Function," *Frontiers in Computational Neuroscience* 11, no. 48 (June 12, 2017), https://doi.org/10.3389/fncom.2017.00048.

31. See John Green, "Compatibilism," CrashCourse, YouTube video, August 22, 2016, https://www.youtube.com/watch?v=KETTtiprINU; McKenna and Coates, "Compatibilism."

32. This is a form of compatibilism, the philosophical school of thought that sees free will as meaningfully possible even in a universe that obeys some kind of deterministic rules. See Green, "Compatibilism"; McKenna and Coates, "Compatibilism," *Stanford Encyclopedia*.

33. For more on the relationship between the hemispheres of our brains, as revealed by the past several decades of neuroscience, see Ned Herrmann, "Is It True That Creativity Resides in the Right Hemisphere of the Brain?," *Scientific American*, January 26, 1998, https://www.scientificamerican.com/article/is-it-true-that-creativit; Dina A. Lienhard, "Roger Sperry's Split Brain Experiments (1959–1968)," *Embryo Project Encyclopedia*, December 27, 2017, https://embryo.asu.edu/pages/roger-sperrys-split-brain-experiments-1959-1968; David Wolman, "The Split Brain: A Tale of Two Halves," *Nature* 483, no. 7389 (March 14, 2012): 260–63, https://www.nature.com/news/the-split-brain-a-tale-of-two-halves-1.10213.

34. Stella de Bode and Susan Curtiss, "Language After Hemispherectomy," *Brain and Cognition* 43, nos. 1–3 (June–August 2000): 135–205.

35. Dana Boatman et al., "Language Recovery After Left Hemispherectomy in Children with Late-Onset Seizures," *Annals of Neurology* 46, no. 4 (April 1999): 579–86, https://www.academia.edu/21485724/Language_recovery_after_left_hemispherectomy_in_children_with_late-onset_seizures.

36. Jing Zhou et al., "Axon Position Within the Corpus Callosum Determines Contralateral Cortical Projection," *Proceedings of the National Academy of Sciences* 110, no. 29 (July 16, 2013): E2714–E2723, https://doi.org/10.1073/pnas.1310233110.

37. Benedict Carey, "Decoding the Brain's Cacophony," *New York Times*, October 31, 2011, https://www.nytimes.com/2011/11/01/science/telling-the-story-of-the-brains-cacophony-of-competing-voices.html; Michael Gazzaniga, "The Split Brain in Man," *Scientific American* 217, no. 2 (August 1967): 24–29, https://doi.org/10.1038%2Fscientificamerican0867-24; Alan Alda and David Huntley, "Pieces of Mind," *Scientific American Frontiers* (PBS, 1997), ctshad, YouTube video, https://www.youtube.com/watch?v=lfGwsAdS9Dc.

38. Michael S. Gazzaniga, "Principles of Human Brain Organization Derived from Split-Brain Studies," *Neuron* 14, no. 2 (February 1995): 217–28, https://doi.org/10.1016/0896-6273(95)90280-5; Roger W. Sperry, "Consciousness, Personal Identity, and the Divided Brain," in *The Dual Brain: Hemispheric Specialization in Humans*, ed. D. F. Benson and Eran Zaidel (New York: Guilford Publications, 1985), 11–25; William Hirstein, "Self-Deception and Confabulation," *Philosophy of Science* 67, no. 3 (September 2000): S418–S429, www.jstor.org/stable/188684.

39. Gazzaniga, "Principles of Human Brain Organization"; Sperry, "Consciousness, Personal Identity, and the Divided Brain"; Hirstein, "Self-Deception and Confabulation."

40. Richard Apps et al., "Cerebellar Modules and Their Role as Operational Cerebellar Processing Units," *Cerebellum* 17, no. 5 (June 6, 2018): 654–682, https://doi.org/10.1007/s12311-018-0952-3; Jan Voogd, "What We Do Not Know About Cerebellar Systems Neuroscience," *Frontiers in Systems Neuroscience* 8, no. 227 (December 18, 2014), https://doi.org/10.3389/fnsys.2014.00227.

41. For more on the society-of-mind theory and how it relates to current neuroscience, see Marvin Minsky, "The Society of Mind Theory Developed from Teaching," Web of Stories—Life Stories of Remarkable People, YouTube video, October 5, 2016, https://www .youtube.com/watch?v=HU2SZEW4EWg; Marvin Minsky, "Biological Plausibility of the Society of Mind Theory," Web of Stories—Life Stories of Remarkable People, YouTube video, October 31, 2016, https://www.youtube.com/watch?v=e02WbBd0F70; Michael N. Shadlen and Adina L. Roskies, "The Neurobiology of Decision-Making and Responsibility: Reconciling Mechanism and Mindedness," *Frontiers in Neuroscience* 6, no. 56 (April 23, 2012), https://doi.org/10.3389/fnins.2012.00056; Johannes Friedrich and Máté Lengyel, "Goal-Directed Decision Making with Spiking Neurons," *Journal of Neuroscience* 36, no. 5 (February 3, 2016): 1529–46, https://doi.org/10.1523/JNEUROSCI.2854-15.2016; Marvin Minsky, *The Society of Mind* (New York: Simon & Schuster, 1986).

42. Sarvi Sharifi et al., "Neuroimaging Essentials in Essential Tremor: A Systematic Review," *NeuroImage Clinical* 5 (May 2014): 217–31, https://doi.org/10.1016/j.nicl.2014.05.003; Rick C. Helmich, David E. Vaillancourt, and David J. Brooks, "The Future of Brain Imaging in Parkinson's Disease," *Journal of Parkinson's Disease* 8, no. s1 (2018): S47–S51, https://doi .org/10.3233/JPD-181482.

43. For more on the philosophy behind identity and change, see Andre Gallois, "Identity Over Time," in *Stanford Encyclopedia of Philosophy*, ed. Edward N. Zalta (Winter 2016), https:// plato.stanford.edu/entries/identity-time.

44. Duncan Graham-Rowe, "World's First Brain Prosthesis Revealed," *New Scientist*, March 2003, https://www.newscientist.com/article/dn3488.

45. Robert E. Hampson et al., "Developing a Hippocampal Neural Prosthetic to Facilitate Human Memory Encoding and Recall," *Journal of Neural Engineering* 15, no. 3 (March 28, 2018), https://doi.org/10.1088/1741-2552/aaaed7.

46. For one less technical and one more technical explanation of mitochondrial turnover, see Jon Lieff, "Dynamic Relationship of Mitochondria and Neurons," jonlieffmd.com, February 2, 2014, https://jonlieffmd.com/tag/dynamic-of-fission-and-fusion; Thomas Misgeld and Thomas L. Schwarz, "Mitostasis in Neurons: Maintaining Mitochondria in an Extended Cellular Architecture," *Neuron* 96, no. 3 (November 1, 2017): 651–66, https://www.ncbi.nlm .nih.gov/pmc/articles/PMC5687842.

47. Samuel F. Bakhoum and Duane A. Compton, "Kinetochores and Disease: Keeping Microtubule Dynamics in Check!," *Current Opinion in Cell Biology* 24, no. 1 (February 2012): 64–70, https://www.ncbi.nlm.nih.gov/pmc/articles/PMC3294090/#R13; Vincent Meininger and Stephane Binet, "Characteristics of Microtubules at the Different Stages of Neuronal Differentiation and Maturation," *International Review of Cytology* 114 (1989): 21–79, https:// doi.org/10.1016/S0074-7696(08)60858-X.

48. Laurie D. Cohen et al., "Metabolic Turnover of Synaptic Proteins: Kinetics, Interdependencies and Implications for Synaptic Maintenance," *PLoS One* 8, no. 5: e63191 (May 2, 2013), https://doi.org/10.1371/journal.pone.0063191.

49. K. H. Huh and R. J. Wenthold, "Turnover Analysis of Glutamate Receptors Identifies a Rapidly Degraded Pool of the N-methyl-D-aspartate Receptor Subunit, NR1, in Cultured Cerebellar Granule Cells," *Journal of Biological Chemistry* 274, no. 1 (January 1, 1999): 151–57, https://www.ncbi.nlm.nih.gov/pubmed/9867823.

50. Erin N. Star, David J. Kwiatkowski, and Venkatesh N. Murthy, "Rapid Turnover of Actin in Dendritic Spines and its Regulation by Activity," *Nature Neuroscience* 5, no. 3 (March 2002): 239–46, https://www.ncbi.nlm.nih.gov/pubmed/11850630.

51. "Female Reproductive System," Cleveland Clinic, accessed March 10, 2023, https:// my.clevelandclinic.org/health/articles/9118-female-reproductive-system; Robert D. Martin, "The Macho Sperm Myth," *Aeon*, August 23, 2018, https://aeon.co/essays/the-idea -that-sperm-race-to-the-egg-is-just-another-macho-myth.

52. Timothy G. Jenkins et al., "Sperm Epigenetics and Aging," *Translational Andrology and Urology* 7, suppl. 3 (July 2018): S328–S335, https://doi.org/10.21037/tau.2018.06.10; Ida Donkin and Romain Barrès, "Sperm Epigenetics and the Influence of Environmental

Factors," *Molecular Metabolism* 14 (August 2018): 1–11, https://doi.org/10.1016/j.molmet.2018.02.006.

53. Holly C. Betts et al., "Integrated Genomic and Fossil Evidence Illuminates Life's Early Evolution and Eukaryote Origin," *Nature Ecology & Evolution* 2 (August 20, 2018): 1556–62, https://doi.org/10.1038/s41559-018-0644-x; Elizabeth Pennisi, "Life May Have Originated on Earth 4 Billion Years Ago, Study of Controversial Fossils Suggests," *Science*, December 18, 2017, https://www.sciencemag.org/news/2017/12/life-may-have-originated-earth-4-billion-years-ago-study-controversial-fossils-suggests; Michael Marshall, "Timeline: The Evolution of Life," *New Scientist*, July 14, 2009, https://institutions.newscientist.com/article/dn17453-timeline-the-evolution-of-life.

54. Without taking into account the odds of our parents meeting and conceiving us, the odds of any two particular reproductive cells meeting if they were going to have a child are about 1 in 2 quintillion. The same is true of both of them, and your grandparents, and all the way back. For a given ancestor even several centuries ago, you probably didn't get any of their genes yourself, but if any link in the genetic chain had been broken, your intermediate ancestors may not have been born, or might have died early of heritable diseases, or gone on to themselves marry different people. Every branch of your family tree rests on this knife-edge of contingency.

So (very roughly speaking) the genetic and epigenetic odds of all human ancestry leading up to you is 1 in 2,000,000,000,000,000,000 raised to the power of the number of ancestors you have—but not just "unique ancestors." Your 100th great-grandparents probably lived sometime between Buddha and Jesus. Just multiplying out the size of your family tree shows that you had about 1.3 nonillion 100th great-grandparents, when the human population was fewer than 400 million. The answer to this mystery is inbreeding. As you traced your family tree back, you'd find that your ancestors were all distant cousins of each other, so that a single person may be your ancestor in many different ways. For a normal count of ancestors, we would focus on these unique individuals. But for estimating the unlikeliness of our own ancestry, we're interested in the relationships, so we should count the same person twice if they're both your eighth and ninth great-grandfather, because it was still up to chance that that person played both roles.

Since there have been roughly 10,000 generations of Homo sapiens, your precise identity relies on some $2^{2^{10,000}}$) ancestry events—about $4 \times 10^{3,010}$. And so, 2 quintillion raised to this number as an exponent gives us the denominator of the overall odds, which is about $10^{10^{3,011}}$. This number is vastly larger than even a googolplex, and has more zeros than there are atoms in the known universe. And that is only for humanity. Oxford biologist Richard Dawkins estimates that it's 300 million generations back to the protostomes, some of our first sexually reproducing ancestors. There is no rigorous data going all the way to the dawn of life, nearly four billion years ago, but the most likely estimates are somewhere on the order of one trillion generations. Let it suffice to say that the odds ratio of this entire genealogy becomes so stupendously large that we can't really grasp it intellectually. We're all standing on the summit of a mountain of happenstance and good fortune that reaches so far down in space and time that we cannot see its base.

For more on human evolution and the evidence underlying these estimates, see Max Ingman et al., "Mitochondrial Genome Variation and the Origin of Modern Humans," *Nature* 408 (December 7, 2000): 713, https://www.eva.mpg.de/fileadmin/content_files/staff/paabo/pdf/Ingman_MitNat_2000.pdf; Donn Devine, "How Long Is a Generation?," Ancestry.ca, March 10, 2023, https://web.archive.org/web/20200111102741/https://www.ancestry.ca/learn/learningcenters/default.aspx?section=lib_Generation; Adam Rutherford, "Ant and Dec's DNA Test Merely Tells Us That We're All Inbred," *Guardian*, November 12, 2019, https://www.theguardian.com/commentisfree/2019/nov/12/ant-and-dec-dna-test-all-inbred-historical-connections; Alva Noë, "DNA, Genealogy and the Search for Who We Are," NPR, January 29, 2016, https://www.npr.org/sections/13.7/2016/01/29/464805509/dna-genealogy-and-the-search-for-who-we-are; Alison Jolly, *Lucy's Legacy: Sex and Intelligence in Human Evolution* (Cambridge, MA: Harvard University Press, 1999), 55, https://books.google.co.uk/books?id=7mSMa2Zl_YkC; Simon Conway Morris, "The Fossils of the Burgess Shale and the Cambrian 'Explosion': Their Implications for Evolution," in *Killers in the Brain: Essays in Science and Technology from the Royal Institution*, ed. Peter Day (Oxford, UK: Oxford University Press, 1999), 22, https://books.google.co.uk/books?id=v3Eo4UqbbYsC; Richard Dawkins and Yan Wong, *The Ancestor's Tale: A Pilgrimage to the Dawn of Evolution* (New York: Houghton Mifflin Harcourt, 2004), 379; John Carl Villanueva, "How Many Atoms Are in the Universe?," *Universe Today*, July 30, 2009, https://www.universetoday.com/36302/atoms-in-the-universe; Tim Urban, "Meet Your Ancestors (All of Them)," *Wait But Why*, December 18, 2013, https://waitbutwhy.com/2013/12/your-ancestor-is-jellyfish.html.

55. For some scientific perspectives on fine-tuning, see Leonard Susskind, "Is the Universe Fine-Tuned for Life and Mind?," Closer to Truth, YouTube video, January 8, 2013, https://www.youtube.com/watch?v=2cT4zZIHR3s; Martin J. Rees, "Why Cosmic Fine-Tuning Demands Explanation," Closer to Truth, YouTube video, January 23, 2017, https://www.youtube.com/watch?v=E0zdXj6fSGY; Simon Friedrich, "Fine-Tuning," in *Stanford Encyclopedia of Philosophy*, ed. Edward N. Zalta (Winter 2018), https://plato.stanford.edu/entries/fine-tuning/#FineTuneCondEarlUniv.

56. According to the Standard Model, there are six kinds of quarks, six antiquarks, six leptons, six antileptons, eight gluons, photons, W+ bosons, W- bosons, Z bosons, and the Higgs boson. For more on the Standard Model's so-called particle zoo, see Julian Huguet, "Subatomic Particles Explained in Under 4 Minutes," *Seeker*, YouTube video, December 18, 2014, https://www.youtube.com/watch?v=eD7hXLRqWWM; Peter Kalmus, "The Physics of Elementary Particles: Part I," *Plus Magazine*, April 21, 2015, https://plus.maths.org/content/physics-elementary-particles; Guido Altarelli and James Wells, "Gauge Theories and the Standard Model," in *Collider Physics Within the Standard Model*, vol. 937 of *Lecture Notes in Physics*, ed. James Wells (Cham, Switzerland: Springer Open, 2017), https://doi.org/10.1007/978-3-319-51920-3_1.

57. For helpful explainers on abiogenesis and the 1952 Miller-Urey experiment that demonstrated its viability, see Paul Anderson, "Abiogenesis," Bozeman Science, YouTube video, June 25, 2011, https://www.youtube.com/watch?v=W3ceg—uQKM; "What Was the Miller-Urey Experiment," Stated Clearly, YouTube video, October 27, 2015, https://www.youtube.com/watch?v=NNijmxsKGbc.

58. Friedrich, "Fine-Tuning," *Stanford Encyclopedia*; Luke A. Barnes, "The Fine-Tuning of the Universe for Intelligent Life," *Publications of the Astronomical Society of Australia* 29, no. 4 (2012): 529–64, doi:10.1071/AS12015.

59. Friedrich, "Fine-Tuning," *Stanford Encyclopedia*; Lawrence J. Hall et al., "The Weak Scale from BBN," *Journal of High Energy Physics* 2014, no. 12, article 134 (2014), doi.org/10.1007/JHEP12(2014)134; Bernard J. Carr and Martin J. Rees, "The Anthropic Principle and the Structure of the Physical World," *Nature* 278 (April 12, 1979): 605–12, https://www.nature.com/articles/278605a0.

60. Craig J. Hogan, "Why the Universe Is Just So," *Reviews of Modern Physics* 72, no. 4 (October 1, 2000): 1149–61, https://journals.aps.org/rmp/abstract/10.1103/RevModPhys.72.1149; Craig J. Hogan, "Quarks, Electrons, and Atoms in Closely Related Universes," in *Universe or Multiverse*, ed. Bernard Carr (Cambridge, UK: Cambridge University Press, 2007), 221–30.

61. Hogan, "Why the Universe Is Just So"; Hogan, "Quarks, Electrons, and Atoms," 221–30.

62. Hogan, "Quarks, Electrons, and Atoms," 224.

63. Hogan, "Quarks, Electrons, and Atoms," 224–25.

64. Hogan, "Quarks, Electrons, and Atoms," 224–25.

65. For more in-depth exploration of evidence on fine-tuning of physical laws, including the fine-tunedness of gravity as necessary for the formation of heavy elements, see Friedrich, "Fine-Tuning," *Stanford Encyclopedia*; Carr and Rees, "Anthropic Principle and the Structure of the Physical World," 605–12.

66. Michael Brooks, "Gravity Mysteries: Why Is Gravity Fine-Tuned?," *New Scientist*, June 10, 2009, https://www.newscientist.com/article/mg20227123-000.

67. Tim Radford, "'Just Six Numbers: The Deep Forces That Shape the Universe by Martin Rees—Review," *Guardian*, June 8, 2012, https://www.theguardian.com/science/2012/jun

/08/just-six-numbers-martin-rees-review; Brooks, "Gravity Mysteries: Why Is Gravity Fine-Tuned?"

68. Friedrich, "Fine-Tuning," *Stanford Encyclopedia*; Max Tegmark and Martin J. Rees, "Why Is the Cosmic Microwave Background Fluctuation Level 10^{-5}?," *Astronomical Journal* 499, no. 2 (June 1, 1998): 526–32, https://iopscience.iop.org/article/10.1086/305673/pdf.

69. Friedrich, "Fine-Tuning," *Stanford Encyclopedia*; Tegmark and Rees, "Why Is the Cosmic Microwave Background Fluctuation Level 10^{-5}?"

70. Martin J. Rees, *Just Six Numbers: The Deep Forces That Shape the Universe* (New York: Weidenfeld & Nicholson, 1999), 104.

71. Rees, *Just Six Numbers*, 104.

72. Friedrich, "Fine-Tuning," *Stanford Encyclopedia*; Roger Penrose, *The Road to Reality: A Complete Guide to the Laws of the Universe* (New York: Vintage, 2004), 729–30.

73. Tony Padilla, "How Many Particles in the Universe?," Numberphile, YouTube video, July 10, 2017, https://www.youtube.com/watch?v=lpj0E0a0mlU; Villanueva, "How Many Atoms Are in the Universe?"; Jacob Aron, "Number of Ways to Arrange 128 Balls Exceeds Atoms in Universe," *New Scientist*, January 28, 2016, https://www.newscientist.com/article/2075593.

74. Luke Barnes (L.A. Barnes), "The Fine-Tuning of the Universe for Intelligent Life," *Publications of the Astronomical Society of Australia* 29, no. 4 (June 7, 2012): 531, https://www.publish.csiro.au/as/pdf/AS12015.

75. Brad Lemley, "Why Is There Life?," *Discover*, November 2000, http://discovermagazine.com/2000/nov/cover.

76. For more on the anthropic principle, including a fascinating conversation between atheist biologist Richard Dawkins and Catholic astronomer Fr. George Coyne, see Roberto Trotta, "What Is the Anthropic Principle?," *Physics World*, YouTube video, January 17, 2014, https://www.youtube.com/watch?v=dWkJ8Pl-8l8; Richard Dawkins and George Coyne, "The Anthropic Principle," jlcamelo, July 9, 2010, https://youtu.be/lm9ZtYkd kEQ?t=102; Joe Rogan and Nick Bostrom, "Joe Rogan Experience #1350—Nick Bostrom," PowerfulJRE, September 11, 2019, https://web.archive.org/web/20190918171740/https://www.youtube.com/watch?v=5c4cv7rVlE8; Christopher Smeenk and George Ellis, "Philosophy of Cosmology," in *Stanford Encyclopedia of Philosophy*, ed. Edward N. Zalta (Winter 2017), https://plato.stanford.edu/entries/cosmology.

77. Lemley, "Why Is There Life?"

78. For an accessible explainer on GANs and some fascinating pieces on emerging technologies that will enable AI re-creations of the dead, see Robert Miles, "Generative Adversarial Networks (GANs)," Computerphile, YouTube video, October 25, 2017, https://www.youtube.com/watch?v=Sw9r8CL98N0; Michael Kammerer, "I Trained an AI to Imitate My Own Art Style. This Is What Happened," *Towards Data Science*, March 28, 2019, https://towardsdatascience.com/i-trained-an-ai-to-imitate-my-own-art-style-this-is-what-happened-461785b9a15b; Ana Santos Rutschman, "Artificial Intelligence Can Now Emulate Human Behaviors—Soon It Will Be Dangerously Good," *The Conversation*, April 5, 2019, https://theconversation.com/artificial-intelligence-can-now-emulate-human-behaviors-soon-it-will-be-dangerously-good-114136; Catherine Stupp, "Fraudsters Used AI to Mimic CEO's Voice in Unusual Cybercrime Case," *Wall Street Journal*, August 30, 2019, https://www.wsj.com/articles/fraudsters-use-ai-to-mimic-ceos-voice-in-unusual-cyber crime-case-11567157402; David Nield, "New AI Generates Freakishly Realistic People Who Don't Actually Exist," *Science Alert*, February 19, 2019, https://www.sciencealert.com/ai-is-getting-creepily-good-at-generating-faces-for-people-who-don-t-actually-exist; Alec Radford et al., "Better Language Models and Their Implications," OpenAI, February 14, 2019, https://openai.com/blog/better-language-models; Richard Socher, "Introducing a Conditional Transformer Language Model for Controllable Generation," Salesforce Einstein, accessed March 10, 2023, https://blog.einstein.ai/introducing-a-cond itional-transformer-language-model-for-controllable-generation.

79. Jeffrey Grubb, "Google Duplex; A.I. Assistant Calls Local Businesses to Make Appointments," Jeff Grubb's Game Mess, YouTube video, May 8, 2018, https://www.youtube.com /watch?v=D5VN56jQMWM.

80. Monkeypaw Productions and BuzzFeed, "You Won't Believe What Obama Says in This Video," BuzzFeedVideo, YouTube video, April 17, 2018, https://www.youtube.com/watch?v=cQ54GDm1eL0; "Could Deepfakes Weaken Democracy?," *Economist*, YouTube video, October 22, 2019, https://www.youtube.com/watch?v=_m2dRDQEC1A; Kristin Houser, "This 'RoboTrump' AI Mimics the President's Writing Style," *Futurism*, October 23, 2019, https://futurism.com/robotrump-ai-text-generator-trump.

81. "No Country for Old Actors," Ctrl Shift Face, YouTube video, November 13, 2019, https://www.youtube.com/watch?v=Ow_uufCxm1A.

82. Casey Newton, "Speak, Memory," *The Verge*, October 6, 2016, https://www.theverge.com/a/luka-artificial-intelligence-memorial-roman-mazurenko-bot.

83. For a clear and engaging look at the science behind the uncanny valley, see Michael Stevens, "Why Are Things Creepy?," Vsauce, YouTube video, July 2, 2013, https://www.youtube.com/watch?v=PEikGKDVsCc.

84. For more on realistic android development and Moravec's paradox, see Tim Hornyak, "Insanely Humanlike Androids Have Entered the Workplace and Soon May Take Your Job," CNBC, October 31, 2019, https://www.cnbc.com/2019/10/31/human-like-androids-have-entered-the-workplace-and-may-take-your-job.html; Jade Tan-Holmes, "Moravec's Paradox—Why Are Machines So Smart, Yet So Dumb?," Up and Atom, YouTube video, July 8, 2019, https://www.youtube.com/watch?v=hcfVRkC3Dp0; J. C. Eccles, "Evolution of Consciousness," *Proceedings of the National Academy of Sciences of the United States of America* 89, no. 16 (August 15, 1992): 7320–24, https://www.ncbi.nlm.nih.gov/pmc/articles/PMC49701; Hans Moravec, *Mind Children* (Cambridge, MA: Harvard University Press, 1988), 1–50.

85. John Hawks, "How Has the Human Brain Evolved?," *Scientific American Mind* 24, no. 3 (July 1, 2013): 76, https://doi.org/10.1038/scientificamericanmind0713-76b.

86. "ASIMO World2001-2/3," jnomw, YouTube video, October 28, 2006, https://www.youtube.com/watch?v=Ph_B_5hKRIE.

87. "More Parkour Atlas," BostonDynamics, YouTube video, September 24, 2019, https://www.youtube.com/watch?v=_sBBaNYex3E.

88. "Sophia the Robot by Hanson Robotics," Hanson Robotics Limited, YouTube video, September 5, 2018, https://www.youtube.com/watch?v=BhU9hOo5Cuc; "Meet Little Sophia, Hanson Robotics' Newest Robot," Hanson Robotics Limited, YouTube video, February 11, 2019, https://www.youtube.com/watch?v=7cGRPvN5430; "Ameca Expressions with GPT3 / 4," Engineered Arts, YouTube video, March 31, 2023, https://www.youtube.com/watch?v=yUszJyS3d7A.

89. Anders Sandberg and Nick Bostrom, *Whole Brain Emulation: A Roadmap*, technical report 2008-3, Future of Humanity Institute, Oxford University (2008), 13, https://www.fhi.ox.ac.uk/brain-emulation-roadmap-report.pdf.

90. Sandberg and Bostrom, *Whole Brain Emulation*, 80–81.

91. For further resources related to mind uploading and brain emulation, see S. A. Graziano, "How Close Are We to Uploading Our Minds?," TED-Ed, YouTube video, October 29, 2019, https://www.youtube.com/watch?v=2DWnvx1NYUA; Trace Dominguez, "How Close Are We to Downloading the Human Brain?," *Seeker*, YouTube video, September 13, 2018, https://www.youtube.com/watch?v=DE5e5zF6a-8; Michio Kaku, "Could We Transport Our Consciousness Into Robots?," *Big Think*, YouTube video, May 31, 2011, https://www.youtube.com/watch?v=tT1vxEpE1aI; Matt O'Dowd, "Computing a Universe Simulation," PBS Space Time, YouTube video, October 10, 2018, https://www.youtube.com/watch?v=0GLgZvTCbaA; Riken, "All-Atom Molecular Dynamics Simulation of the Bacterial Cytoplasm," rikenchannel, YouTube video, June 6, 2017, https://www.youtube.com/watch?v=5JcFgj2gHx8; "Scientists Create First Billion-Atom Biomolecular Simulation," Los Alamos National Lab, YouTube video, April 22, 2019, https://www.youtube.com/watch?v=jmeik65RkJw; "Matrioshka Brains," Isaac Arthur, YouTube video, June 23, 2016, https://www.youtube.com/watch?v=Ef-mxjYkllw; Simon Makin, "The Four Biggest Challenges in Brain Simulation," *Nature* 571, S9 (July 25, 2019), https://www.nature.com/articles/d41586-019-02209-z; Egidio D'Angelo et al., "Realistic Modeling of Neurons and

Networks: Towards Brain Simulation," *Functional Neurology* 28, no. 3 (July–September 2013): 153–66, https://www.ncbi.nlm.nih.gov/pmc/articles/PMC3812748; Elon Musk and Neuralink, "An Integrated Brain-Machine Interface Platform with Thousands of Channels," Neuralink working paper, July 17, 2019, https://doi.org/10.1101/703801; Fujitsu, "Supercomputer Used to Simulate 3,000-Atom Nano Device," Phys.org, January 14, 2014, https://phys.org/news/2014-01-supercomputer-simulate-atom-nano-device.html; Anders Sandberg, "Ethics of Brain Emulations," *Journal of Experimental & Theoretical Artificial Intelligence* 26, no. 3 (April 14, 2014): 439–57, https://doi.org/10.1080/0952813X.2014.895113; Ray Kurzweil, *How to Create a Mind: The Secret of Human Thought Revealed* (New York: Viking, 2012).

92. Michael Merzenich, "Growing Evidence of Brain Plasticity," TED video, February 2004, https://www.ted.com/talks/michael_merzenich_on_the_elastic_brain.

CHAPTER 4: LIFE IS GETTING EXPONENTIALLY BETTER

1. World Bank Development Research Group, "Poverty Headcount Ratio at $2.15 a Day (2017 PPP) (% of population)," World Bank, accessed March 25, 2023, https://data.worldbank.org/indicator/SI.POV.DDAY; World Bank, "Population, Total for World (SPPOPTOTLWLD)," retrieved from FRED, Federal Reserve Bank of St. Louis, updated July 4, 2023, https://fred.stlouisfed.org/series/SPPOPTOTLWLD.

2. UNESCO Institute for Statistics, "Literacy Rate, Adult Total (% of People Ages 15 and Above)," retrieved from Worldbank.org, October 24 2022, https://data.worldbank.org/indicator/SE.ADT.LITR.ZS.

3. *Progress on Household Drinking Water, Sanitation and Hygiene 2000–2020: Five Years into the SDGs* (Geneva: World Health Organization and United Nations Children's Fund, 2021), 9, https://apps.who.int/iris/rest/bitstreams/1369501/retrieve.

4. World Bank Development Research Group, "Poverty Headcount Ratio at $2.15 a day"; World Bank, "Population, Total for World (SPPOPTOTLWLD)."

5. UNESCO Institute for Statistics, "Literacy Rate, Adult Total."

6. *Progress on Household Drinking Water, Sanitation and Hygiene 2000–2020*, 9.

7. Ray Kurzweil, *The Age of Spiritual Machines: When Computers Exceed Human Intelligence* (New York: Viking, 1999).

8. Ray Kurzweil, *The Singularity Is Near* (New York: Viking, 2005).

9. Peter H. Diamandis and Steven Kotler, *Abundance: The Future Is Better Than You Think* (New York: Simon & Schuster, 2012).

10. Steven Pinker, *Enlightenment Now: The Case for Reason, Science, Humanism, and Progress* (New York: Penguin, 2018).

11. Andrew Evans, "The First Thanksgiving Travel," *National Geographic*, November 24, 2011, https://www.nationalgeographic.com/travel/digital-nomad/2011/11/24/the-first-thanksgiving-travel.

12. Henry Fairlie, "Henry Fairlie on What Europeans Thought of Our Revolution," *New Republic*, July 3, 2014, https://newrepublic.com/article/118527/american-revolution-what-did-europeans-think; Peter Stanford, "The Street of Ships in New York," *Boating* 21, no. 1 (January 1967): 48, https://books.google.com/books?id=VsMYj3YIKLcC.

13. Vaclav Smil, "Crossing the Atlantic," *IEEE Spectrum*, March 28, 2018, https://spectrum.ieee.org/transportation/marine/crossing-the-atlantic.

14. Frank W. Geels, *Technological Transitions and System Innovations: A Co-evolutionary and Socio-Technical Analysis* (Cambridge, UK: Edward Elgar, 2005), 135, https://books.google.com/books?id=SDfrb7TNX5oC.

15. Steven Ujifusa, *A Man and His Ship: America's Greatest Naval Architect and His Quest to Build the S.S. United States* (New York: Simon & Schuster, 2012), 152, https://books.google.com/books?id=H6KB4q7M938C.

16. John C. Spychalski, "Transportation," in *The Columbia History of the 20th Century*, ed. Richard W. Bulliet (New York: Columbia University Press, 1998), 409, https://books.google.com/books?id=9QsqpvR0nq0C.

17. Jason Paur, "Oct. 4, 1958: 'Comets' Debut Trans-Atlantic Jet Age," *Wired*, October 4, 2010, https://www.wired.com/2010/10/1004first-transatlantic-jet-service-boac.

18. Howard Slutsken, "What It Was Really Like to Fly on Concorde," CNN, March 2, 2019, https://www.cnn.com/travel/article/concorde-flying-what-was-it-like/index.html.

19. "London to New York Flight Duration," Finance.co.uk, accessed April 14, 2023, https://www.finance.co.uk/travel/flight-times-and-durations-calculator/london-to-new-york.

20. For a helpful essay on the role of conflict in storytelling and evidence for negativity bias in the news, see Jerry Flattum, "What Is a Story? Conflict—The Foundation of Storytelling," *Script*, March 18, 2013, https://scriptmag.com/features/conflict-the-foundation-of-story telling; Stuart Soroka, Patrick Fournier, and Lilach Nir, "Cross-National Evidence of a Negativity Bias in Psychophysiological Reactions to News," *Proceedings of the National Academy of Sciences* 116, no. 38 (September 17, 2019): 18888–92, https://doi.org/10.1073/pnas.1908369116.

21. For more on how social media algorithms emphasize conflict and fuel polarization, see "Jonathan Haidt: How Social Media Drives Polarization," Amanpour and Company, You-Tube video, December 4, 2019, https://www.youtube.com/watch?v=G9ofYEfewNE; Eli Pariser, "How News Feed Algorithms Supercharge Confirmation Bias," *Big Think*, You-Tube video, December 18, 2018, https://www.youtube.com/watch?v=prx9bxzns3g; Jeremy B. Merrill and Will Oremus, "Five Points for Anger, One for a 'Like': How Facebook's Formula Fostered Rage and Misinformation," *Washington Post*, October 26, 2021, https://www.washingtonpost.com/technology/2021/10/26/facebook-angry-emoji-algorithm; Damon Centola, "Why Social Media Makes Us More Polarized and How to Fix It," *Scientific American*, October 15, 2020, https://www.scientificamerican.com/article/why-social-media-makes-us-more-polarized-and-how-to-fix-it.

22. Max Roser, "Economic Growth," Our World in Data, accessed October 11, 2021, https://ourworldindata.org/economic-growth; Ryland Thomas and Nicholas Dimsdale, "A Millennium of UK Data," Bank of England OBRA dataset, 2017, https://www.bankofengland.co.uk/-/media/boe/files/statistics/research-datasets/a-millennium-of-macroeconomic-data-for-the-uk.xlsx; "Inflation Calculator," Bank of England, accessed April 14, 2023, https://www.bankofengland.co.uk/monetary-policy/inflation/inflation-calculator; Stephen Broadberry et al., *British Economic Growth, 1270–1870* (Cambridge, UK: Cambridge University Press, 2015).

23. Roser, "Economic Growth"; Thomas and Dimsdale, "A Millennium of UK Data"; "Inflation Calculator," Bank of England; Broadberry et al., *British Economic Growth*.

24. Roser, "Economic Growth"; Thomas and Dimsdale, "A Millennium of UK Data"; "Inflation Calculator," Bank of England; Broadberry et al., *British Economic Growth*.

25. Roser, "Economic Growth"; Thomas and Dimsdale, "A Millennium of UK Data"; "Inflation Calculator," Bank of England; Broadberry et al., *British Economic Growth*.

26. Peter Nowak, "The Rise of Mean World Syndrome in Social Media," *Globe and Mail*, November 6, 2014, https://www.theglobeandmail.com/life/relationships/the-rise-of-mean-world-syndrome-in-social-media/article21481089.

27. For more on this research, see Paula McGrath, "Why Good Memories Are Less Likely to Fade," BBC News, May 4, 2014, https://www.bbc.com/news/health-27193607; Colin Allen, "Past Perfect: Why Bad Memories Fade," *Psychology Today*, June 3, 2003, https://www.psychologytoday.com/us/articles/200306/past-perfect-why-bad-memories-fade; W. Richard Walker, John J. Skowronski, and Charles P. Thompson, "Life Is Pleasant—and Memory Helps to Keep It That Way!," *Review of General Psychology* 7, no. 2 (June 2003): 203–10, https://www.apa.org/pubs/journals/releases/gpr-72203.pdf.

28. Walker, Skowronski, and Thompson, "Life Is Pleasant."

29. Timothy D. Ritchie et al., "A Pancultural Perspective on the Fading Affect Bias in Autobiographical Memory," *Memory* 23, no. 2 (February 14, 2014): 278–90, https://doi.org/10.1080/09658211.2014.884138.

30. John Tierney, "What Is Nostalgia Good For? Quite a Bit, Research Shows," *New York Times*, July 8, 2013, https://www.nytimes.com/2013/07/09/science/what-is-nostalgia-good-for-quite-a-bit-research-shows.html.

31. Mike Mariani, "How Nostalgia Made America Great Again," *Nautilus*, April 20, 2017, https://nautil.us/how-nostalgia-made-america-great-again-236556.

32. Ed O'Brien and Nadav Klein, "The Tipping Point of Perceived Change: Asymmetric Thresholds in Diagnosing Improvement Versus Decline," *Journal of Personality and Social Psychology* 112, no. 2 (February 2017): 161–85, https://doi.org/10.1037/pspa0000070.

33. Art Markman, "How Do You Decide Things Are Getting Worse?," *Psychology Today*, February 7, 2017, https://www.psychologytoday.com/nz/blog/ulterior-motives/201702/how-do-you-decide-things-are-getting-worse.

34. Robert P. Jones et al., *The Divide over America's Future: 1950 or 2050? Findings from the 2016 American Values Survey*, Public Religion Research Institute, October 25, 2016, https://www.prri.org/wp-content/uploads/2016/10/PRRI-2016-American-Values-Survey.pdf; Mariani, "How Nostalgia Made America Great Again."

35. Pete Etchells, "Declinism: Is the World Actually Getting Worse?," *Guardian*, January 16, 2015, https://www.theguardian.com/science/head-quarters/2015/jan/16/declinism-is-the-world-actually-getting-worse.

36. Martijn Lampert, Anne Blanksma Çeta, and Panos Papadongonas, *Increasing Knowledge and Activating Millennials for Making Poverty History*, Glocalities global survey report, July 2018, https://xs.motivaction.nl/fileArchive/?f=116335&o=5880&key=4444.

37. Ipsos, "Perceptions Are Not Reality," Ipsos *Perils of Perception* blog, July 8, 2013, https://www.ipsos.com/ipsos-mori/en-uk/perceptions-are-not-reality.

38. Jamiles Lartey et al., "Ahead of Midterms, Most Americans Say Crime Is Up. What Does the Data Say?," Marshall Project, November 5, 2022, https://www.themarshallproject.org/2022/11/05/ahead-of-midterms-most-americans-say-crime-is-up-what-does-the-data-say; Dara Lind, "The US Is Safer Than Ever—and Americans Don't Have any Idea," *Vox*, April 7, 2016, https://www.vox.com/2015/5/4/8546497/crime-rate-america.

39. Julia Belluz, "You May Think the World Is Falling Apart. Steven Pinker Is Here to Tell You It Isn't," *Vox*, September 10, 2016, https://www.vox.com/2016/8/16/12486586/2016-worst-year-ever-violence-trump-terrorism.

40. For accessible video explainers of the Kahneman-Tversky research, as well as the original published work, see "Thinking, Fast and Slow by Daniel Kahneman: Animated Book Summary," FightMediocrity, YouTube video, June 5, 2015, https://www.youtube.com/watch?v=uqXVAo7dVRU; "Thinking Fast and Slow by Daniel Kahneman #2—Heuristics and Biases: Animated Book Summary," One Percent Better, YouTube video, November 12, 2016, https://www.youtube.com/watch?v=Q_wBt5aSRYY; "Kahneman and Tversky: How Heuristics Impact Our Judgment," Intermittent Diversion, YouTube video, June 7, 2018, https://www.youtube.com/watch?v=3IjIVD-KYF4; Richard H. Thaler et al., "The Effect of Myopia and Loss Aversion on Risk Taking: An Experimental Test," *Quarterly Journal of Economics* 112, no. 2 (May 1997): 647–61, https://www.jstor.org/stable/2951249; Daniel Kahneman and Amos Tversky, "The Psychology of Preferences," *Scientific American* 246, no. 1 (January 1981): 160–73; Daniel Kahneman, Paul Slovic, and Amos Tversky, eds., *Judgment Under Uncertainty: Heuristics and Biases* (Cambridge, UK: Cambridge University Press, 1982); Amos Tversky and Daniel Kahneman, "Judgment Under Uncertainty: Heuristics and Biases," *Science* 185, no. 4157 (September 27, 1974): 1124–31, http://doi.org/10.1126/science.185.4157.1124; Daniel Kahneman and Amos Tversky, "On the Study of Statistical Intuitions," *Cognition* 11, no. 2 (March 1982): 123–41; Daniel Kahneman and Amos Tversky, "Variants of Uncertainty," *Cognition* 11, no. 2 (March 1982): 143–57.

41. Daniel Kahneman, *Thinking, Fast and Slow* (New York: Farrar, Straus and Giroux, 2011), 7; Daniel Kahneman and Amos Tversky, "On the Psychology of Prediction," *Psychological Review* 80, no. 4 (1973): 237–51, https://doi.org/10.1037/h0034747.

42. Tversky and Kahneman, "Judgment Under Uncertainty," 1125–26.

43. Tversky and Kahneman, "Judgment Under Uncertainty," 1127–28.

44. Max Roser and Esteban Ortiz-Ospina, "Literacy," Our World in Data, September 20, 2018, https://ourworldindata.org/literacy; Eltjo Buringh and Jan Luiten van Zanden, "Charting the 'Rise of the West': Manuscripts and Printed Books in Europe, a Long-Term

Perspective from the Sixth Through Eighteenth Centuries," *Journal of Economic History* 69, no. 2 (June 2009): 409–45, https://doi.org/10.1017/S0022050709000837.

45. Franz H. Bäuml, "Varieties and Consequences of Medieval Literacy and Illiteracy," *Speculum* 55, no. 2 (April 1980): 237–65, https://www.jstor.org/stable/2847287; Denise E. Murray, "Changing Technologies, Changing Literacy Communities?," *Language Learning & Technology* 4, no. 2 (September 2000): 39–53, https://scholarspace.manoa.hawaii.edu/bitstream/10125/25099/1/04_02_murray.pdf.

46. Roser and Ortiz-Ospina, "Literacy"; Buringh and van Zanden, "Charting the 'Rise of the West,'" 409–45.

47. Roser and Ortiz-Ospina, "Literacy"; Sevket Pamuk and Jan Luiten van Zanden, "Standards of Living," in *The Cambridge Economic History of Modern Europe: Volume 1: 1700–1870*, ed. Stephen Broadberry and Kevin H. O'Rourke (New York: Cambridge University Press, 2010), 229.

48. Christelle Garrouste, *100 Years of Educational Reforms in Europe: A Contextual Database*, JRC Scientific and Technical Reports, European Union, 2010, https://publications.jrc.ec.europa.eu/repository/bitstream/JRC57357/reqno_jrc57357.pdf.

49. Roser and Ortiz-Ospina, "Literacy"; Jan Luiten van Zanden et al., eds., *How Was Life?: Global Well-being Since 1820* (Paris: OECD Publishing, 2014), https://doi.org/10.1787/9789264214262-en.

50. Van Zanden et al., *How Was Life?*; United Nations Educational, Scientific and Cultural Organization, *Literacy, 1969–1971: Progress Achieved in Literacy Throughout the World* (Paris: UNESCO, 1972), https://unesdoc.unesco.org/in/rest/annotationSVC/DownloadWatermarkedAttachment/attach_import_c0206949-c3f1-4eac-a189-9c5bcfdac220.

51. Roser and Ortiz-Ospina, "Literacy"; Roy Carr-Hill and José Pessoa, *International Literacy Statistics: A Review of Concepts, Methodology and Current Data* (Montreal: UNESCO Institute for Statistics, 2008), http://uis.unesco.org/sites/default/files/documents/international-literacy-statistics-a-review-of-concepts-methodology-and-current-data-en_0.pdf; van Zanden et al., *How Was Life?*; United Nations Educational, Scientific and Cultural Organization, *Literacy, 1969–1971*; Friedrich Huebler and Weixin Lu, *Adult and Youth Literacy: National, Regional and Global Trends, 1985–2015*, UIS Information Paper (Montreal: UNESCO Institute for Statistics, 2013), http://uis.unesco.org/sites/default/files/documents/adult-and-youth-literacy-national-regional-and-global-trends-1985-2015-en_0.pdf.

52. UNESCO Institute for Statistics, "Literacy Rate, Adult Total"; Central Intelligence Agency, "Field Listing—Literacy," CIA World Factbook, accessed October 11, 2021, https://www.cia.gov/the-world-factbook/field/literacy.

53. National Assessment of Adult Literacy, *A First Look at the Literacy of America's Adults in the 21st Century* (Washington, DC: National Center for Education Statistics, 2005): 5, https://nces.ed.gov/NAAL/PDF/2006470.PDF.

54. Madeline Goodman et al., *Literacy, Numeracy, and Problem Solving in Technology-Rich Environments Among U.S. Adults: Results from the Program for the International Assessment of Adult Competencies 2012: First Look* (NCES 2014-008) (Washington, DC: US Department of Education, 2013), 14, https://nces.ed.gov/pubs2014/2014008.pdf.

55. Roser and Ortiz-Ospina, "Literacy"; Bas van Leeuwen and Jieli van Leeuwen-Li, "Education Since 1820," in *How Was Life?: Global Well-Being Since 1820*, ed. Jan Luiten van Zanden et al. (Paris: OECD Publishing, 2014), http://dx.doi.org/10.1787/9789264214262-9-en; UNESCO Institute for Statistics, "Literacy Rate, Adult Total"; Tom Snyder, ed., *120 Years of American Education: A Statistical Portrait* (Washington, DC: National Center for Education Statistics, 1993), excerpted at http://nces.ed.gov/naal/lit_history.asp.

56. Roser and Ortiz-Ospina, "Literacy"; van Leeuwen and van Leeuwen-Li, "Education Since 1820"; UNESCO Institute for Statistics, "Literacy Rate, Adult Total"; Snyder, ed., *120 Years of American Education*; Buringh and van Zanden, "Charting the 'Rise of the West,'" 409–45; Pamuk and van Zanden, "Standards of Living," 229; "Illiteracy, 1870–2010, All Countries," Montevideo-Oxford Latin American Economic History Data Base, accessed October 31, 2021, http://moxlad.cienciassociales.edu.uy/en; UNESCO Institute for

Statistics, "Literacy Rate, Adult Total (% of People Ages 15 and Above)—Argentina, Brazil," retrieved from Worldbank.org, September 2021, https://data.worldbank.org/indicator/SE.ADT.LITR.ZS?locations=AR-BR.

57. Hannah Ritchie et al., "Global Education," Our World in Data, 2016, accessed October 29, 2021, https://ourworldindata.org/global-education; Jong-Wha Lee and Hanol Lee, "Human Capital in the Long Run," *Journal of Development Economics* 122 (September 2016): 147–69, https://doi.org/10.1016/j.jdeveco.2016.05.006, data available at http://www.barrolee.com/Lee_Lee_LRdata_dn.htm.

58. Richie et al., "Global Education"; Lee and Lee, "Human Capital in the Long Run"; Vito Tanzi and Ludger Schuknecht, *Public Spending in the 20th Century: A Global Perspective* (Cambridge, UK: Cambridge University Press, 2000), https://www.google.com/books/edition/Public_Spending_in_the_20th_Century/kHl6xCgd3aAC?gbpv=1.

59. Richie et al., "Global Education"; Lee and Lee, "Human Capital in the Long Run"; Robert J. Barro and Jong Wha Lee, "A New Data Set of Educational Attainment in the World, 1950–2010," *Journal of Development Economics* 104 (September 2013): 184–98, https://doi.org/10.1016/j.jdeveco.2012.10.001, data available at http://www.barrolee.com/data/yrsch2.htm.

60. "Human Development Index and Its Components," United Nations Development Programme, accessed March 19, 2023, available for download at https://hdr.undp.org/sites/default/files/2021-22_HDR/HDR21-22_Statistical_Annex_HDI_Table.xlsx.

61. "Human Development Index and Its Components," United Nations Development Programme.

62. National Center for Education Statistics, "Expenditures of Educational Institutions Related to the Gross Domestic Product, by Level of Institution: Selected Years, 1929–30 Through 2020–21," in *Digest of Education Statistics: 2021*, US Department of Education, March 2023, https://nces.ed.gov/programs/digest/d21/tables/dt21_106.10.asp; "Consumer Price Index, 1913–," Federal Reserve Bank of Minneapolis, accessed April 28, 2023, https://www.minneapolisfed.org/about-us/monetary-policy/inflation-calculator/consumer-price-index-1913-; US Bureau of Labor Statistics, "Consumer Price Index for All Urban Consumers: All Items in U.S. City Average (CPIAUCSL)," retrieved from FRED, Federal Reserve Bank of St. Louis, updated April 12, 2023, https://fred.stlouisfed.org/series/CPIAUCSL.

63. National Center for Education Statistics, "Expenditures of Educational Institutions Related to the Gross Domestic Product"; "Consumer Price Index, 1913–," Federal Reserve Bank of Minneapolis; US Bureau of Economic Analysis, "Population (B230RC0A052NBEA)," retrieved from FRED, Federal Reserve Bank of St. Louis, updated January 26, 2023, https://fred.stlouisfed.org/series/B230RC0A052NBEA; US Bureau of Labor Statistics, "Consumer Price Index for All Urban Consumers."

64. The United Nations Development Programme publishes the most comprehensive and authoritative data, but it reaches back to only 1990. The 1870–2017 dataset from Our World in Data draws on older sources with less consistent methodology. To preserve relative changes while maximizing commensurability across the period in question, the UNDP data for 1990–2021 is used and the prior OWD data is rescaled so that it reaches consistency with the UN data in 1990. See "Human Development Insights," United Nations Development Programme Human Development Reports, accessed March 22, 2023, https://hdr.undp.org/data-center/country-insights#/ranks; Roser and Ortiz-Ospina, "Global Education"; Lee and Lee, "Human Capital in the Long Run"; Barro and Lee, "New Data Set of Educational Attainment in the World, 1950–2010."

65. For more information on the role of waterborne diseases in public health across history, see "What Exactly Is Typhoid Fever?," *Seeker*, YouTube video, August 20, 2019, https://www.youtube.com/watch?v=NllKW2CYU68; "The Pandemic the World Has Forgotten," *Seeker*, YouTube video, September 8, 2020, https://www.youtube.com/watch?v=hj95IZMlZWw; "The Story of Cholera," Global Health Media Project, YouTube video, December 10, 2011, https://www.youtube.com/watch?v=jG1VNSCsP5Q; Theodore H. Tulchinsky

and Elena A. Varavikova, "A History of Public Health," *The New Public Health*, October 10, 2014, 1–42, https://doi.org/10.1016/B978-0-12-415766-8.00001-X.

66. Suzanne Spellen, "From Pakistan to Brooklyn: A Quick History of the Bathroom," *Brownstoner*, November 28, 2016, https://www.brownstoner.com/architecture/victorian-bathroom-history-plumbing-brooklyn-architecture-interiors; Anthony Mitchell Sammarco, *The Great Boston Fire of 1872* (Charleston, SC: Arcadia, 1997), 30, https://books.google.com/books?id=v3lzw5d2K8wC; Price V. Fishback, *Soft Coal, Hard Choices: The Economic Welfare of Bituminous Coal Miners, 1890–1930* (New York: Oxford University Press, 1992), 170, https://www.google.com/books/edition/Soft_Coal_Hard_Choices/EjnnCwAAQBAJ; Stanley Lebergott, *Wealth and Want* (Princeton, NJ: Princeton University Press, 1975), 7, https://www.google.com/books/edition/Wealth_and_Want/Lrx9BgAAQBAJ; "Historical Census of Housing Tables: Sewage Disposal," US Census Bureau, 1990, revised October 8, 2021, https://www.census.gov/data/tables/time-series/dec/coh-sewage.html; Gary M. Walton and Hugh Rockoff, *History of the American Economy*, 13th ed. (Boston: Cengage Learning, 2017), 377.

67. Walton and Rockoff, *History of the American Economy*, 377; US Census Bureau, "Historical Census of Housing Tables: Sewage Disposal."

68. See, for example, Susan Carpenter, "After Two Years of Eco-Living, What Works and What Doesn't," *Los Angeles Times*, March 11, 2014, https://www.latimes.com/home/la-hm-realist-main-20101016-story.html.

69. Jane Otai, "Happy #WorldToiletDay! Here's What It's Like To Live Without One," NPR, November 19, 2015, https://www.npr.org/sections/goatsandsoda/2015/11/19/456495448/happy-worldtoiletday-here-s-what-it-s-like-to-live-without-one.

70. WHO/UNICEF Joint Monitoring Programme (JMP) for Water Supply, Sanitation and Hygiene, "People Using Safely Managed Sanitation Services (% of Population)," retrieved from worldbank.org, accessed October 12, 2021, https://data.worldbank.org/indicator/SH.STA.SMSS.ZS.

71. Stanley Lebergott, *Pursuing Happiness: American Consumers in the Twentieth Century* (Princeton, NJ: Princeton University Press, 1993), 102, https://www.google.com/books/edition/Pursuing_Happiness/bD0ABAAAQBAJ; US Census Bureau, "Historical Census of Housing Tables: Sewage Disposal"; Marc Jeuland et al., "Water and Sanitation: Economic Losses from Poor Water and Sanitation—Past, Present, and Future," in *How Much Have Global Problems Cost the World?*, ed. Bjørn Lomborg (Cambridge, UK: Cambridge University Press, 2013), 333; David A. Raglin, "Plumbing and Kitchen Facilities in Housing Units," 2015 American Community Survey Research and Evaluation Report Memorandum Series No. ACS15-RER-06, US Census Bureau, May 29, 2015, 3, https://www.census.gov/content/dam/Census/library/working-papers/2015/acs/2015_Raglin_01.pdf; Katie Meehan et al., *Plumbing Poverty in U.S. Cities: A Report on Gaps and Trends in Household Water Access, 2000 to 2017*, King's College London, September 27, 2021, 4, https://kclpure.kcl.ac.uk/portal/files/159767495/Plumbing_Poverty_in_US_Cities.pdf; US Census Bureau, "Total Households (TTLHH)," retrieved from FRED, Federal Reserve Bank of St. Louis, updated November 21, 2022, https://fred.stlouisfed.org/series/TTLHH; WHO/UNICEF Joint Monitoring Programme (JMP) for Water Supply, Sanitation, and Hygiene, "People Using at Least Basic Sanitation Services (% of Population)," retrieved from Worldbank.org, accessed April 28, 2023, https://data.worldbank.org/indicator/SH.STA.BASS.ZS; WHO/UNICEF Joint Monitoring Programme (JMP) for Water Supply, Sanitation, and Hygiene, "Improved Sanitation Facilities (% of Population with Access)," retrieved from Worldbank.org, updated January 25, 2018, accessed April 28, 2023, originally available at https://databank.worldbank.org/source/millennium-development-goals/Series/SH.STA.ACSN; World Health Organization and United Nations Children's Fund, *Progress on Sanitation and Drinking Water—2015 Update and MDG Assessment*, World Health Organization, 2015, 14, https://data.unicef.org/wp-content/uploads/2015/12/Progress-on-Sanitation-and-Drinking-Water_234.pdf; *Progress on Household Drinking Water, Sanitation and Hygiene 2000–2020: Five Years into the SDGs* (Geneva: World Health Organization and the

United Nations Children's Fund, 2021), 7, https://washdata.org/sites/default/files/2021-07/jmp-2021-wash-households.pdf.

72. Elizabeth Nix, "How Edison, Tesla and Westinghouse Battled to Electrify America," History.com, October 24, 2019, https://www.history.com/news/what-was-the-war-of-the-currents; Harold D. Wallace Jr., "Power from the People: Rural Electrification Brought More than Lights," National Museum of American History, February 12, 2016, https://americanhistory.si.edu/blog/rural-electrification; Stanley Lebergott, *The American Economy: Income, Wealth and Want* (Princeton, NJ: Princeton University Press, 1976), 334, https://www.google.com/books/edition/The_American_Economy/HYV9BgAAQBAJ.

73. For more detail on America's rural electrification, see "On the Line—Rural Electrification Administration Film," Russell Library Audiovisual Collections, YouTube video, June 4, 2018, https://www.youtube.com/watch?v=DbAM-CwOxu0; General Electric, "More Power to the American Farmer—1946," miSci: Museum of Innovation and Science, YouTube video, June 13, 2012, https://www.youtube.com/watch?v=aY5eFQTYkaw; US Department of Agriculture, Rural Electrification Administration, *Rural Lines, USA: The Story of the Rural Electrification Administration's First Twenty-five Years, 1935–1960*, US Department of Agriculture Miscellaneous Publication No. 811 (1960), https://www.google.com/books/edition/Rural_Lines_USA/IBkuAAAAYAAJ; US Census Bureau, *Historical Statistics of the United States: Colonial Times to 1970*, part 1 (Washington, DC: US Census Bureau, 1975), 827, https://www.census.gov/history/pdf/histstats-colonial-1970.pdf.

74. US Census Bureau, *Historical Statistics of the United States: Colonial Times to 1970*, part 1.

75. Lily Odarno, "Closing Sub-Saharan Africa's Electricity Access Gap: Why Cities Must Be Part of the Solution," World Resources Institute, August 14, 2019, https://www.wri.org/blog/2019/08/closing-sub-saharan-africa-electricity-access-gap-why-cities-must-be-part-solution; Giacomo Falchetta et al., "Satellite Observations Reveal Inequalities in the Progress and Effectiveness of Recent Electrification in Sub-Saharan Africa," *One Earth* 2, no. 4 (April 24, 2020): 364–79, https://doi.org/10.1016/j.oneear.2020.03.007.

76. IEA, IRENA, UNSD, World Bank, and WHO, "Access to Electricity (% of Population)," from *Tracking SDG 7: The Energy Progress Report* (Washington, DC: World Bank, 2023), https://data.worldbank.org/indicator/EG.ELC.ACCS.ZS.

77. Daron Acemoğlu and James A. Robinson, *Why Nations Fail: The Origins of Power, Prosperity, and Poverty* (New York: Crown, 2012).

78. Lebergott, *The American Economy: Income, Wealth and Want*, 334; US Census Bureau, *Historical Statistics of the United States: Colonial Times to 1970*, part 1; IEA, IRENA, UNSD, World Bank, and WHO, "Access to Electricity (% of Population)"; IEA, IRENA, UNSD, World Bank, and WHO, "Access to Electricity (% of Population)—United States," from *Tracking SDG 7: The Energy Progress Report* (Washington, DC: World Bank, 2023), https://data.worldbank.org/indicator/EG.ELC.ACCS.ZS?locations=US; World Bank, "Access to Electricity (% of Population)," Sustainable Energy or All (SE4ALL) database, SE4ALL Global Tracking Framework, retrieved from Worldbank.org, accessed April 28, 2023, originally available at https://data.worldbank.org/indicator/EG.ELC.ACCS.ZS.

79. US Census Bureau, "Selected Communications Media: 1920 to 1998," *Statistical Abstract of the United States: 1999* (Washington, DC: US Census Bureau, 1999): 885, https://www.census.gov/history/pdf/radioownership1920-1998.pdf.

80. US Census Bureau, "Selected Communications Media: 1920 to 1998."

81. "Kathleen Hall Jamieson on Talk Radio's History and Impact," PBS.org, February 13, 2004, https://web.archive.org/web/20170301204706/https://www.pbs.org/now/politics/talkradiohistory.html; "'Radio' Listening Dominates Audio In-Cat," Edison Research, January 11, 2019, https://www.edisonresearch.com/am-fm-radio-still-dominant-audio-in-car; Aniko Bodroghkozy, ed., *A Companion to the History of American Broadcasting* (Hoboken, NJ: Wiley, 2018).

82. Ezra Klein, "Something Is Breaking American Politics, but It's Not Social Media," *Vox*, April 12, 2017, https://www.vox.com/policy-and-politics/2017/4/12/15259438/social-media-political-polarization; Jeffrey M. Berry and Sarah Sobieraj, "Understanding the Rise of

Talk Radio," *PS: Political Science and Politics* 44, no. 4 (October 2011): 762–67, https://www.jstor.org/stable/41319965.

83. "Audio and Podcasting Fact Sheet," Pew Research Center, June 29, 2021, https://www.pewresearch.org/journalism/fact-sheet/audio-and-podcasting.

84. For a closer look at how television was invented and its earliest development, see "The Invention of Television (1929)," British Pathé, YouTube video, April 13, 2014, https://www.youtube.com/watch?v=nwJ2bMATIAM; "The Origins of Television," Nirali Pathak, YouTube video, January 2, 2012, https://www.youtube.com/watch?v=uM7ZD5f9Pb8; "Evolution of Television 1920–2020," Captain Gizmo, YouTube video, December 18, 2018, https://www.youtube.com/watch?v=PveVwQhNnq8; Albert Abramson, *The History of Television, 1880–1941* (Jefferson, NC: McFarland, 1987).

85. For more on the state of television in 1939 and the technology being put on hold due to World War II, see "What Television Was Like in 1939," Smithsonian Channel, YouTube video, December 30, 2013, https://www.youtube.com/watch?v=wj_Mcpff-Ks; BBC, "Close Down of Television Service for the Duration of the War," History of the BBC, accessed April 28, 2023, https://www.bbc.com/historyofthebbc/anniversaries/september/closedown-of-television; "Television Facts and Statistics—1939 to 2000," TVhistory.tv, accessed March 31, 2022, https://web.archive.org/web/20220331223237/http://www.tvhistory.tv/facts-stats.htm.

86. Cobbett S. Steinberg, *TV Facts*, Facts on File (1980), 142, available in part at "Television Facts and Statistics—1939 to 2000," TVhistory.tv, accessed March 31, 2022, https://web.archive.org/web/20220331223237/http://www.tvhistory.tv/facts-stats.htm.

87. Steinberg, *TV Facts*, 142.

88. Lance Venta, "Infinite Dial: Mean Number of Radios In Home Drops in Half Since 2008," Radio Insight, March 20, 2020, https://radioinsight.com/headlines/184900/infinite-dial-mean-number-of-radios-in-home-drops-in-half-since-2008; "Mobile Fact Sheet," Pew Research Center, April 7, 2021, https://www.pewresearch.org/internet/fact-sheet/mobile; US Census Bureau, "Selected Communications Media: 1920 to 1998," 885; Douglas B. Craig, *Fireside Politics: Radio and Political Culture in the United States, 1920–1940* (Baltimore: Johns Hopkins University Press, 2000; paperback, 2006), 12, https://www.google.com/books/edition/Fireside_Politics/haWh203m7aIC; US Census Bureau, "Utilization of Selected Media: 1980 to 2005," *Statistical Abstract of the United States: 2008* (Washington, DC: US Census Bureau, 2007): 704, https://www.census.gov/prod/2007pubs/08abstract/infocomm.pdf; US Census Bureau, "Utilization and Number of Selected Media: 2000 to 2009," *Statistical Abstract of the United States: 2012* (Washington, DC: US Census Bureau, 2011): 712, https://www.google.com/books/edition/Statistical_Abstract_of_the_United_State/pW9NAQAAMAAJ.

89. Steinberg, *TV Facts*, 142; US Census Bureau, "Total Households (TTLHH)"; Jack W. Plunkett, *Plunkett's Entertainment & Media Industry Almanac 2006* (Houston, TX: Plunkett Research, 2006), 35.

90. At the time of writing in 2023, the latest official estimates by Nielsen cover the first half of 2021. See Plunkett, *Plunkett's Entertainment & Media Industry Almanac 2006*, 35; "Nielsen Estimates 121 Million TV Homes in the U.S. for the 2020–2021 TV Season," Nielsen Company, August 28, 2020, https://www.nielsen.com/us/en/insights/article/2020/nielsen-estimates-121-million-tv-homes-in-the-u-s-for-the-2020-2021-tv-season.

91. Rick Porter, "TV Long View: Five Years of Network Ratings Declines in Context," *Hollywood Reporter*, September 21, 2019, https://www.hollywoodreporter.com/live-feed/five-years-network-ratings-declines-explained-1241524; Sapna Maheshwari and John Koblin, "Why Traditional TV Is in Trouble," *New York Times*, May 13, 2018, https://www.nytimes.com/2018/05/13/business/media/television-advertising.html.

92. For more on the transformative impact of the Altair 8800, see Jason Fitzpatrick, "The Computer That Changed Everything (Altair 8800)," Computerphile, YouTube video, May 15, 2015, https://www.youtube.com/watch?v=cwEmnfy2BhI; "The PC That Started Microsoft & Apple! (Altair 8800)," ColdFusion, YouTube video, March 18, 2016, https://www.youtube.com/watch?v=X5lpOskKF9I.

93. US Census Bureau, *Historical Statistics of the United States, Colonial Times to 1957* (Washington, DC: US Government Publishing Office, 1960), 491, https://www.google.co.uk/books/edition/Historical_Statistics_of_the_United_Stat/hyI1AAAAIAAJ; US Census Bureau, "Total Households (TTLHH)"; Steinberg, *TV Facts*, 142; Plunkett, *Plunkett's Entertainment & Media Industry Almanac 2006*, 35; "Nielsen: 109.6 Million TV Households in the U.S.," HispanicAd.com, July 30, 2004, http://hispanicad.com/blog/news-article/had/research/nielsen-1096-million-tv-households-us; "Late News," *AdAge*, August 29, 2005, https://adage.com/article/late-news/late-news/104427; TVTechnology, "Nielsen Reports Slight Increase in TV Households," TV Tech, August 28, 2006, https://www.tvtechnology.com/news/nielsen-reports-slight-increase-in-tv-households; Variety Staff, "TV Nation: 112.8 Million Strong," *Variety*, August 23, 2007, https://variety.com/2007/tv/opinion/tv-nation-1128-22127/?jwsource=cl; "Nielsen Reports Growth of 4.4% in Asian and 4.3% in Hispanic U.S. Households for 2008-2009 Television Season," Nielsen Company, August 28, 2008, https://www.nielsen.com/wp-content/uploads/sites/3/2019/04/press_release34.pdf; "114.9 Million U.S. Television Homes Estimated for 2009–2010 Season," Nielsen Company, August 29, 2009, https://www.nielsen.com/us/en/insights/article/2009/1149-million-us-television-homes-estimated-for-2009-2010-season; "Number of U.S. TV Households Climbs by One Million for 2010–11 TV Season," Nielsen Company, August 27, 2010, https://www.nielsen.com/us/en/insights/article/2010/number-of-u-s-tv-households-climbs-by-one-million-for-2010-11-tv-season; Cynthia Littleton, "Nielsen Tackles Web Viewing," *Variety*, February 25, 2013, https://variety.com/2013/digital/news/nielsen-tackles-web-viewing-1118066529; "Nielsen Estimates 115.6 Million TV Homes in the U.S., Up 1.2%," Nielsen Company, May 7, 2013, https://www.nielsen.com/us/en/insights/article/2013/nielsen-estimates-115-6-million-tv-homes-in-the-u-s-up-1-2; "Nielsen Estimates More Than 116 Million TV Homes in the U.S.," Nielsen Company, August 29, 2014, https://www.nielsen.com/us/en/insights/article/2014/nielsen-estimates-more-than-116-million-tv-homes-in-the-us; "Nielsen Estimates 116.4 Million TV Homes in the U.S. for the 2015–16 TV Season," Nielsen Company, August 28, 2015, https://www.nielsen.com/us/en/insights/article/2015/nielsen-estimates-116-4-million-tv-homes-in-the-us-for-the-2015-16-tv-season; "Nielsen Estimates 118.4 Million TV Homes in the U.S. for the 2016–17 TV Season," Nielsen Company, August 26, 2016, https://www.nielsen.com/us/en/insights/article/2016/nielsen-estimates-118-4-million-tv-homes-in-the-us-for-the-2016-17-season; "Nielsen Estimates 119.6 Million TV Homes in the U.S. for the 2017–18 TV Season," Nielsen Company, August 25, 2017, https://www.nielsen.com/us/en/insights/article/2017/nielsen-estimates-119-6-million-us-tv-homes-2017-2018-tv-season; "Nielsen Estimates 119.9 Million TV Homes in the U.S. for the 2018–2019 TV Season," Nielsen Company, September 7, 2018, https://www.nielsen.com/us/en/insights/article/2018/nielsen-estimates-119-9-million-tv-homes-in-the-us-for-the-2018-19-season; "Nielsen Estimates 120.6 Million TV Homes in the U.S. for the 2019–2020 TV Season," Nielsen Company, August 27, 2019, https://www.nielsen.com/us/en/insights/article/2019/nielsen-estimates-120-6-million-tv-homes-in-the-u-s-for-the-2019-202-tv-season; "Nielsen Estimates 121 Million TV Homes in the U.S. for the 2020–2021 TV Season."

94. For more extensive histories of the personal computer revolution and its significance, see Carrie Anne Philbin, "The Personal Computer Revolution: Crash Course Computer Science #25," CrashCourse, YouTube video, August 23, 2017, https://www.youtube.com/watch?v=M5BZou6C01w; "History of Apple I and Steve Jobs' Personal Computer," *TechCrunch*, YouTube video, April 17, 2017, https://www.youtube.com/watch?v=LTJPdHeibOQ; "History of Microsoft—1975," jonpaulmoen, YouTube video, December 18, 2010, https://www.youtube.com/watch?v=BLaMbaVT22E; Leo Rowe, "History of Personal Computers Part 1," (from *Triumph of the Nerds*, PBS, 1996), Chasing 80, YouTube video, December 17, 2013, https://www.youtube.com/watch?v=AIBr-kPgYuU; Gerard O'Regan, *A Brief History of Computing*, 2nd ed. (London: Springer, 2008).

95. "1984 Apple's Macintosh Commercial," Mac History, YouTube video, February 1, 2012, https://www.youtube.com/watch?v=VtvjbmoDx-I; "Computer and Internet Use in the

United States: 1984 to 2009," US Census Bureau, February 2010, https://www.census.gov /data/tables/time-series/demo/computer-internet/computer-use-1984-2009.html.

96. "Internet Host Count History," Internet Systems Consortium, accessed May 18, 2012, https://web.archive.org/web/20120518101749/http://www.isc.org/solutions/survey/history; 3way Labs, "Internet Domain Survey, January, 2019," Internet Systems Consortium, accessed April 28, 2023, http://ftp.isc.org/www/survey/reports/current.

97. "Internet Host Count History," Internet Systems Consortium; 3way Labs, "Internet Domain Survey, January, 2019."

98. Arielle Sumits, "The History and Future of Internet Traffic," Cisco, August 28, 2015, https://blogs.cisco.com/sp/the-history-and-future-of-internet-traffic.

99. "QuickFacts United States," US Census Bureau, accessed April 28, 2023, https://www .census.gov/quickfacts/fact/table/US/HCN010217; Michael Martin, "Computer and Internet Use in the United States: 2018," American Community Survey Reports ACS-49, US Census Bureau, April 2021, https://www.census.gov/newsroom/press-releases/2021/com puter-internet-use.html.

100. Estimated based primarily on data from the International Telecommunication Union. While the ITU does not directly collect data on the percentage of households with a computer, smartphone, or tablet (which would be the optimal metric), its estimates of the percentage of households with internet access at home is the best available proxy. We know all these households have some kind of functional computer, and this metric leaves out a small additional number of households with a computer but no internet access on any device, so I use that conservative figure here. See "Statistics—Individuals Using the Internet," International Telecommunication Union, January 31, 2023, https://www.itu.int/en/ITU-D/Sta tistics/Pages/stat/default.aspx; "Key ICT Indicators for Developed and Developing Countries," *ITU World Telecommunication/ICT Indicators Database*.

101. Data includes both PCs and computer-containing devices such as tablets and smartphones. See "Computer and Internet Access in the United States: 2012—Table 4: Households with a Computer and Internet Use: 1984 to 2012," US Census Bureau, February 3, 2014, https://www2.census.gov/programs-surveys/demo/tables/computer-internet/2012/com puter-use-2012/table4.xls; Thom File and Camille Ryan, "Computer and Internet Use in the United States: 2013," American Community Survey Reports ACS-28, US Census Bureau, November 2014, 3, https://www.census.gov/content/dam/Census/library/publica tions/2017/acs/acs-37.pdf; Camille Ryan and Jamie M. Lewis, "Computer and Internet Use in the United States: 2015," American Community Survey Reports ACS-37, US Census Bureau, September 2017, 4, https://www.census.gov/content/dam/Census/library/publica tions/2017/acs/acs-37.pdf; Martin, "Computer and Internet Use in the United States: 2018"; International Telecommunication Union, "Key ICT Indicators for Developed and Developing Countries, the World and Special Regions (Totals and Penetration Rates)," from *ITU World Telecommunication/ICT Indicators Database*, International Telecommunication Union, November 2020, https://www.itu.int/en/ITU-D/Statistics/Documents/facts /ITU_regional_global_Key_ICT_indicator_aggregates_Nov_2020.xlsx; "Statistics—Indi viduals Using the Internet," International Telecommunication Union; International Tele communication Union, "Key ICT Indicators for the World and Special Regions (Totals and Penetration Rates)," ITU World Telecommunication/ICT Indicators database, November 2022, updated February 15, 2023, https://www.itu.int/en/ITU-D/Statistics/Docu ments/facts/ITU_regional_global_Key_ICT_indicator_aggregates_Nov_2022_revised _15Feb2023.xlsx.

102. "Drug Discovery and Development Process," Novartis, YouTube video, January 14, 2011, https://www.youtube.com/watch?v=3Gl0gAcW8rw; Robert Gaynes, "The Discovery of Penicillin—New Insights After More than 75 Years of Clinical Use," *Emerging Infectious Diseases* 23, no. 5 (May 2017): 849–53, http://dx.doi.org/10.3201/eid2305.161556; Sabrina Barr, "Penicillin Allergy: How Common Is It and What Are the Symptoms," *Independent*, October 23, 2018, https://www.independent.co.uk/life-style/health-and-families/penicil lin-allergy-how-common-symptoms-antibiotic-drug-bacteria-a8597246.html.

103. Dan Usher, *Political Economy* (Malden, MA: Wiley, 2003), 5, https://books.google.com /books?id=-2210y5aPZgC; Max Roser, Hannah Ritchie, and Bernadeta Dadonaite, "Child and Infant Mortality," Our World in Data, November 2019, https://ourworldindata.org /child-mortality; Anthony A. Volk and Jeremy A. Atkinson, "Infant and Child Death in the Human Environment of Evolutionary Adaptation," *Evolution and Human Behavior* 34, no. 3 (May 2013): 182–92, https://doi.org/10.1016/j.evolhumbehav.2012.11.007Get.

104. Mattias Lindgren, "Life Expectancy at Birth," *Gapminder*, accessed April 28, 2023, https:// www.gapminder.org/data/documentation/gd004.

105. Toshiko Kaneda, Charlotte Greenbaum, and Carl Haub, *2021 World Population Data Sheet*, Population Reference Bureau, August 2021, https://www.prb.org/wp-content/uploads /2021/08/letter-booklet-2021-world-population.pdf; Lindgren, "Life Expectancy at Birth."

106. Terry Grossman and I describe the first, second, and third bridges to life extension in more detail in our 2009 book *Transcend*. That was a book about health and wellness, so I describe the fourth bridge—extending our consciousness into digital mediums where it can be backed up and expanded—in more detail throughout this book. See Ray Kurzweil and Terry Grossman, *Transcend: Nine Steps to Living Well Forever* (Emmaus, PA: Rodale, 2009).

107. Lindgren, "Life Expectancy at Birth."

108. "Products," Sequencing.com, accessed March 25, 2023, https://web.archive.org/web/2023 0315065708/https://sequencing.com/products/purchase-kit; Elizabeth Pennisi, "A $100 Ge-nome? New DNA Sequencers Could Be a 'Game Changer' for Biology, Medicine," *Science*, June 15, 2022, https://www.science.org/content/article/100-genome-new-dna-sequencers -could-be-game-changer-biology-medicine; Kris A. Wetterstrand, "The Cost of Sequencing a Human Genome," National Human Genome Research Institute, accessed April 28, 2023, https://www.genome.gov/about-genomics/fact-sheets/Sequencing-Human-Genome-cost; Kris A. Wetterstrand, "DNA Sequencing Costs: Data," National Human Genome Research Institute, November 1, 2021, https://www.genome.gov/about-genomics/fact-sheets/DNA -Sequencing-Costs-Data; Andrew Carroll and Pi-Chuan Chang, "Improving the Accuracy of Genomic Analysis with DeepVariant 1.0," *Google AI Blog*, Google Research, September 18, 2020, https://ai.googleblog.com/2020/09/improving-accuracy-of-genomic-analysis.html.

109. For more details on current clinical applications of artificial intelligence, see Bernard Marr, "How Is AI Used in Healthcare—5 Powerful Real-World Examples that Show the Latest Advances," *Forbes*, July 27, 2018, https://www.forbes.com/sites/bernardmarr/2018 /07/27/how-is-ai-used-in-healthcare-5-powerful-real-world-examples-that-show-the -latest-advances/#55fa6ef05dfb; Giovanni Briganti and Olivier Le Moine, "Artificial Intel-ligence in Medicine: Today and Tomorrow," *Frontiers in Medicine* 7, article 27 (February 5, 2020), https://doi.org/10.3389/fmed.2020.00027.

110. See chapter 6 for a much more detailed discussion of the factors that cause human life ex-pectancy to currently max out at 120. As I explain in that chapter, research over the past few decades has identified the specific biochemical processes that cause aging, and as of 2023 there is active research working toward addressing all of them. This doesn't have to immediately totally cure aging to allow radical life extension—the tipping point will be when, every year, medicine adds at least one additional year to our life expectancy, allowing people to get ahead of the curve, so to speak, and achieve "longevity escape velocity."

111. Due to the disruptions of the COVID-19 pandemic, 2020 is the most recent year for which good projected UK life expectancy tables exist at the time of writing in 2023. See "English Life Tables," Office for National Statistics (United Kingdom), September 1, 2015, https:// www.ons.gov.uk/file?uri=%2fpeoplepopulationandcommunity%2fbirthsdeathsandmarri ages%2flifeexpectancies%2fdatasets%2f2englishlifetables%2fcurrent/eolselt1to17_tcm77 -414359.xls; "National Life Tables, United Kingdom, Period Expectation of Life, Based on Data for the Years 2018–2020," Office for National Statistics (United Kingdom), Septem-ber 23, 2021, https://www.ons.gov.uk/file?uri=%2Fpeoplepopulationandcommunity%2Fbirth sdeathsandmarriages%2Flifeexpectancies%2Fdatasets%2Fnationallifetablesunitedking domreferencetables%2Fcurrent%2Fnationallifetables3yruk.xlsx.

112. Due to the disruptions of the COVID-19 pandemic, 2020 is the most recent year for which good projected US life expectancy tables exist at the time of writing in 2023. See United

Nations Department of Economic and Social Affairs, Population Division, *World Population Prospects 2019—Special Aggregates, Online Edition*, rev. 1. (United Nations, 2019), https://population.un.org/wpp/Download/Files/3_Indicators%20(Special%20Aggregates) /EXCEL_FILES/5_Geographical/Mortality/WPP2019_SA5_MORT_F16_1_LIFE_EX PECTANCY_BY_AGE_BOTH_SEXES.XLSX.

113. Max Roser and Esteban Ortiz-Ospina, "Global Extreme Poverty," Our World in Data, March 27, 2017, https://ourworldindata.org/extreme-poverty; François Bourguignon and Christian Morrisson, "Inequality Among World Citizens: 1820–1992," *American Economic Review* 92, no. 4 (September 2002): 727–44, https://doi.org/10.1257/00028280260344443; *PovcalNet: An Online Analysis Tool for Global Poverty Monitoring*, World Bank, March 17, 2020, http://iresearch.worldbank.org/PovcalNet/home.aspx.

114. For some clear and compelling summaries and visualizations of the global development and industrialization process, see Council on Foreign Relations, "Global Development Explained | World101," CFR Education, YouTube video, June 18, 2019, https://www.you tube.com/watch?v=Po0o3Gk9FPQ; "Hans Rosling's 200 Countries, 200 Years, 4 Minutes—the Joy of Stats—BBC Four," BBC, YouTube video, November 26, 2010, https:// www.youtube.com/watch?v=jbkSRLYSojo; "The History of International Development | Max Roser | EAGxOxford 2016," Centre for Effective Altruism, YouTube video, April 16, 2017, https://www.youtube.com/watch?v=XbBn8OEqL4k; Max Roser, "The Short History of Global Living Conditions and Why It Matters That We Know It," Our World in Data, accessed October 29, 2021, https://ourworldindata.org/a-history-of-global-living-conditions -in-5-charts; Bourguignon and Morrisson, "Inequality Among World Citizens: 1820–1992."

115. For deeper detail on the rapid (and sometimes horribly mismanaged) agricultural transformations of China and India, see Yi Wen, "China's Rapid Rise: From Backward Agrarian Society to Industrial Powerhouse in Just 35 Years," Federal Reserve Bank of St. Louis, April 12, 2016, https://www.stlouisfed.org/publications/regional-economist/april-2016/chi nas-rapid-rise-from-backward-agrarian-society-to-industrial-powerhouse-in-just-35 -years; Xiao-qiang Jiao, Nyamdavaa Mongol, and Fu-suo Zhang, "The Transformation of Agriculture in China: Looking Back and Looking Forward," *Journal of Integrative Agriculture* 17, no. 4 (April 2018): 755–64, https://doi.org/10.1016/S2095-3119(17)61774-X; Amarnath Tripathi and A. R. Prasad, "Agricultural Development in India Since Independence: A Study on Progress, Performance, and Determinants," *Journal of Emerging Knowledge on Emerging Markets* 1, no. 1 (November 2009): 63–92, https://digitalcommons.kennesaw.edu /cgi/viewcontent.cgi?article=1007&context=jekem; M. S. Swaminathan, *50 Years of Green Revolution: An Anthology of Research Papers* (Singapore: World Scientific, 2017); Francine R. Frankel, *India's Green Revolution: Economic Gains and Political Costs* (Princeton, NJ: Princeton University Press, 1971).

116. World Bank Development Research Group, "Poverty Headcount Ratio at $2.15 a Day."

117. World Bank, "Regional Aggregation Using 2011 PPP and $1.9/Day Poverty Line," Povcal-Net: The On-line Tool for Poverty Measurement Developed by the Development Research Group of the World Bank, September 1, 2022, https://web.archive.org/web/202209010 35616/http://iresearch.worldbank.org/PovcalNet/povDuplicateWB.aspx.

118. "Regional Aggregation Using 2011 PPP," World Bank; Eric W. Sievers, *The Post-Soviet Decline of Central Asia: Sustainable Development and Comprehensive Capital* (London: Routledge Curzon, 2003).

119. David Hulme, "The Making of the Millennium Development Goals: Human Development Meets Results-Based Management in an Imperfect World," (BWPI working paper 16, Brooks World Poverty Institute, December 2007), https://sustainabledevelopment.un.org /content/documents/773bwpi-wp-1607.pdf.

120. For more on the Millennium Development Goals and their impact, see United Nations Department of Economic and Social Affairs, *The Millennium Development Goals Report 2015* (New York: United Nations, April 2016), https://www.un.org/millenniumgoals/2015 _MDG_Report/pdf/MDG%202015%20rev%20(July%201).pdf; Hannah Ritchie and Max Roser, "Now It Is Possible to Take Stock—Did the World Achieve the Millennium Development Goals?," Our World in Data, September 20, 2018, https://ourworldindata.org

/millennium-development-goals; Charles Kenny, "MDGs to SDGs: Have We Lost the Plot?," Center for Global Development, May 27, 2015, https://www.cgdev.org/publication /mdgs-sdgs-have-we-lost-plot.

121. World Bank Development Research Group, "Poverty Headcount Ratio at $2.15 a Day (2017 PPP) (% of population) – United States," World Bank, accessed April 28, 2023, https://data.worldbank.org/indicator/SI.POV.DDAY?locations=US.

122. Emily A. Shrider et al., *Income and Poverty in the United States: 2020*, Current Population Reports P60–273, US Census Bureau, September 2021, 56, https://www.census.gov/con tent/dam/Census/library/publications/2021/demo/p60-273.pdf.

123. David R. Dickens Jr. and Christina Morales, "Income Distribution and Poverty in Nevada," in *The Social Health of Nevada: Leading Indicators and Quality of Life in the Silver State*, ed. Dmitri N. Shalin (Las Vegas: UNLV Center for Democratic Culture Publications, 2006): 1–24, https://digitalscholarship.unlv.edu/cgi/viewcontent.cgi?article=1019 &context=social_health_nevada_reports; Shrider et al., *Income and Poverty in the United States: 2020*, 56.

124. Shrider et al., *Income and Poverty in the United States: 2020*, 14, 17.

125. For a closer look at how the US poverty line is determined and updated, see Institute for Research on Poverty, "What Are Poverty Thresholds and Poverty Guidelines?," University of Wisconsin–Madison, accessed April 28, 2023, https://www.irp.wisc.edu/resources /what-are-poverty-thresholds-and-poverty-guidelines/#:~:text=9902; Institute for Research on Poverty, "How Is Poverty Measured?," University of Wisconsin–Madison, accessed April 28, 2023, https://www.irp.wisc.edu/resources/how-is-poverty-measured.

126. John Creamer et al., US Census Bureau, *Poverty in the United States: 2021* (Washington, DC: US Government Publishing Office, September 2022), 25, https://www.census.gov /content/dam/Census/library/publications/2022/demo/p60-277.pdf.

127. Creamer et al., *Poverty in the United States: 2021*, 25, 36.

128. Creamer et al., *Poverty in the United States: 2021*, 25.

129. "MIT OpenCourseWare at 20," MIT OpenCourseWare, YouTube video, April 12, 2021, https://www.youtube.com/watch?v=0aAEamhJHUI.

130. Creamer et al., *Poverty in the United States: 2021*, 25; World Bank Development Research Group, "Poverty Headcount Ratio at $2.15 a Day—United States."

131. Max Roser and Esteban Ortiz-Ospina, "Global Extreme Poverty—World Population Living in Extreme Poverty, World, 1820 to 2015," Our World in Data, March 27, 2017, updated 2019, https://ourworldindata.org/grapher/world-population-in-extreme-poverty -absolute; Bourguignon and Morrisson, "Inequality Among World Citizens: 1820–1992"; World Bank Development Research Group, "Poverty Headcount Ratio at $2.15 a Day"; World Bank, "Regional Aggregation Using 2011 PPP and $1.9/Day Poverty Line"; World Bank, Development Research Group, "Poverty Headcount Ratio at $2.15 a day (2017 PPP) (% of population)—United States."

132. Office of the Assistant Secretary for Planning and Evaluation, "HHS Poverty Guidelines for 2023," US Department of Health and Human Services, January 19, 2023, https://aspe .hhs.gov/poverty-guidelines.

133. US Bureau of Economic Analysis, "Personal Income per Capita (A792RC0A052NBEA)," retrieved from FRED, Federal Reserve Bank of St. Louis, updated March 30, 2023, https:// fred.stlouisfed.org/series/A792RC0A052NBEA; "Consumer Price Index, 1913–," Federal Reserve Bank of Minneapolis; US Bureau of Labor Statistics, "Consumer Price Index for All Urban Consumers."

134. US Bureau of Economic Analysis, "Personal Income Per Capita (A792RC0A052NBEA)"; Peter H. Lindert and Jeffrey G. Williamson, "American Incomes 1774–1860" (working paper 18396, National Bureau of Economic Research, September 2012), 33, https://www .nber.org/system/files/working_papers/w18396/w18396.pdf; Alexander Klein, "New State-Level Estimates of Personal Income in the United States, 1880–1910," in *Research in Economic History*, vol. 29, ed. Christopher Hanes and Susan Wolcott (Bingley, UK: Emerald Group, 2013), 220, https://doi.org/10.1108/S0363-3268(2013)0000029008; "Consumer

Price Index, 1800–," Federal Reserve Bank of Minneapolis, accessed April 28, 2023, https://www.minneapolisfed.org/about-us/monetary-policy/inflation-calculator/consumer -price-index-1800-.

135. Dickens and Morales, "Income Distribution and Poverty in Nevada," 1; Creamer et al., *Poverty in the United States: 2021*, 25.

136. Bourguignon and Morrisson, "Inequality Among World Citizens: 1820–1992"; World Bank Development Research Group, "Poverty Headcount Ratio at $2.15 a Day"; World Bank Development Research Group, "Poverty Headcount Ratio at $2.15 a Day (2017 PPP) (% of population) – United States."

137. Jutta Bolt and Jan Luiten van Zanden, *Maddison Project Database*, version 2020, Groningen Growth and Development Centre, November 2, 2020, https://www.rug.nl/ggdc/histori caldevelopment/maddison/data/mpd2020.xlsx; Jutta Bolt and Jan Luiten van Zanden, "Maddison Style Estimates of the Evolution of the World Economy: A New 2020 Update," (working paper WP-15, Maddison Project, October 2020), https://www.rug.nl/ggdc/histo ricaldevelopment/maddison/publications/wp15.pdf; US Bureau of Economic Analysis, "Real Gross Domestic Product per Capita (A939RX0Q048SBEA)," retrieved from FRED, Federal Reserve Bank of St. Louis, April 27, 2023, https://fred.stlouisfed.org/series /A939RX0Q048SBEA; "Consumer Price Index, 1913–," Federal Reserve Bank of Minne-apolis; John J. McCusker, "Colonial Statistics," in *Historical Statistics of the United States: Earliest Times to the Present*, ed. Susan G. Carter et al. (Cambridge, UK: Cambridge Uni-versity Press, 2006), V-671; Richard Sutch, "National Income and Product," in *Historical Statistics of the United States: Earliest Times to the Present*, ed. Susan G. Carter et al. (Cam-bridge, UK: Cambridge University Press, 2006), III-23–25; Leandro Prados de la Esco-sura, "Lost Decades? Economic Performance in Post-Independence Latin America," *Journal of Latin American Studies* 41, no. 2 (May 2009): 279–307, https://www.jstor.org /stable/27744128; Jutta Bolt et al., "Rebasing 'Maddison': New Income Comparisons and the Shape of Long-Run Economic Development" (working paper 10, Maddison Project, Groningen Growth and Development Centre, 2018), https://www.rug.nl/ggdc/html_publi cations/memorandum/gd174.pdf.

138. Bolt and van Zanden, *Maddison Project Database*; Bolt and van Zanden, "Maddison Style Estimates of the Evolution of the World Economy"; US Bureau of Economic Analysis, "Real Gross Domestic Product per Capita (A939RX0Q048SBEA)"; "Consumer Price Index, 1913–," Federal Reserve Bank of Minneapolis; US Bureau of Labor Statistics, "Consumer Price Index for All Urban Consumers"; McCusker, "Colonial Statistics," V-671; Sutch, "National Income and Product"; Prados de la Escosura, "Lost Decades?"; Bolt et al., "Rebasing 'Maddison.'"

139. Bolt and van Zanden, *Maddison Project Database*; Bolt and van Zanden, "Maddison Style Estimates of the Evolution of the World Economy"; US Bureau of Economic Analysis, "Real Gross Domestic Product Per Capita (A939RX0Q048SBEA)"; "Consumer Price Index, 1913," Federal Reserve Bank of Minneapolis; US Bureau of Labor Statistics, "Con-sumer Price Index for All Urban Consumers"; McCusker, "Colonial Statistics," V-671; Sutch, "National Income and Product"; Prados de la Escosura, "Lost Decades?"; Bolt et al., "Rebasing 'Maddison.'"

140. Max Roser, "Working Hours," Our World in Data, 2013, https://ourworldindata.org/work ing-hours; Michael Huberman and Chris Minns, "The Times They Are Not Changin': Days and Hours of Work in Old and New Worlds, 1870–2000," *Explorations in Economic History* 44, no. 4 (July 12, 2007): 548, https://personal.lse.ac.uk/minns/Huberman_Minns _EEH_2007.pdf; University of Groningen and University of California, Davis, "Average Annual Hours Worked by Persons Engaged for United States (AVHWPEUSA065N-RUG)," retrieved from FRED, Federal Reserve Bank of St. Louis, January 21, 2021, https://fred.stlouisfed.org/series/AVHWPEUSA065NRUG.

141. US Census Bureau, "Real Median Personal Income in the United States (MEPAINU-SA672N)," retrieved from FRED, Federal Reserve Bank of St. Louis, updated September 13, 2022, https://fred.stlouisfed.org/series/MEPAINUSA672N; "Consumer Price Index,

1913–," Federal Reserve Bank of Minneapolis; US Bureau of Labor Statistics, "Consumer Price Index for All Urban Consumers."

142. US Bureau of Economic Analysis, "Personal Income Per Capita (A792RC0A052NBEA)"; Lindert and Williamson, "American Incomes 1774–1860"; Klein, "New State-Level Estimates of Personal Income in the United States, 1880–1910," 220; "Consumer Price Index, 1800–," Federal Reserve Bank of Minneapolis; US Bureau of Labor Statistics, "Consumer Price Index for All Urban Consumers."

143. US Bureau of Economic Analysis, "Personal Income Per Capita (A792RC0A052NBEA)"; Huberman and Minns, "The Times They Are Not Changin'," 548; University of Groningen and University of California, Davis, "Average Annual Hours Worked by Persons Engaged for United States; "Consumer Price Index, 1913–," Federal Reserve Bank of Minneapolis.

144. Bolt and van Zanden, *Maddison Project Database*; US Bureau of Economic Analysis, "Population (B230RC0A052NBEA)"; US Bureau of Economic Analysis, "Personal Income (PI)," retrieved from FRED, Federal Reserve Bank of St. Louis, updated February 24, 2023, https://fred.stlouisfed.org/series/PI; US Bureau of Economic Analysis, "Hours Worked by Full-Time and Part-Time Employees (B4701C0A222NBEA)," retrieved from FRED, Federal Reserve Bank of St. Louis, updated October 12, 2022, https://fred.stlouis fed.org/series/B4701C0A222NBEA; Stanley Lebergott, "Labor Force and Employment, 1800–1960," in *Output, Employment, and Productivity in the United States After 1800*, ed. Dorothy S. Brady (Washington, DC: National Bureau of Economic Research, 1966), 118, https://www.nber.org/chapters/c1567.pdf; US Census Bureau, *Statistical Abstract of the United States: 1999* (Washington, DC: US Census Bureau, 1999): 879, https://www.census .gov/prod/99pubs/99statab/sec31.pdf; Stanley Lebergott, "Labor Force, Employment, and Unemployment, 1929–39 Estimating Methods," *Monthly Labor Review* 67, no. 1 (July 1948): 51, https://www.bls.gov/opub/mlr/1948/article/pdf/labor-force-employment-and -unemployment-1929-39-estimating-methods.pdf; Huberman and Minns, "The Times They Are Not Changin'," 548; US Bureau of Labor Statistics, "Employment Level (CE16OV)," retrieved from FRED, Federal Reserve Bank of St. Louis, updated March 10, 2023, https://fred.stlouisfed.org/series/CE16OV; "Consumer Price Index, 1800–," Federal Reserve Bank of Minneapolis; US Bureau of Labor Statistics, "Consumer Price Index for All Urban Consumers."

145. Huberman and Minns, "The Times They Are Not Changin'," 548.

146. For a few quick explainers on the role of the US labor movement in shortening working hours, as well as some detailed contemporaneous sources, see "History of the 40-Hour Workweek," CNBC Make It, YouTube video, May 3, 2017, https://www.youtube.com /watch?v=BcRlq-Hrtc0; "The 40 Hour Work Week," Prosocial Progress Foundation, You-Tube video, December 3, 2017, https://www.youtube.com/watch?v=7KtJNYZySjU; Shana Lebowitz and Marguerite Ward, "Most Americans Support Andrew Yang's Call for a 4-Day Workweek—But Before Any Policy Changes, We Should Understand Why the 5-Day, 40-Hour Workweek Was So Revolutionary," *Business Insider,* June 12, 2020, https://www .businessinsider.com/history-of-the-40-hour-workweek-2015-10; George E. Barnett, "Growth of Labor Organization in the United States, 1897–1914," *Quarterly Journal of Economics* 30, no. 4 (August 1916): 780–95, http://www.jstor.com/stable/1884242; Leo Wolman, *The Growth of American Trade Unions, 1880–1923* (New York: National Bureau of Economic Research, 1924).

147. Huberman and Minns, "The Times They Are Not Changin'," 548.

148. Huberman and Minns, "The Times They Are Not Changin'," 548.

149. Huberman and Minns, "The Times They Are Not Changin'," 548; University of Groningen and University of California, Davis, "Average Annual Hours Worked by Persons Engaged for United States."

150. Throughout this chapter, I focus most heavily on prosperity trends in the US and other OECD nations not out of any belief that they are more important than other countries but because as I write this they have reached steeper points along these exponential trends. In addition, high-quality data is often less available for developing countries, which some-

times limits scholars' ability to measure trends rigorously across the entire globe. See Huberman and Minns, "The Times They Are Not Changin'," 548; University of Groningen and University of California, Davis, "Average Annual Hours Worked by Persons Engaged for United States"; Robert C. Feenstra, Robert Inklaar, and Marcel P. Timmer, *PWT 9.1: Penn World Table Version 9.1*, Groningen Growth and Development Centre, September 26, 2019, https://www.rug.nl/ggdc/productivity/pwt; Robert C. Feenstra, Robert Inklaar, and Marcel P. Timmer, "The Next Generation of the Penn World Table," *American Economic Review* 105, no. 10 (October 2015): 3150–82, http://dx.doi.org/10.1257/aer.20130954; "Kurzarbeit: Germany's Short-Time Work Benefit," International Monetary Fund, June 15, 2020, https://www.imf.org/en/News/Articles/2020/06/11/na061120-kurzarbeit-germanys -short-time-work-benefit.

151. For more on this shift and its probable long-term impacts, see Rani Molla, "Office Work Will Never Be the Same," *Vox*, May 21, 2020, https://www.vox.com/recode/2020/5/21 /21234242/coronavirus-covid-19-remote-work-from-home-office-reopening; Gil Press, "The Future of Work Post-Covid-19," *Forbes*, July 15, 2020, https://www.forbes.com/sites /gilpress/2020/07/15/the-future-of-work-post-covid-19/#3c9ea15e4baf; Nick Routley, "6 Charts That Show What Employers and Employees Really Think About Remote Working," World Economic Forum, June 3, 2020, https://www.weforum.org/agenda/2020/06 /coronavirus-covid19-remote-working-office-employees-employers; Matthew Dey et al., "Ability to Work from Home: Evidence from Two Surveys and Implications for the Labor Market in the COVID-19 Pandemic," *Monthly Labor Review* (US Bureau of Labor Statistics), June 2020, https://doi.org/10.21916/mlr.2020.14.

152. Huberman and Minns, "The Times They Are Not Changin'," 548; University of Groningen and University of California, Davis, "Average Annual Hours Worked by Persons Engaged for United States"; Feenstra, Inklaar, and Timmer, "The Next Generation of the Penn World Table," 3150–82; OECD Statistics, "Average Annual Hours Actually Worked per Worker," Organisation for Economic Co-operation and Development, retrieved October 22, 2021, https://stats.oecd.org/Index.aspx?DataSetCode=ANHRS.

153. International Labour Office, *Marking Progress Against Child Labour: Global Estimates and Trends 2000–2012* (Geneva, Switzerland: International Labour Organization, 2013), 16, http://www.ilo.org/wcmsp5/groups/public/@ed_norm/@ipec/documents/publication /wcms_221513.pdf.

154. International Labour Office, *Marking Progress Against Child Labour*, 3; International Labour Office, *Global Estimates of Child Labour: Results and Trends, 2012–2016* (Geneva, Switzerland: International Labour Organization, 2017), 9, http://www.ilo.org/wcmsp5/groups /public/---dgreports/---dcomm/documents/publication/wcms_575499.pdf; International Labour Office and United Nations Children's Fund, *Child Labour: Global Estimates 2020, Trends and the Road Forward* (Geneva, Switzerland: ILO and UNICEF, 2021), 23, https:// www.ilo.org/wcmsp5/groups/public/---ed_norm/---ipec/documents/publication/wcms _797515.pdf; US Department of Labor, *2021 Findings on the Worst Forms of Child Labor* (Washington, DC: US Department of Labor, 2022), https://www.dol.gov/sites/dolgov/files /ILAB/child_labor_reports/tda2021/2021_TDA_Big_Book.pdf.

155. International Labour Office, *Marking Progress Against Child Labour*, 3; International Labour Office, *Global Estimates of Child Labour: Results and Trends, 2012–2016*, 9; International Labour Office and United Nations Children's Fund, *Child Labour: Global Estimates 2020, Trends and the Road Forward*, 23, 82.

156. United Nations Office on Drugs and Crime, *Global Study on Homicide: Executive Summary* (Vienna: United Nations, July 2019), 26–28, https://www.unodc.org/documents/data-and -analysis/gsh/Booklet1.pdf.

157. In England in 1325, there were 21.4 homicides per 100,000 people per year. By 1575 it was 5.2, by 1675 it was 3.5, and by 1862 it was 1.6, and it has stayed below that ever since. In Italy, it was 71.7 in 1375, 7.0 by 1862, and 0.9 by 2010. See Max Roser and Hannah Ritchie, "Homicides," Our World in Data, December 2019, https://ourworldindata.org/homicides; Manuel Eisner, "From Swords to Words: Does Macro-Level Change in Self-Control

Predict Long-Term Variation in Levels of Homicide?," *Crime and Justice* 43, no. 1 (September 2014): 80–81, https://doi.org/10.1086/677662; UN Office on Drugs and Crime, "Intentional Homicides (per 100,000 People)—France, Netherlands, Sweden, Germany, Switzerland, Italy, United Kingdom, Spain," retrieved from Worldbank.org, accessed March 25, 2023, https://data.worldbank.org/indicator/VC.IHR.PSRC.P5?end=2020&locations= FR-NL-SE-DE-CH-IT-GB-ES&start=2020&view=bar; "Appendix Tables: Homicide in England and Wales," UK Office for National Statistics, February 9, 2023, https://www.ons .gov.uk/file?uri=/peoplepopulationandcommunity/crimeandjustice/datasets/appendixtables homicideinenglandandwales/current/homicideyemarch22appendixtables.xlsx.

158. This chart draws largely from Our World in Data's chart "Long-Term Homicide Rates Across Western Europe, 1250 to 2017." However, the latest version of this chart incorporates World Health Organization data since 1950 that was compiled for public health purposes and significantly underestimates judicially reported homicides. Therefore, the WHO data is not sufficiently commensurable with data for previous years, which largely comes from the archival research of Eisner (2014). Instead, for most of the data since 1990, this chart draws on the UN Office on Drugs and Crime's International Homicide Statistics database. For 2019–2021 UK data, reporting years run April–March, so weighted averages are taken here to approximate the January–December calendar years in the rest of the data. See Roser and Ritchie, "Homicides"; Eisner, "From Swords to Words," 80–81; UN Office on Drugs and Crime, "Intentional Homicides (per 100,000 People)—France, Netherlands, Sweden, Germany, Switzerland, Italy, United Kingdom, Spain"; "Appendix Tables: Homicide in England and Wales," UK Office for National Statistics.

159. Alexia Cooper and Erica L. Smith, *Homicide Trends in the United States, 1980–2008* (Washington, DC: US Department of Justice, Bureau of Justice Statistics, November 2011), 2, https://www.bjs.gov/content/pub/pdf/htus8008.pdf; "Crime in the United States: By Volume and Rate per 100,000 Inhabitants, 1999–2018," Federal Bureau of Investigation, accessed April 28, 2023, https://ucr.fbi.gov/crime-in-the-u.s/2018/crime-in-the-u.s.-2018 /topic-pages/tables/table-1; Federal Bureau of Investigation, "Crime Data Explorer: Expanded Homicide Offense Counts in the United States," US Department of Justice, Federal Bureau of Investigation, accessed April 28, 2023, https://cde.ucr.cjis.gov/LATEST /webapp/#/pages/explorer/crime/shr; Emily J. Hanson, "Violent Crime Trends, 1990–2021," Congressional Research Service report IF12281, December 12, 2022, https://sgp .fas.org/crs/misc/IF12281.pdf; US Bureau of Economic Analysis, "Population (B230R-C0A052NBEA)."

160. For more on the effects of Prohibition on crime and violence in America, see "How Prohibition Created the Mafia," History, YouTube video, February 21, 2019, https://www.you tube.com/watch?v=N-K60XXaPKw; Dave Roos, "How Prohibition Put the 'Organized' in Organized Crime," History.com, February 22, 2019, https://www.history.com/news/prohi bition-organized-crime-al-capone; Bureau of Justice Statistics, "Key Facts at a Glance: Homicide Rate Trends," US Department of Justice, accessed September 29, 2006, https:// web.archive.org/web/20060929061431/http://www.ojp.usdoj.gov/bjs/glance/tables/hmrt tab.htm.

161. For some further explainers on the impact of the war on drugs and its relationship to violence in the United States, see "Why The War on Drugs Is a Huge Failure," Kurzgesagt—In a Nutshell, YouTube video, March 1, 2016, https://www.youtube.com/watch?v= wJUXLqNHCaI; German Lopez, "The War on Drugs, Explained," *Vox*, May 8, 2016, https://www.vox.com/2016/5/8/18089368/war-on-drugs-marijuana-cocaine-heroin-meth; PBS, "Thirty Years of America's Drug War: A Chronology," *Frontline*, accessed April 28, 2023, https://www.pbs.org/wgbh/pages/frontline/shows/drugs/cron; Bureau of Justice Statistics, "Key Facts at a Glance: Homicide Rate Trends."

162. For more on broken windows theory and proactive policing, see George L. Kelling and James Q. Wilson, "Broken Windows: The Police and Neighborhood Safety," *Atlantic*, March 1982, https://www.theatlantic.com/magazine/archive/1982/03/broken-windows /304465/?single_page=true; "Broken Windows Policing," Center for Evidence-Based Crime Policy, accessed April 28, 2023, https://cebcp.org/evidence-based-policing/what

-works-in-policing/research-evidence-review/broken-windows-policing; Shankar Vedan-tam et al., "How a Theory of Crime and Policing Was Born, and Went Terribly Wrong," NPR, November 1, 2016, https://www.npr.org/2016/11/01/500104506/broken-windows -policing-and-the-origins-of-stop-and-frisk-and-how-it-went-wrong; National Academies of Sciences, Engineering, and Medicine, *Proactive Policing: Effects on Crime and Communi-ties* (Washington, DC: National Academies Press, 2018), https://doi.org/10.17226/24928; Kevin Strom, *Research on the Impact of Technology on Policing Strategy in the 21st Century, Final Report* (Research Triangle Park, NC: RTI International, 2017), https://www.ncjrs .gov/pdffiles1/nij/grants/251140.pdf.

163. "Lead for Life—The History of Leaded Gasoline—An Excerpt," from *Late Lessons from Early Warnings*, Jakob Gottschau, YouTube video, September 16, 2013, https://www.you tube.com/watch?v=pqg9jH1xwjI; Jennifer L. Doleac, "New Evidence That Lead Exposure Increases Crime," Brookings Institution, June 1, 2017, https://www.brookings.edu/blog /up-front/2017/06/01/new-evidence-that-lead-exposure-increases-crime.

164. Bureau of Justice Statistics, "Key Facts at a Glance: Homicide Rate Trends"; James Alan Fox and Marianne W. Zawitz, *Homicide Trends in the United States* (Washington, DC: Bu-reau of Justice Statistics, 2010), 9–10, https://www.bjs.gov/content/pub/pdf/htius.pdf; "Crime in the United States: By Volume and Rate per 100,000 Inhabitants, 1999–2018," Federal Bureau of Investigation; Federal Bureau of Investigation, "Crime Data Explorer: Expanded Homicide Offense Counts in the United States"; Hanson, "Violent Crime Trends, 1990–2021"; US Bureau of Economic Analysis, "Population (B230RC0A0 52NBEA)."

165. Ann L. Pastore and Kathleen Maguire, eds., *Sourcebook of Criminal Justice Statistics* (Wash-ington, DC: US Department of Justice, Bureau of Justice Statistics, 2005), 278–79, https:// www.ojp.gov/pdffiles1/Digitization/208756NCJRS.pdf; "Crime in the United States: By Volume and Rate per 100,000 Inhabitants, 1999–2018," Federal Bureau of Investigation; Federal Bureau of Investigation, "Crime Data Explorer: Expanded Homicide Offense Counts in the United States"; Hanson, "Violent Crime Trends, 1990–2021"; US Bureau of Economic Analysis, "Population (B230RC0A052NBEA)."

166. Steven Pinker, *The Better Angels of Our Nature: Why Violence Has Declined* (New York: Pen-guin, 2011), 60–91.

167. Pinker, *Better Angels of Our Nature*, 60.

168. Pinker, *Better Angels of Our Nature*, 49, 53, 63–64.

169. Pinker, *Better Angels of Our Nature*, 52–53.

170. Pinker, *Better Angels of Our Nature*, 193–98.

171. Pinker, *Better Angels of Our Nature*, 175–77, 580–92; Peter Singer, *The Expanding Circle: Ethics, Evolution, and Moral Progress* (Princeton, NJ: Princeton University Press, 1981).

172. For several more detailed summaries of recent applications of AI to materials discovery for solar electricity and energy storage, see Elizabeth Montalbano, "AI Enables Design of Spray-on Coating That Can Generate Solar Energy," *Design News*, December 26, 2019, https://www.designnews.com/materials-assembly/ai-enables-design-spray-coating-can -generate-solar-energy; Shinji Nagasawa, Eman Al-Naamani, and Akinori Saeki, "Com-puter-Aided Screening of Conjugated Polymers for Organic Solar Cell: Classification by Random Forest," *Journal of Physical Chemistry Letters* 9, no. 10 (May 7, 2018): 2639–46, https://doi.org/10.1021/acs.jpclett.8b00635; Geun Ho Gu et al., "Machine Learning for Renewable Energy Materials," *Journal of Materials Chemistry A* 7, no. 29 (April 30, 2019): 17096–117, https://doi.org/10.1039/C9TA02356A; Ziyi Luo et al., "A Survey of Artificial Intelligence Techniques Applied in Energy Storage Materials R&D," *Frontiers in Energy Research* 8, no. 116 (July 3, 2020), https://doi.org/10.3389/fenrg.2020.00116; An Chen, Xu Zhang, and Zhen Zhou, "Machine Learning: Accelerating Materials Development for En-ergy Storage and Conversion," *InfoMat* 2, no. 3 (February 23, 2020): 553–76, https://doi .org/10.1002/inf2.12094; Xinyi Yang et al., "Development Status and Prospects of Artifi-cial Intelligence in the Field of Energy Conversion Materials," *Frontiers in Energy Research* 8, no. 167 (July 31, 2020), https://doi.org/10.3389/fenrg.2020.00167; Teng Zhou, Zhen Song, and Kai Sundmacher, "Big Data Creates New Opportunities for Materials Research: A

Review on Methods and Applications of Machine Learning for Materials Design," *Engineering* 5, no. 6 (December 2019): 1017–26, https://doi.org/10.1016/j.eng.2019.02.011.

173. Hannah Ritchie and Max Roser, "Renewable Energy," Our World in Data, accessed April 28, 2023, https://ourworldindata.org/renewable-energy; International Energy Agency Statistics/OECD, "Electricity Production from Renewable Sources, Excluding Hydroelectric (% of Total)," retrieved from Worldbank.org, 2014, https://data.worldbank.org/indicator /EG.ELC.RNWX.ZS; *BP Statistical Review of World Energy 2021* (London: BP, 2021), 64–65, https://www.bp.com/content/dam/bp/business-sites/en/global/corporate/pdfs/energy -economics/statistical-review/bp-stats-review-2021-full-report.pdf; BP, "Statistical Review of World Energy—All Data, 1965–2021," from *BP Statistical Review of World Energy 2022* (London: BP, 2022), https://www.bp.com/content/dam/bp/business-sites/en/global /corporate/xlsx/energy-economics/statistical-review/bp-stats-review-2022-all-data.xlsx.

174. *BP Statistical Review of World Energy 2022* (London: BP, 2022): 45, 51, https://www.bp.com /content/dam/bp/business-sites/en/global/corporate/pdfs/energy-economics/statistical -review/bp-stats-review-2022-full-report.pdf; *BP Statistical Review of World Energy 2021*, 64–65; *BP Statistical Review of World Energy 2020* (London: BP, 2020), 52–53, 59, 61, https://www.bp.com/content/dam/bp/business-sites/en/global/corporate/pdfs/energy -economics/statistical-review/bp-stats-review-2020-full-report.pdf; BP, "Statistical Review of World Energy—All Data, 1965–2021"; International Energy Agency Statistics/ OECD, "Electric Power Consumption (kWh Per Capita)," retrieved from Worldbank.org, 2014, https://data.worldbank.org/indicator/EG.USE.ELEC.KH.PC; United Nations Department of Economic and Social Affairs, Population Division, "Total Population—Both Sexes," *World Population Prospects 2019*, online ed. rev. 1 (New York: United Nations, 2019), https://population.un.org/wpp/Download/Files/1_Indicators%20(Standard)/EXCEL _FILES/1_Population/WPP2019_POP_F01_1_TOTAL_POPULATION_BOTH _SEXES.xlsx; Ritchie and Roser, "Renewable Energy."

175. "Solar (Photovoltaic) Panel Prices vs. Cumulative Capacity," Our World in Data, accessed March 25, 2023, https://ourworldindata.org/grapher/solar-pv-prices-vs-cumulative-capacity; Gregory F. Nemet, "Interim Monitoring of Cost Dynamics for Publicly Supported Energy Technologies," *Energy Policy* 37, no. 3 (March 2009): 825–35, https://doi.org/10.1016 /j.enpol.2008.10.031; J. Doyne Farmer and François Lafond, "How Predictable Is Technological Progress?," *Research Policy* 45, no. 3 (April 2016): 647–65, https://doi.org/10.1016 /j.respol.2015.11.001; "IRENASTAT Online Data Query Tool," International Renewable Energy Agency, accessed March 25, 2023, https://www.irena.org/Data/Downloads/IRE NASTAT; IRENA, *Renewable Power Generation Costs in 2021* (Abu Dhabi: International Renewable Energy Agency, 2022), https://www.irena.org/-/media/Files/IRENA/Agency /Publication/2022/Jul/IRENA_Power_Generation_Costs_2021.pdf?rev=34c22a 4b244d434da0accde7de7c73d8; "Consumer Price Index, 1913–," Federal Reserve Bank of Minneapolis; US Bureau of Labor Statistics, "Consumer Price Index for All Urban Consumers."

176. Molly Cox, "Key 2020 US Solar PV Cost Trends and a Look Ahead," Greentech Media, December 17, 2020, https://www.greentechmedia.com/articles/read/key-2020-us-solar-pv -cost-trends-and-a-look-ahead.

177. "Solar (Photovoltaic) Panel Prices vs. Cumulative Capacity," Our World in Data; Nemet, "Interim Monitoring of Cost Dynamics for Publicly Supported Energy Technologies," 825–35; Farmer and Lafond, "How Predictable Is Technological Progress?"; "IRENA-STAT Online Data Query Tool"; IRENA, *Renewable Power Generation Costs in 2021*; François Lafond et al., "How Well Do Experience Curves Predict Technological Progress? A Method for Making Distributional Forecasts," *Technological Forecasting & Social Change* 128 (March 2018): 104–17, https://doi.org/10.1016/j.techfore.2017.11.001; Sandra Enkhardt, "Global Solar Capacity Additions Hit 268 GW in 2022, Says BNEF," *PV Magazine*, December 23, 2022, https://www.pv-magazine.com/2022/12/23/global-solar-capacity-additions -hit-268-gw-in-2022-says-bnef.

178. "Solar (Photovoltaic) Panel Prices vs. Cumulative Capacity," Our World in Data; Nemet, "Interim Monitoring of Cost Dynamics for Publicly Supported Energy Technologies,"

825–35; Farmer and Lafond, "How Predictable Is Technological Progress?"; "IRENA-STAT Online Data Query Tool"; IRENA, *Renewable Power Generation Costs in 2021*; Lafond et al., "How Well Do Experience Curves Predict Technological Progress?," 104–17; Enkhardt, "Global Solar Capacity Additions Hit 268 GW in 2022, Says BNEF."

179. Hannah Ritchie and Max Roser, "Renewable Energy—Renewable Energy Generation, World," Our World in Data, accessed April 28, 2023, https://ourworldindata.org/grapher /modern-renewable-energy-consumption?country=~OWID_WRL; Ritchie and Roser, "Renewable Energy—Solar Power Generation"; "World Electricity Generation by Fuel, 1971–2017," International Energy Agency, November 26, 2019, https://www.iea.org/data -and-statistics/charts/world-electricity-generation-by-fuel-1971-2017; *BP Statistical Review of World Energy 2022*, 45, 51; BP, "Statistical Review of World Energy—All Data, 1965– 2021"; "Share of Low-Carbon Sources and Coal in World Electricity Generation, 1971– 2021," International Energy Agency, April 19, 2021, https://www.iea.org/data-and-statistics /charts/share-of-low-carbon-sources-and-coal-in-world-electricity-generation-1971-2021.

180. Ryan Wiser et al., "Land-Based Wind Market Report: 2022 Edition," US Department of Energy, August 2022, 50, https://doi.org/10.2172/1882594; "Consumer Price Index, 1913–," Federal Reserve Bank of Minneapolis; US Bureau of Labor Statistics, "Consumer Price Index for All Urban Consumers."

181. Our World in Data, "Wind Power Generation," Our World in Data, accessed March 25, 2023, https://ourworldindata.org/grapher/wind-generation?tab=chart; *BP Statistical Review of World Energy 2022*; "Yearly Electricity Data," Ember, March 28, 2023, https:// ember-climate.org/data-catalogue/yearly-electricity-data; Charles Moore, "European Electricity Review 2022," Ember, February 1, 2022, https://ember-climate.org/insights /research/european-electricity-review-2022.

182. "Renewable Energy Generation, World," Our World in Data; *BP Statistical Review of World Energy 2022*.

183. "Renewable Energy Generation, World," Our World in Data; *BP Statistical Review of World Energy 2022*, 45, 51; "World Electricity Generation by Fuel, 1971–2017," International Energy Agency.

184. For a closer look at the history of the Magna Carta and its effect on the American founding, see "800 Years of Magna Carta," British Library, YouTube video, March 10, 2015, https:// www.youtube.com/watch?v=RQ7vUkbtlQA; Nicholas Vincent, "Consequences of Magna Carta," British Library, March 13, 2015, https://www.bl.uk/magna-carta/articles/con sequences-of-magna-carta; Dave Roos, "How Did Magna Carta Influence the U.S. Constitution?," History.com, September 30, 2019, https://www.history.com/news/magna-carta -influence-us-constitution-bill-of-rights.

185. For further information on Gutenberg's printing press and its impact on European civilization, see Dave Roos, "7 Ways the Printing Press Changed the World," History.com, September 3, 2019, https://www.history.com/news/printing-press-renaissance; Jeremiah Dittmar, "Information Technology and Economic Change: The Impact of the Printing Press," VoxEU, February 11, 2011, https://voxeu.org/article/information-technology-and -economic-change-impact-printing-press; Patrick McGrady, "The Medieval Invention That Changed the Course of History: The Machine That Made Us," Timeline—World History Documentaries, YouTube video, August 25, 2018, https://www.youtube.com /watch?v=uQ88yC35NjI; Jeremiah Dittmar and Skipper Seabold, "Gutenberg's Moving Type Propelled Europe Towards the Scientific Revolution," *LSE Business Review*, March 19, 2019, https://blogs.lse.ac.uk/businessreview/2019/03/19/gutenbergs-moving-type-pro pelled-europe-towards-the-scientific-revolution.

186. For an excellent summary on the origins and early evolution of Parliament in England, see Gwilym Dodd, "The Birth of Parliament," BBC, February 17, 2011, https://www.bbc .co.uk/history/british/middle_ages/birth_of_parliament_01.shtml.

187. For a range of additional resources on the English Civil War and England's 1689 Bill of Rights, see John Green, "English Civil War: Crash Course European History #14," Crash-Course, YouTube video, August 6, 2019, https://www.youtube.com/watch?v=dyk3bI _Y68Y; Avalon Project at Yale Law School, "English Bill of Rights 1689," Lillian Goldman

Law Library, accessed April 28, 2023, https://avalon.law.yale.edu/17th_century/england
.asp; Geoffrey Lock, "The 1689 Bill of Rights," *Political Studies* 37, no. 4 (December 1,
1989), https://doi.org/10.1111/j.1467-9248.1989.tb00288.x; Peter Ackroyd, *Civil War: The
History of England*, vol. 3 (New York: St. Martin's, 2014).

188. Neil Johnston, "The History of the Parliamentary Franchise" (research paper 13/14, UK
House of Commons Library, March 1, 2013), http://researchbriefings.files.parliament.uk
/documents/RP13-14/RP13-14.pdf; Roser and Ortiz-Ospina, "Literacy"; Pamuk and van
Zanden, "Standards of Living," 229.

189. Dalibor Rohac, "Mechanism Design in the Venetian Republic," Cato Institute, July 17,
2013, https://www.cato.org/publications/commentary/mechanism-design-venetian-repub
lic; Thomas F. Madden, *Venice: A New History* (New York: Penguin, 2012).

190. Anna Grześkowiak-Krwawicz, *Queen Liberty: The Concept of Freedom in the Polish-
Lithuanian Commonwealth* (Leiden, Netherlands: Brill, 2012).

191. Steve Umhoefer, "Mark Pocan Says Less than 25 Percent of Population Could Vote When
Constitution Was Written," Politifact, April 16, 2015, https://www.politifact.com/fact
checks/2015/apr/16/mark-pocan/mark-pocan-says-less-25-percent-population-could-v.

192. For more information on early American suffrage, from simplified to more detailed, see
"Who Voted in Early America?," Constitutional Rights Foundation, accessed April 28,
2023, https://www.crf-usa.org/bill-of-rights-in-action/bria-8-1-b-who-voted-in-early-america
#.UW36ebWsiSo; Donald Ratcliffe, "The Right to Vote and the Rise of Democracy, 1787–
1828," *Journal of the Early Republic* 33, no. 2 (Summer 2013): 219–54, https://www.jstor
.org/stable/24768843.

193. For more detail on the liberal movements that flourished and faltered in Europe during the
mid-nineteenth century, see John Green, "Revolutions of 1848: Crash Course European
History #26," CrashCourse, YouTube video, November 19, 2019, https://www.youtube
.com/watch?v=cXTaP1BD1YY; "Alexander II—History of Russia in 100 Minutes (Part 17
of 36)," Smart History of Russia, YouTube video, July 21, 2017, https://www.youtube.com
/watch?v=cqGBRn7oBEg; "Alexander III—History of Russia in 100 Minutes (Part 18 of
36)," Smart History of Russia, YouTube video, July 21, 2017, https://www.youtube.com
/watch?v=XGCzmjwfSSs; Mike Rapport, *1848: Year of Revolution* (New York: Basic Books,
2009); Paul Bushkovitch, *A Concise History of Russia* (New York: Cambridge University
Press, 2011).

194. "People Living in Democracies and Autocracies, World," Our World in Data, accessed
March 29, 2023, https://ourworldindata.org/grapher/people-living-in-democracies-autoc
racies?country=~OWID_WRL; "The V-Dem Dataset (v13)," V-Dem (Varieties of Democracy), accessed March 25, 2023, https://v-dem.net/data/the-v-dem-dataset; Bastian Herre,
"Scripts and Datasets on Democracy," GitHub, accessed March 25, 2023, https://github
.com/owid/notebooks/tree/main/BastianHerre/democracy; Anna Lührmann et al., "Regimes of the World (RoW): Opening New Avenues for the Comparative Study of Political
Regimes," *Politics and Governance* 6, no. 1 (March 19, 2018): 60–77, https://doi.org/10.17645
/pag.v6i1.1214.

195. "People Living in Democracies and Autocracies, World," Our World in Data; "The V-Dem
Dataset (v13)," V-Dem; Lührmann et al., "Regimes of the World (RoW): Opening New
Avenues."

196. "People Living in Democracies and Autocracies, World," Our World in Data; "The V-Dem
Dataset (v13)," V-Dem; Herre, "Scripts and Datasets on Democracy"; Lührmann et al.,
"Regimes of the World (RoW): Opening New Avenues."

197. "People Living in Democracies and Autocracies, World," Our World in Data; "The V-Dem
Dataset (v13)," V-Dem; Herre, "Scripts and Datasets on Democracy"; Lührmann et al.,
"Regimes of the World (RoW): Opening New Avenues."

198. "People Living in Democracies and Autocracies, World," Our World in Data; "The V-Dem
Dataset (v13)," V-Dem; Herre, "Scripts and Datasets on Democracy"; Lührmann et al.,
"Regimes of the World (RoW): Opening New Avenues."

199. Bradley Honigberg, "The Existential Threat of AI-Enhanced Disinformation Operations,"
Just Security, July 8, 2022, https://www.justsecurity.org/82246/the-existential-threat-of

-ai-enhanced-disinformation-operations; Tiffany Hsu and Stuart A. Thompson, "Disinformation Researchers Raise Alarms About A.I. Chatbots," *New York Times*, February 13, 2023, https://www.nytimes.com/2023/02/08/technology/ai-chatbots-disinformation.html.

200. This chart includes data from two useful datasets. Our World in Data has estimates covering a very long period, from 1800 to 2022. These are based primarily on the level of authoritarianism in a society and include some loose approximations, especially for the period before World War II. The Economist Intelligence Unit assesses whether countries are democracies based on much more granular information about a range of factors in each nation, such as political participation and civil liberties. Yet this data stretches back to only 2006. Because the two datasets use different methodology and criteria, and because there is not a consistent relationship between them for the years they overlap, it would not be suitable to splice them into a single set. Instead, both are presented: the OWD data gives a longer view of overall trends, while the EIU data offers a more precise picture of contemporary democratization. Note that the retreat of democracy since 2015 is largely driven by erosions of democracy in India, which slipped into being considered an "electoral autocracy." See "People Living in Democracies and Autocracies, World," Our World in Data; "The V-Dem Dataset (v13)," V-Dem; Herre, "Scripts and Datasets on Democracy"; Lührmann et al., "Regimes of the World (RoW): Opening New Avenues"; Laza Kekic, "The World in 2007: The Economist Intelligence Unit's Index of Democracy," *Economist*, November 15, 2006, 6, https://www.economist.com/media/pdf/DEMOCRACY_INDEX _2007_v3.pdf; *The Economist Intelligence Unit's Index of Democracy 2008* (London: Economist Intelligence Unit, 2008), 2, https://graphics.eiu.com/PDF/Democracy%20Index%202008 .pdf; *Democracy Index 2010: Democracy in Retreat* (London: Economist Intelligence Unit, 2010), 1, https://graphics.eiu.com/PDF/Democracy_Index_2010_web.pdf; *Democracy Index 2011: Democracy Under Stress* (London: Economist Intelligence Unit, 2011), 2, https:// www.eiu.com/public/topical_report.aspx?campaignid=DemocracyIndex2011; *Democracy Index 2012: Democracy at a Standstill* (London: Economist Intelligence Unit, 2012), 2, https://web.archive.org/web/20170320185156/http://pages.eiu.com/rs/eiu2/images/Democ racy-Index-2012.pdf; *Democracy Index 2013: Democracy in Limbo* (London: Economist Intelligence Unit, 2013), 2, https://www.eiu.com/public/topical_report.aspx?campaignid= Democracy0814; *Democracy Index 2014: Democracy and Its Discontents* (London: Economist Intelligence Unit, 2014), 2, https://www.eiu.com/public/topical_report.aspx?campaignid= Democracy0115; *Democracy Index 2015: Democracy in an Age of Anxiety* (London: Economist Intelligence Unit, 2015), 1, https://web.archive.org/web/20160305143559/http://www .yabiladi.com/img/content/EIU-Democracy-Index-2015.pdf; *Democracy Index 2016: Revenge of the "Deplorables,"* (London: Economist Intelligence Unit, 2016), 2, https://www .eiu.com/public/topical_report.aspx?campaignid=DemocracyIndex2016; *Democracy Index 2017: Free Speech Under Attack* (London: Economist Intelligence Unit, 2017), 2, https:// www.eiu.com/public/topical_report.aspx?campaignid=DemocracyIndex2017; *Democracy Index 2018: Me Too?* (London: Economist Intelligence Unit, 2018), 2, https://www.eiu.com /public/topical_report.aspx?campaignid=democracy2018; *Democracy Index 2019: A Year of Democratic Setbacks and Popular Protest* (London: Economist Intelligence Unit, 2019), 3, https://www.eiu.com/public/topical_report.aspx?campaignid=democracyindex2019; *Democracy Index 2020: In Sickness and in Health?* (London: Economist Intelligence Unit, 2020), 3, https://pages.eiu.com/rs/753-RIQ-438/images/democracy-index-2020.pdf; *Democracy Index 2021: The China Challenge* (London: Economist Intelligence Unit, 2020), 4, https://www .eiu.com/n/campaigns/democracy-index-2021; *Democracy Index 2022: Frontline Democracy and the Battle for Ukraine* (London: Economist Intelligence Unit, 2020), 3, https://www.eiu .com/n/campaigns/democracy-index-2022.

201. See the appendix for the sources used for all the historical cost-of-computation calculations in this book.

202. See the appendix for the sources used for all the historical cost-of-computation calculations in this book.

203. Joshua Ho and Andrei Frumusanu, "Understanding Qualcomm's Snapdragon 810: Performance Review," Anandtech, February 12, 2015, https://www.anandtech.com/show/8933

/snapdragon-810-performance-preview/5. See the appendix for the sources used for all the historical cost-of-computation calculations in this book.

204. See the appendix for the sources used for all the historical cost-of-computation calculations in this book.

205. Paul E. Ceruzzi, *A History of Modern Computing*, 2nd ed. (Cambridge, MA: MIT Press, 1990), 73, https://www.google.com/books/edition/A_History_of_Modern_Computing/x1 YESXanrgQC; "Reference / FAQ / Products and Services," IBM, April 28, 2023, https:// www.ibm.com/ibm/history/reference/faq_0000000011.html; "Consumer Price Index, 1913–," Federal Reserve Bank of Minneapolis.

206. Kyle Wiggers, "Apple Unveils the A16 Bionic, Its Most Powerful Mobile Chip Yet," *Tech-Crunch*, September 7, 2022, https://techcrunch.com/2022/09/07/apple-unveils-new-mobile -chips-including-the-a16-bionic; Nick Guy and Roderick Scott, "Which iPhone Should I Buy?," *New York Times*, October 28, 2022, https://www.nytimes.com/wirecutter/reviews /the-iphone-is-our-favorite-smartphone.

207. Gordon Moore, "Cramming More Components onto Integrated Circuits," *Electronics* 38, no. 8 (April 19, 1965), https://archive.computerhistory.org/resources/access/text/2017/03 /102770822-05-01-acc.pdf; "1965: 'Moore's Law' Predicts the Future of Integrated Cir-cuits," Computer History Museum, accessed April 28, 2023, https://www.computerhis tory.org/siliconengine/moores-law-predicts-the-future-of-integrated-circuits; Fernando J. Corbató et al., *The Compatible Time-Sharing System: A Programmer's Guide* (Cambridge, MA: MIT Press, 1990), http://www.bitsavers.org/pdf/mit/ctss/CTSS_ProgrammersGuide.pdf.

208. Robert W. Keyes, "Physics of Digital Devices," *Reviews of Modern Physics* 61, no. 2 (April 1, 1989): 279–98, https://doi.org/10.1103/RevModPhys.61.279.

209. "Wikipedia: Size Comparisons," *Wikipedia: The Free Encyclopedia*, Wikimedia Foundation, accessed April 28, 2023, https://en.wikipedia.org/wiki/Wikipedia:Size_comparisons# Wikipedia.

210. I describe in more detail the conversion of many physical goods into information technolo-gies in chapter 6.

211. K. Eric Drexler, *Radical Abundance: How a Revolution in Nanotechnology Will Change Civili-zation* (New York: PublicAffairs, 2013), 168–72.

212. For a deeper look at cultured meat and the current impact of meat production, see "The Meat of the Future: How Lab-Grown Meat Is Made," *Eater*, YouTube video, October 2, 2015, https://www.youtube.com/watch?v=u468xY1T8fw; "Inside the Quest to Make Lab Grown Meat," *Wired*, YouTube video, February 16, 2018, https://www.youtube.com /watch?v=QO9SS1NS6MM; Mark Post, "Cultured Beef for Food-Security and the Envi-ronment: Mark Post at TEDxMaastricht," TEDx Talks, YouTube video, May 11, 2014, https://www.youtube.com/watch?v=FITvEUSJ8TM; Julian Huguet, "This Breakthrough in Lab-Grown Meat Could Make It Look Like Real Flesh," *Seeker*, YouTube video, No-vember 14, 2019, https://www.youtube.com/watch?v=1lUuDi_s_Zo; "How Close Are We to Affordable Lab-Grown Meat?," PBS Terra, YouTube video, August 25, 2022, https:// www.youtube.com/watch?v=M-weFARkGi4; "Can Lab-Grown Steak be the Future of Meat? | Big Business | Business Insider," Insider Business, YouTube video, July 17, 2022, https://www.youtube.com/watch?v=UQejwvnog0M; Leah Douglas, "Lab-Grown Meat Moves Closer to American Dinner Plates," Reuters, January 23, 2023, https://www .reuters.com/business/retail-consumer/lab-grown-meat-moves-closer-american-dinner -plates-2023-01-23; "Yearly Number of Animals Slaughtered for Meat, World, 1961 to 2020," Our World in Data, accessed March 25, 2023, https://ourworldindata.org/grapher /animals-slaughtered-for-meat; "Global Meat Production, 1961 to 2020," Our World in Data, accessed March 25, 2023, https://ourworldindata.org/grapher/global-meat -production; FAOSTAT, Food and Agriculture Organization of the United Nations, ac-cessed March 25, 2023, http://www.fao.org/faostat/en/#data; Xiaoming Xu et al., "Global Greenhouse Gas Emissions from Animal-Based Foods Are Twice Those of Plant-Based Foods," *Nature Food* 2 (September 13, 2021): 724–32, https://www.fao.org/3/cb7033en /cb7033en.pdf.

213. "GLEAM v3.0 Dashboard—Emissions—Global Emissions from Livestock in 2015," Food and Agriculture Organization of the United Nations, accessed March 29, 2023, https://foodandagricultureorganization.shinyapps.io/GLEAMV3_Public.

214. For more on the metaverse concept, see John Herrman and Kellen Browning, "Are We in the Metaverse Yet?," *New York Times*, July 10, 2021, https://www.nytimes.com/2021/07/10/style/metaverse-virtual-worlds.html; Rabindra Ratan and Yiming Lei, "The Metaverse: From Science Fiction to Virtual Reality," *Big Think*, August 13, 2021, https://bigthink.com/the-future/metaverse; Casey Newton, "Mark in the Metaverse," *The Verge*, July 22, 2021, https://www.theverge.com/22588022/mark-zuckerberg-facebook-ceo-metaverse-interview.

215. Hannah Ritchie, "Half of the World's Habitable Land Is Used for Agriculture," Our World in Data, November 11, 2019, https://ourworldindata.org/global-land-for-agriculture; Erle C. Ellis et al., "Anthropogenic Transformation of the Biomes, 1700 to 2000, *Global Ecology and Biogeography* 19, no. 5 (September 2010): 589–606, https://doi.org/10.1111/j.1466-8238.2010.00540.x; "Food and Agriculture Data," FAOSTAT, Food and Agriculture Organization of the United Nations, accessed April 28, 2023, http://www.fao.org/faostat/en/#home.

216. Ritchie, "Half of the World's Habitable Land Is Used for Agriculture"; Ellis et al., "Anthropogenic Transformation of the Biomes"; "Food and Agriculture Data," FAOSTAT.

217. Jackson Burke, "As Working from Home Becomes More Widespread, Many Say They Don't Want to Go Back," CNBC, April 24, 2020, https://www.cnbc.com/2020/04/24/as-working-from-home-becomes-more-widespread-many-say-they-dont-want-to-go-back.html.

218. For a closer look at how materials with nanoscale features are being used to increase solar photovoltaic efficiency, see Maren Hunsberger, "Carbon Nanotubes Might Be the Secret Boost Solar Energy Has Been Looking For," *Seeker*, YouTube video, September 16, 2019, https://www.youtube.com/watch?v=EwiDGxkD9_c; Matt Ferrell, "How Carbon Nanotubes Might Boost Solar Energy—Explained," *Undecided with Matt Ferrell*, YouTube video, July 7, 2020, https://www.youtube.com/watch?v=lnZpaunXhGc; David Grossman, "Carbon Nanotubes Could Increase Solar Efficiency to 80 Percent," *Popular Mechanics*, July 25, 2019, https://www.popularmechanics.com/science/green-tech/a28506867/carbon-nanotubes-solar-efficiency; Nasim Tavakoli and Esther Alarcon-Llado, "Combining 1D and 2D Waveguiding in an Ultrathin GaAs NW/Si Tandem Solar Cell," *Optics Express* 27, no. 12 (June 10, 2019): A909–A923, https://doi.org/10.1364/OE.27.00A909.

219. Mark Hutchins, "A Quantum Dot Solar Cell with 16.6% Efficiency," *PV Magazine*, February 19, 2020, https://www.pv-magazine.com/2020/02/19/a-quantum-dot-solar-cell-with-16-6-efficiency; C. Jackson Stolle, Taylor B. Harvey, and Brian A. Korgel, "Nanocrystal Photovoltaics: A Review of Recent Progress," *Current Opinion in Chemical Engineering* 2, no. 2 (May 2013): 160–67, https://doi.org/10.1016/j.coche.2013.03.001.

220. Qiulin Tan et al., "Nano-Fabrication Methods and Novel Applications of Black Silicon," *Sensors and Actuators A: Physical* 295 (August 15, 2019): 560–73, https://doi.org/10.1016/j.sna.2019.04.044.

221. Stephen Y. Chou and Wei Ding, "Ultrathin, High-Efficiency, Broad-Band, Omni-Acceptance, Organic Solar Cells Enhanced by Plasmonic Cavity with Subwavelength Hole Array," *Optics Express* 21, no. S1 (January 14, 2013): A60–A76, https://doi.org/10.1364/OE.21.000A60.

222. David L. Chandler, "Solar Power Heads in a New Direction: Thinner," MIT News, June 26, 2013, http://news.mit.edu/2013/thinner-solar-panels-0626; Marco Bernardi, Maurizia Palummo, and Jeffrey C. Grossman, "Extraordinary Sunlight Absorption and One Nanometer Thick Photovoltaics Using Two-Dimensional Monolayer Materials," *Nano Letters* 13, no. 8 (June 10, 2013): 3664–70, https://doi.org/10.1021/nl401544y.

223. Andy Extance, "The Dawn of Solar Windows," *IEEE Spectrum*, January 24, 2018, https://spectrum.ieee.org/energy/renewables/the-dawn-of-solar-windows; Glenn McDonald, "This Liquid Coating Turns Windows Into Solar Panels," *Seeker*, September 1, 2017,

https://www.seeker.com/earth/energy/clear-liquid-coating-turns-windows-into-solar
-panels.

224. "Renewable Energy Generation, World," Our World in Data; *BP Statistical Review of World Energy 2022*, 45, 51; "World Electricity Generation by Fuel, 1971–2017," International Energy Agency.

225. "Renewable Energy Generation, World," Our World in Data; *BP Statistical Review of World Energy 2022*, 45, 51; "World Electricity Generation by Fuel, 1971–2017," International Energy Agency.

226. "Renewable Energy Generation, World," Our World in Data; *BP Statistical Review of World Energy 2022*, 45, 51; "World Electricity Generation by Fuel, 1971–2017," International Energy Agency.

227. "Lazard's Levelized Cost of Energy Analysis—Version 15.0," Lazard, October 2021, 9, https://www.lazard.com/media/sptlfats/lazards-levelized-cost-of-energy-version-150-vf .pdf; Mark Bolinger et al., "Levelized Cost-Based Learning Analysis of Utilityscale Wind and Solar in the United States," *iScience* 25, no. 6 (May 2022): 4, https://doi.org/10.1016 /j.isci.2022.104378; Jeffrey Logan et al., *Electricity Generation Baseline Report* (technical report NREL/TP-6A20-67645, National Renewable Energy Laboratory, January 2017), 6, https://www.nrel.gov/docs/fy17osti/67645.pdf; "Lazard's Levelized Cost of Energy Analysis—Version 11.0," Lazard, 2017, 2, 10, https://www.lazard.com/media/450337/laz ard-levelized-cost-of-energy-version-110.pdf; Center for Sustainable Systems, "Wind Energy Factsheet" (pub. no. CSS07-09, University of Michigan, August 2019), http://css .umich.edu/sites/default/files/Wind%20Energy_CSS07-09_e2019.pdf; "Renewable Energy Generation, World," Our World in Data; *BP Statistical Review of World Energy 2022*, 45, 51; "World Electricity Generation by Fuel, 1971–2017," International Energy Agency.

228. Alexis De Vos, "Detailed Balance Limit of the Efficiency of Tandem Solar Cells," *Journal of Physics D: Applied Physics* 13, no. 5 (1980): 845, https://doi.org/10.1088/0022-3727/13/5 /018; "Best Research-Cell Efficiency Chart," National Renewable Energy Laboratory, accessed April 28, 2023, https://www.nrel.gov/pv/cell-efficiency.html.

229. Marcelo De Lellis, "The Betz Limit Applied to Airborne Wind Energy," *Renewable Energy* 127 (November 2018): 32–40, https://doi.org/10.1016/j.renene.2018.04.034.

230. Ritchie and Roser, "Renewable Energy—Renewable Energy Generation, World"; Ritchie and Roser, "Renewable Energy—Solar Power Generation"; "World Electricity Generation by Fuel, 1971–2017," International Energy Agency; *BP Statistical Review of World Energy 2022*, 45, 51; BP, "Statistical Review of World Energy—All Data, 1965–2021"; "Share of Low-Carbon Sources and Coal in World Electricity Generation, 1971–2021," International Energy Agency.

231. Science on a Sphere, "Energy on a Sphere," National Oceanic and Atmospheric Administration, accessed May 30, 2021, http://web.archive.org/web/20210530160109/https://sos .noaa.gov/datasets/energy-on-a-sphere.

232. Jeff Tsao, Nate Lewis, and George Crabtree, "Solar FAQs" (working draft, US Department of Energy, April 20, 2006), 9–12, https://web.archive.org/web/20200424084337 /https://www.sandia.gov/~jytsao/Solar%20FAQs.pdf.

233. *BP Statistical Review of World Energy 2022*, 8.

234. Ritchie and Roser, "Renewable Energy—Renewable Energy Generation, World"; Ritchie and Roser, "Renewable Energy—Solar Power Generation"; "World Electricity Generation by Fuel, 1971–2017," International Energy Agency; *BP Statistical Review of World Energy 2022*, 45, 51; BP, "Statistical Review of World Energy—All Data, 1965–2021"; "Share of Low-Carbon Sources and Coal in World Electricity Generation, 1971–2021," International Energy Agency.

235. Will de Freitas, "Could the Sahara Turn Africa into a Solar Superpower?," World Economic Forum, January 17, 2020, https://www.weforum.org/agenda/2020/01/solar-panels -sahara-desert-renewable-energy.

236. For a general overview of emerging energy storage technologies, see "Fact Sheet | Energy Storage (2019)," Environmental and Energy Study Institute, February 22, 2019, https:// www.eesi.org/papers/view/energy-storage-2019.

237. Andy Colthorpe, "Behind the Numbers: The Rapidly Falling LCOE of Battery Storage," *Energy Storage News*, May 6, 2020, https://www.energy-storage.news/behind-the-numbers -the-rapidly-falling-lcoe-of-battery-storage; "Levelized Costs of New Generation Resources in the Annual Energy Outlook 2022," US Energy Information Administration, March 2022, https://www.eia.gov/outlooks/aeo/pdf/electricity_generation.pdf.

238. For more detail on the rapid growth of energy storage capacity for US utilities, see "Battery Storage in the United States: An Update on Market Trends," US Energy Information Administration, August 16, 2021, https://www.eia.gov/analysis/studies/electricity/batterystor age; *Energy Storage Grand Challenge: Energy Storage Market Report*, US Department of Energy technical report DOE/GO-102020-5497 (December 2020), https://www.energy.gov/sites /default/files/2020/12/f81/Energy%20Storage%20Market%20Report%202020_0.pdf.

239. "Lazard's Levelized Cost of Storage Analysis—Version 7.0," Lazard, 2021, 6, https://web .archive.org/web/20220729095608/https://www.lazard.com/media/451882/lazards -levelized-cost-of-storage-version-70-vf.pdf; "Lazard's Levelized Cost of Storage Analysis— Version 6.0," Lazard, 2020, 6, https://web.archive.org/web/20221006123556/https://www .lazard.com/media/451566/lazards-levelized-cost-of-storage-version-60-vf2.pdf; "Lazard's Levelized Cost of Storage Analysis—Version 5.0," Lazard, 2019, 4, https://web.archive .org/web/20221104121921/https://www.lazard.com/media/451087/lazards-levelized -cost-of-storage-version-50-vf.pdf; "Lazard's Levelized Cost of Storage Analysis—Version 4.0," Lazard, 2018, 11, https://www.lazard.com/media/sckbar5m/lazards-levelized-cost-of -storage-version-40-vfinal.pdf; "Lazard's Levelized Cost of Storage Analysis—Version 3.0," Lazard, 2017, 12, https://www.scribd.com/document/413797533/Lazard-Levelized -Cost-of-Storage-Version-30; "Lazard's Levelized Cost of Storage Analysis—Version 2.0," Lazard, 2016, 11, https://web.archive.org/web/20221104121905/https://www.lazard.com /media/438042/lazard-levelized-cost-of-storage-v20.pdf; "Lazard's Levelized Cost of Stor age Analysis—Version 1.0," Lazard, 2015, 9, https://web.archive.org/web/20221105052132 /https://www.lazard.com/media/2391/lazards-levelized-cost-of-storage-analysis-10.pdf; "Consumer Price Index, 1913–," Federal Reserve Bank of Minneapolis; US Bureau of Labor Statistics, "Consumer Price Index for All Urban Consumers."

240. US Energy Information Administration, *Electric Power Annual 2021* (Washington, DC: US Department of Energy, November 2022), 64, https://web.archive.org/web/202302011 94905/http://www.eia.gov/electricity/annual/pdf/epa.pdf; US Energy Information Admin istration, *Electric Power Annual 2020* (Washington, DC: US Department of Energy, Octo ber 2021), 64, https://web.archive.org/web/20220301172156/http://www.eia.gov/electricity /annual/pdf/epa.pdf.

241. "Key Facts from JMP 2015 Report," World Health Organization, 2015, https://web.ar chive.org/web/20211209095710/https://www.who.int/water_sanitation_health/publica tions/JMP-2015-keyfacts-en-rev.pdf.

242. "Key Facts from JMP 2015 Report," World Health Organization, 2015; *Progress on House hold Drinking Water, Sanitation and Hygiene 2000–2020*, 7–8.

243. 2019 is the most recent year for which good data is available at the time of writing, likely because of the disruptions of COVID-19 on health metrics collection. See Institute for Health Metrics and Evaluation, "GBD Results Tool," Global Heath Data Exchange, ac cessed April 28, 2023, http://ghdx.healthdata.org/gbd-results-tool; World Health Organi zation, "Diarrhoeal Disease," World Health Organization, May 2, 2017, https://www.who .int/news-room/fact-sheets/detail/diarrhoeal-disease.

244. For more on these technologies, including an amusing video of Bill Gates and Jimmy Fal lon drinking water extracted from raw sewage by the Janicki Omni Processor, see Bill Gates, "Janicki Omniprocessor," *GatesNotes*, YouTube video, January 5, 2015, https://www .youtube.com/watch?v=bVzppWSIFU0; "Bill Gates and Jimmy Drink Poop Water," *The Tonight Show Starring Jimmy Fallon*, YouTube video, January 22, 2015, https://www.you tube.com/watch?v=FHgsL0dpQ-U; Stephen Beacham, "How the LifeStraw Is Eradicating an Ancient Disease," CNET, April 9, 2020, https://www.cnet.com/news/how-the -lifestraw-is-eradicating-an-ancient-disease; "Lifestraw Challenge: Drinking Pee, Back wash & More!," Vat19, YouTube video, June 2, 2017, https://www.youtube.com/watch?v=_

mkUTSGCF3I; Rebecca Paul, "6 Water-purifying Devices for Clean Drinking Water in the Developing World," *Inhabitat*, November 8, 2013, https://inhabitat.com/6-water -purifying-devices-for-clean-drinking-water-in-the-developing-world.

245. See "Roving Blue O-Pen Silver Advanced Portable Water Purification Device," Roving Blue Inc., YouTube video, November 5, 2018, https://www.youtube.com/watch?v=XeT p1iKQW28; Laurel Wilson, "Church Volunteers Install Water Systems in Other Countries," *Bowling Green Daily News*, December 27, 2014, https://www.bgdailynews.com/news /church-volunteers-install-water-systems-in-other-countries/article_969f45ad-7694-54af -8cb3-155e67ca54ad.html.

246. Aimee M. Gall et al., "Waterborne Viruses: A Barrier to Safe Drinking Water," *PLoS Pathogens* 11, no. 6, article e1004867 (June 25, 2015), https://doi.org/10.1371/journal .ppat.1004867.

247. "Microfiber Matters," Minnesota Pollution Control Agency, February 4, 2019, https:// www.pca.state.mn.us/featured/microfiber-matters.

248. For more on Dean Kamen's Slingshot technology, see "Slingshot Water Purifier," Atlas Initiative Group, YouTube video, February 11, 2012, https://www.youtube.com/watch?v= Uk_T9MiZKRs; Tom Foster, "Pure Genius: How Dean Kamen's Invention Could Bring Clean Water to Millions," *Popular Science*, June 16, 2014, https://www.popsci.com/article /science/pure-genius-how-dean-kamens-invention-could-bring-clean-water-millions.

249. For a long but clear and entertaining video on how Stirling engines work and why they offer some important advantages, see "Stirling Engines—The Power of the Future?," Lindybeige, YouTube video, November 28, 2016, https://www.youtube.com/watch?v=vGlDsFAOWXc.

250. For more detail on scientific evidence about the birth of agriculture, see Rhitu Chatterjee, "Where Did Agriculture Begin? Oh Boy, It's Complicated," NPR, July 15, 2016, https:// www.npr.org/sections/thesalt/2016/07/15/485722228/where-did-agriculture-begin-oh -boy-its-complicated; Ainit Snir et al., "The Origin of Cultivation and Proto-Weeds, Long Before Neolithic Farming," *PLOS One* 10, no. 7 (July 22, 2015), https://doi.org/10.1371 /journal.pone.0131422.

251. David A. Pietz, Dorothy Zeisler-Vralsted, *Water and Human Societies* (Cham, Switzerland: Springer International Publishing, 2021), 55–57.

252. National Agricultural Statistics Service, "Corn, Grain—Yield, Measured in Bu / Acre," Quick Stats, US Department of Agriculture, accessed April 28, 2023, https://quickstats .nass.usda.gov/results/FBDE769A-0982-37DB-BA6D-A312ABDAA2B6; National Agricultural Statistics Service, "Corn and Soybean Production Down in 2022, USDA Reports Corn Stocks Down, Soybean Stocks Down from Year Earlier Winter Wheat Seedings Up for 2023," US Department of Agriculture, January 12, 2023, https://www.nass.usda.gov /Newsroom/2023/01-12-2023.php.

253. Hannah Ritchie and Max Roser, "Crop Yields," Our World in Data, updated September 2019, https://ourworldindata.org/crop-yields; Hannah Ritchie and Max Roser, "Land Use," Our World in Data, September 2019, https://ourworldindata.org/land-use; "Food and Agriculture Data," FAOSTAT.

254. Lebergott, "Labor Force and Employment, 1800–1960," 119; Organisation for Economic Co-operation and Development, "Employment by Economic Activity: Agriculture: All Persons for the United States (LFEAAGTTUSQ647S)," retrieved from FRED, Federal Reserve Bank of St. Louis, updated April 20, 2023, https://fred.stlouisfed.org/series /LFEAAGTTUSQ647S; US Bureau of Labor Statistics, "Civilian Labor Force Level (CL-F16OV)," retrieved from FRED, Federal Reserve Bank of St. Louis, updated April 7, 2023, https://fred.stlouisfed.org/series/CLF16OV.

255. For more on the current vertical agriculture industry and its near-term prospects, see "Why Vertical Farming Is the Future of Food," RealLifeLore2, YouTube video, May 17, 2020, https://www.youtube.com/watch?v=IBleQycVanU; "This Farm of the Future Uses No Soil and 95% Less Water," Seeker Stories, YouTube video, July 5, 2016, https://www .youtube.com/watch?v=-_tvJtUHnmU; Stuart Oda, "Are Indoor Vertical Farms the Future of Agriculture?," TED, YouTube video, February 7, 2020, https://www.youtube.com /watch?v=z9jXW9r1xr8; David Roberts, "This Company Wants to Build a Giant Indoor

Farm Next to Every Major City in the World," *Vox*, April 11, 2018, https://www.vox.com /energy-and-environment/2017/11/8/16611710/vertical-farms; Selina Wang, "This High-Tech Vertical Farm Promises Whole Foods Quality at Walmart Prices," *Bloomberg*, September 6, 2017, https://www.bloomberg.com/news/features/2017-09-06/this-high-tech -vertical-farm-promises-whole-foods-quality-at-walmart-prices.

256. For further short explainers of vertical farming technologies, see Kyree Leary, "Crops Are Harvested Without Human Input, Teasing the Future of Agriculture," *Futurism*, February 26, 2018, https://futurism.com/automated-agriculture-uk; "Growing Up: How Vertical Farming Works," B1M, YouTube video, March 6, 2019, https://www.youtube.com/watch ?v=QT4TWbPLrN8.

257. William Park, "How Far Can Vertical Farming Go?," BBC, January 11, 2023, https://www .bbc.com/future/article/20230106-what-if-all-our-food-was-grown-in-indoor-vertical -farms; Ian Frazier, "The Vertical Farm," *New Yorker*, January 1, 2017, https://www.new yorker.com/magazine/2017/01/09/the-vertical-farm.

258. For more on water-efficient vertical farming operations like Gotham Greens and Aero-Farms, see Brian Heater, "Gotham Greens Just Raised $310M to Expand Its Greenhouses Nationwide," *TechCrunch*, September 12, 2022, https://techcrunch.com/2022/09/12/gotham -greens-just-raised-310m-to-expand-its-greenhouses-nationwide; "Our Farms," Gotham Greens, accessed March 31, 2023, https://www.gothamgreens.com/our-farms; "This Future Farm Uses No Soil and 95% Less Water," FutureWise, YouTube video, June 26, 2018, https://www.youtube.com/watch?v=SHkwXRMLcmE.

259. Laura Reiley, "Indoor Farming Looks Like It Could Be the Answer to Feeding a Hot and Hungry Planet. It's Not That Easy," *Washington Post*, November 19, 2019, https://www .washingtonpost.com/business/2019/11/19/indoor-farming-is-one-decades-hottest-trends -regulations-make-success-elusive.

260. Ritchie, "Half of the World's Habitable Land Is Used for Agriculture"; Ellis et al., "Anthropogenic Transformation of the Biomes"; "Food and Agriculture Data," FAOSTAT.

261. For a closer look at the early history of 3D printing, see Drew Turney, "History of 3D Printing: It's Older Than You Think," *Design and Make with Autodesk*, August 31, 2021, https://www.autodesk.com/redshift/history-of-3d-printing; Leo Gregurić, "History of 3D Printing: When Was 3D Printing Invented?," *All3DP*, December 10, 2018, https://web.ar chive.org/web/20211227053912/https://all3dp.com/2/history-of-3d-printing-when -was-3d-printing-invented/.

262. For more on the process of 3D printing itself, see "How Does 3D Printing Work? | The Deets," *Digital Trends*, YouTube video, September 22, 2019, https://www.youtube.com /watch?v=dGajFRaS834; Rebecca Matulka and Matty Green, "How 3D Printers Work," Department of Energy, June 19, 2014, https://www.energy.gov/articles/how-3d-printers -work.

263. For a wide sampling of views from 3D-printing experts on trends for the industry, and photos showing improving resolution in manufactured feature sizes, see Michael Petch, "80 Additive Manufacturing Experts Predict the 3D Printing Trends to Watch in 2020," 3DPrintingIndustry.com, January 15, 2020, https://3dprintingindustry.com/news/80-additive -manufacturing-experts-predict-the-3d-printing-trends-to-watch-in-2020-167177; Leo Gregurić, "The Smallest 3D Printed Things," *All3DP*, January 30, 2019, https://all3dp .com/2/the-smallest-3d-printed-things.

264. "How It Works," FitMyFoot, accessed June 29, 2022, https://web.archive.org/web /20220629040739/https://fitmyfoot.com/pages/how-it-works.

265. For a sampling of 3D printing being used to customize objects to people's bodies, see "IKEA Partnering with eSports Academy to Scan Bodies and 3D-Print Chairs," NowThis, September 15, 2018, https://nowthisnews.com/videos/future/ikea-and-esports-academy -are-making-3d-printed-chairs; Bianca Britton, "The 3D-Printed Wheelchair: A Revolution in Comfort?," CNN, January 24, 2017, https://money.cnn.com/2017/01/24/technol ogy/3d-printed-wheelchair-benjamin-hubert-layer/index.html; Clare Scott, "Knife Maker Points to 3D Printing as Alternative Method of Craftsmanship," 3DPrint.com, October 16, 2018, https://3dprint.com/227502/knife-maker-uses-3d-printing.

266. For more on applications of 3D printing in medical implants, see "3D Printed Implants Could Help Patients with Bone Cancer," Insider, YouTube video, November 7, 2017, https://www.youtube.com/watch?v=jcp-aaa1PBk; "3D Printed Lattices Improve Orthopaedic Implants," Renishaw, YouTube video, November 28, 2019, https://www.youtube.com/watch?v=2rm_3rUl3QE; AMFG, "Application Spotlight: 3D Printing for Medical Implants," Autonomous Manufacturing Ltd., August 15, 2019, https://amfg.ai/2019/08/15/application-spotlight-3d-printing-for-medical-implants.

267. *The Carbon Footprint of Global Trade: Tackling Emissions from International Freight Transport* (Brussels: International Transport Forum, 2016), 2, https://www.itf-oecd.org/sites/default/files/docs/cop-pdf-06.pdf.

268. Dean Takahashi, "Dyndrite Launches GPU-Powered Improvements for Better 3D Printing Speed and Quality," *VentureBeat*, November 18, 2019, https://venturebeat.com/2019/11/18/dyndrite-launches-gpu-powered-improvements-for-better-3d-printing-speed-and-quality; Agiimaa Kruchkin, "Innovation in Creation: Demand Rises While Prices Drop for 3D Printing Machines," *Manufacturing Tomorrow*, February 16, 2016, https://www.manufacturingtomorrow.com/article/2016/02/innovation-in-creation-demand-rises-while-prices-drop-for-3d-printing-machines/7631.

269. "Profiles of 15 of the World's Major Plant and Animal Fibres," Food and Agriculture Organization of the United Nations, accessed April 28, 2023, http://www.fao.org/natural-fibres-2009/about/15-natural-fibres/en.

270. See "Cytosurge FluidFM µ3Dprinter, World's First 3D Printer at Sub-Micron Direct Metal Printing," Charbax, YouTube video, May 21, 2017, https://www.youtube.com/watch?v=n9oO6EiBt40; Sam Davies, "Nanofabrica Announces Commercial Launch of Micro-Level Resolution Additive Manufacturing Technology," *TCT Magazine*, March 14, 2019, https://www.tctmagazine.com/additive-manufacturing-3d-printing-news/nanofabrica-micro-level-resolution-additive-manufacturing.

271. For more on 3D-printed fabric, see Zachary Hay, "3D Printed Fabric: The Most Promising Projects," *All3DP*, November 7, 2019, https://all3dp.com/2/3d-printed-fabric-most-promising-project; Roni Jacobson, "The Shattering Truth of 3D-Printed Clothing," *Wired*, May 12, 2017, https://www.wired.com/2017/05/the-shattering-truth-of-3d-printed-clothing.

272. Danny Paez, "An Incredible New 3D Printer Is 100X Faster Than What Was Possible: Video," *Inverse*, January 26, 2019, https://www.inverse.com/article/52721-high-speed-3d-printing-mass-production; Mark Zastrow, "3D Printing Gets Bigger, Faster and Stronger," *Nature* 578, no. 7793 (February 5, 2020), https://doi.org/10.1038/d41586-020-00271-6; "Prediction 5: 3D Printing Reaches the 'Plateau of Productivity,'" Deloitte, 2019, https://www.deloitte.co.uk/tmtpredictions/predictions/3d-printing.

273. For more on the process of 3D printing for organ generation, see Amanda Deisler, "This 3D Bioprinted Organ Just Took Its First 'Breath,'" *Seeker*, YouTube video, May 3, 2019, https://www.youtube.com/watch?v=V0rIP_u1JPQ; NIH Research Matters, "3D-Printed Scaffold Engineered to Grow Complex Tissues," National Institutes of Health, April 7, 2020, https://www.nih.gov/news-events/nih-research-matters/3d-printed-scaffold-engineered-grow-complex-tissues; Luis Diaz-Gomez et al., "Fiber Engraving for Bioink Bioprinting Within 3D Printed Tissue Engineering Scaffolds," *Bioprinting* 18, article e00076 (June 2020), https://doi.org/10.1016/j.bprint.2020.e00076Get; Luke Dormehl, "Ceramic Ink Could Let Doctors 3D Print Bones Directly into a Patient's Body," *Digital Trends*, January 30, 2021, https://www.digitaltrends.com/news/ceramic-ink-3d-printed-bones.

274. For more information on the work of United Therapeutics, including an illuminating interview with its CEO, my friend Martine Rothblatt, see CNBC Squawk Box, "Watch CNBC's Full Interview with United Therapeutics CEO Martine Rothblatt," CNBC, June 25, 2019, https://www.cnbc.com/video/2019/06/25/watch-cnbcs-full-interview-with-united-therapeutics-ceo-martine-rothblatt.html; Antonio Regalado, "Inside the Effort to Print Lungs and Breathe Life into Them with Stem Cells," *MIT Technology Review*, June 28, 2018, https://www.technologyreview.com/2018/06/28/240446/inside-the-effort-to-print-lungs-and-breathe-life-into-them-with-stem-cells.

275. For further explanations of why transplanted organs often fail, see "Why Do Organ Transplants Fail So Often," Julia Wilde, YouTube video, July 12, 2015, https://www.youtube.com/watch?v=LQ0K02m6_KM; Amy Shira Teitel, "Your Body Is Designed to Attack a New Organ, Now We Know Why," *Seeker*, YouTube video, July 22, 2017, https://www.youtube.com/watch?v=yfDL9PWubCs; "Lowering Rejection Risk in Organ Transplants," Mayo Clinic, YouTube video, March 18, 2014, https://www.youtube.com/watch?v=bUz3X9ZYd5s.

276. Elizabeth Ferrill and Robert Yoches, "IP Law and 3D Printing: Designers Can Work Around Lack of Cover," *Wired*, September 2013, https://www.wired.com/insights/2013/09/ip-law-and-3d-printing-designers-can-work-around-lack-of-cover; Michael K. Henry, "How 3D Printing Challenges Existing Intellectual Property Law," Henry Patent Law Firm, August 13, 2018, https://henry.law/blog/3d-printing-challenges-patent-law.

277. Jake Hanrahan, "3D-Printed Guns Are Back, and This Time They Are Unstoppable," *Wired*, May 20, 2019, https://www.wired.co.uk/article/3d-printed-guns-blueprints.

278. For more information on how this works, see Innovative Manufacturing and Construction Research Centre, "Future of Construction Process: 3D Concrete Printing," Concrete Printing, YouTube video, May 30, 2010, https://www.youtube.com/watch?v=EfbhdZKPHro; Nathalie Labonnote et al., "Additive Construction: State-of-the-Art, Challenges and Opportunities," *Automation in Construction* 72, no. 3 (December 2016): 347–66, https://doi.org/10.1016/j.autcon.2016.08.026.

279. For a helpful overview of recent advances in this area, see Sriram Renganathan, "3D Printed House/Construction Materials: What Are They?," *All3DP*, April 23, 2019, https://all3dp.com/2/3d-printing-in-construction-what-are-3d-printed-houses-made-of.

280. Melissa Goldin, "Chinese Company Builds Houses Quickly with 3D Printing," *Mashable*, April 28, 2014, https://mashable.com/2014/04/28/3d-printing-houses-china.

281. For impressive time-lapse video of constructing the frame of an entire building via 3D printing, see "The Biggest 3D Printed Building," Apis Cor, YouTube video, October 24, 2019, https://www.youtube.com/watch?v=69HrqNnrfh4.

282. Rick Stella, "It's Hideous, But This 3D-Printed Villa in China Can Withstand a Major Quake," *Digital Trends*, July 11, 2016, https://www.digitaltrends.com/home/3d-printed-chinese-villas-huashang-tenda.

283. Emma Bowman, "3D-Printed Homes Level Up with a 2-Story House in Houston," NPR, January 16, 2023, https://www.npr.org/2023/01/16/1148943607/3d-printed-homes-level-up-with-a-2-story-house-in-houston; "Habitat for Humanity Home Completed," Alquist, accessed December 6, 2022, https://web.archive.org/web/20221206002108/https://www.alquist3d.com/habitat; "How Concrete Homes Are Built with a 3D Printer," Insider Art, YouTube video, June 28, 2022, https://www.youtube.com/watch?v=vL2KoMNzGTo.

284. Wetterstrand, "The Cost of Sequencing a Human Genome"; Kris A. Wetterstrand, "DNA Sequencing Costs: Data," National Human Genome Research Institute, November 19, 2021, https://www.genome.gov/about-genomics/fact-sheets/DNA-Sequencing-Costs-Data; National Research Council Committee on Mapping and Sequencing the Human Genome, *Mapping and Sequencing the Human Genome*, chap. 5 (Washington, DC: National Academies Press, 1988), https://www.ncbi.nlm.nih.gov/books/NBK218256; E. Y. Chan (Applied Biosystems), email to author, October 7, 2008.

285. For a deeper dive on cancer immunotherapy, see "How Does Cancer Immunotherapy Work?," MD Anderson Cancer Center, YouTube video, April 20, 2017, https://www.youtube.com/watch?v=CwaMZCu4kpI; "Tumour Immunology and Immunotherapy," Nature Video, YouTube video, September 17, 2015, https://www.youtube.com/watch?v=K09xzIQ8zsg; Alex D. Waldman, Jill M. Fritz, and Michael J. Lenardo, "A Guide to Cancer Immunotherapy: From T Cell Basic Science to Clinical Practice," *Nature Reviews Immunology* 20 (2020): 651–68, https://doi.org/10.1038/s41577-020-0306-5.

286. For further explanations of CAR-T cell therapy and its exciting promise for cancer treatment, see "CAR T-Cell Therapy: How Does It Work?," Dana-Farber Cancer Institute, YouTube video, August 31, 2017, https://www.youtube.com/watch?v=OadAW99s4Ik; Carl

June, "A 'Living Drug' That Could Change the Way We Treat Cancer," TED, YouTube video, October 2, 2019, https://www.youtube.com/watch?v=7qWvVcBZzRg.

287. For a representative sampling of recent research on iPS therapies, see Krista Conger, "Old Human Cells Rejuvenated with Stem Cell Technology," Stanford Medicine, March 24, 2020, https://med.stanford.edu/news/all-news/2020/03/old-human-cells-rejuvenated-with -stem-cell-technology.html; Qiliang Zhou et al., "Trachea Engineering Using a Centrifugation Method and Mouse-Induced Pluripotent Stem Cells," Tissue Engineering Part C: Methods 24, no. 9 (September 14, 2018): 524–33, https://doi.org/10.1089/ten.TEC.2018 .0115; Kazuko Kikuchi et al., "Craniofacial Bone Regeneration Using iPS Cell-Derived Neural Crest Like Cells," Journal of Hard Tissue Biology 27, no. 1 (January 1, 2018): 1–10, https://doi.org/10.2485/jhtb.27.1; "The World's First Allogeneic iPS-Derived Retina Cell Transplant," Japan Agency for Medical Research and Development, September 20, 2018, https://www.amed.go.jp/en/seika/fy2018-05.html; Hiroo Kimura et al., "Stem Cells Purified from Human Induced Pluripotent Stem Cell-Derived Neural Crest-Like Cells Promote Peripheral Nerve Regeneration," Scientific Reports 8, no. 1, article 10071 (July 3, 2018), https://doi.org/10.1038/s41598-018-27952-7; Suman Kanji and Hiranmoy Das, "Advances of Stem Cell Therapeutics in Cutaneous Wound Healing and Regeneration," Mediators of Inflammation, article 5217967 (October 29, 2017), https://doi.org/10.1155/2017/5217967; David Cyranoski, "'Reprogrammed' Stem Cells Approved to Mend Human Hearts for the First Time," Nature 557, no. 7707 (May 29, 2018): 619–20, https://doi.org/10.1038/d41586 -018-05278-8; Yue Yu et al., "Application of Induced Pluripotent Stem Cells in Liver Diseases," Cell Medicine 7, no. 1 (April 22, 2014): 1–13, https://doi.org/10.3727/215517914X68 0056; Susumu Tajiri et al., "Regenerative Potential of Induced Pluripotent Stem Cells Derived from Patients Undergoing Haemodialysis in Kidney Regeneration," Scientific Reports 8, no. 1, article 14919 (October 8, 2018), https://doi.org/10.1038/s41598-018-33256-7; Sharon Begley, "Cancer-Causing DNA Is Found in Some Sem Cells Being Used in Patients," STAT News, April 26, 2017, https://www.statnews.com/2017/04/26/stem-cells-cancer -mutations.

288. For more detail on other factors that influenced the evolution of the human immune system, see Jorge Domínguez-Andrés and Mihai G. Netea, "Impact of Historic Migrations and Evolutionary Processes on Human Immunity," Trends in Immunology 40, no. 12 (November 27, 2019): P1105–P1119, https://doi.org/10.1016/j.it.2019.10.001.

289. For a clear explainer of how type 1 diabetes works in the body, see "Type 1 Diabetes | Nucleus Health," Nucleus Medical Media, YouTube video, January 10, 2012, https://www .youtube.com/watch?v=jxbbBmbvu7I.

290. For a nontechnical explainer on this research and links to key studies, see Todd B. Kashdan, "Why Do People Kill Themselves? New Warning Signs," Psychology Today, May 15, 2014, https://www.psychologytoday.com/us/blog/curious/201405/why-do-people-kill-them selves-new-warning-signs.

CHAPTER 5: THE FUTURE OF JOBS: GOOD OR BAD?

1. For further information on this rapid evolution, see Alex Davies, "An Oral History of the Darpa Grand Challenge, the Grueling Robot Race That Launched the Self-Driving Car," Wired, August 3, 2017, https://www.wired.com/story/darpa-grand-challenge-2004-oral -history; Joshua Davies, "Say Hello to Stanley," Wired, January 1, 2006, https://www .wired.com/2006/01/stanley; Ronan Glon and Stephen Edelstein, "The History of Self-Driving Cars," Digitaltrends, July 31, 2020, https://www.digitaltrends.com/cars/history -of-self-driving-cars-milestones.

2. Kristen Korosec, "Waymo's Driverless Taxi Service Can Now Be Accessed on Google Maps," TechCrunch, June 3, 2021, https://techcrunch.com/2021/06/03/waymos-driverless -taxi-service-can-now-be-accessed-on-google-maps; Rebecca Bellan, "Waymo Launches Robotaxi Service in San Francisco," TechCrunch, August 24, 2021, https://techcrunch .com/2021/08/24/waymo-launches-robotaxi-service-in-san-francisco; Jonathan M. Gitlin,

"Self-Driving Waymo Trucks to Haul Loads Between Houston and Fort Worth," *Ars Technica*, June 10, 2021, https://arstechnica.com/cars/2021/06/self-driving-waymo-trucks-to-haul-loads-between-houston-and-fort-worth; Dug Begley, "More Computer-Controlled Trucks Coming to Test-Drive I-45 Between Dallas and Houston," *Houston Chronicle*, August 25, 2022, https://www.houstonchronicle.com/news/houston-texas/transportation/article/More-computer-controlled-trucks-coming-to-17398269.php.

3. "Next Stop for Waymo One: Los Angeles," Waymo, October 19, 2022, https://blog.waymo.com/2022/10/next-stop-for-waymo-one-los-angeles.html.

4. Aaron Pressman, "Google's Waymo Reaches 20 Million Miles of Autonomous Driving," *Fortune*, January 7, 2020, https://fortune.com/2020/01/07/googles-waymo-reaches-20-million-miles-of-autonomous-driving.

5. Will Knight, "Waymo's Cars Drive 10 Million Miles a Day in a Perilous Virtual World," *MIT Technology Review*, October 10, 2018, https://www.technologyreview.com/s/612251/waymos-cars-drive-10-million-miles-a-day-in-a-perilous-virtual-world; Alexis C. Madrigal, "Inside Waymo's Secret World for Training Self-Driving Cars," *Atlantic*, August 23, 2017, https://www.theatlantic.com/technology/archive/2017/08/inside-waymos-secret-testing-and-simulation-facilities/537648; John Krafcik, "Waymo Livestream Unveil: The Next Step in Self-Driving," Waymo, YouTube video, March 27, 2018, https://www.youtube.com/watch?v=-EBcpIvPWnY; Mario Herger, "2020 Disengagement Reports from California," The Last Driver License Holder, February 9, 2021, https://thelastdriverlicenseholder.com/2021/02/09/2020-disengagement-reports-from-california; "Off Road, but Not Offline: How Simulation Helps Advance Our Waymo Driver," Waymo, April 28, 2020, https://blog.waymo.com/2020/04/off-road-but-not-offline—simulation27.html; Kris Holt, "Waymo's Autonomous Vehicles Have Clocked 20 Million Miles on Public Roads," *Engadget*, August 19, 2021, https://www.engadget.com/waymo-autonomous-vehicles-update-san-francisco-193934150.html.

6. Technically, "deep" neural networks may have as few as three layers, but advances in computing power over the past decade have made much deeper networks practical. A key element of AlphaGo was a thirteen-layer neural network, which it used in 2015–16 to surpass the best human Go players. In order for this network to be useful, it needed massive amounts of data, so researchers trained it by simulating up to 1,000 games per second per computer processing unit. In 2017, AlphaGo Zero simulated about 29 million games with a 79-layer network, and beat the original AlphaGo 100 games to 0. Some AI projects now use neural networks with more than 100 layers. More layers doesn't necessarily translate to more intelligence, but it's useful to think of layers as adding subtlety and abstraction. If a subject is complex enough and you have enough data, a network with more layers can often discover hidden patterns that a shallower network would miss. This is a very important idea in AI more broadly, and a central reason why so many fields like health care and materials science are starting to turn into information technologies. Namely, as computation becomes cheaper, it becomes practical to use deeper neural networks. As data collection and storage becomes cheaper, it becomes practical to feed those deep networks enough data to harness their potential in more areas. And as deep learning is applied more broadly, those fields benefit from exponentially increasing intelligence. See "AlphaGo," Google DeepMind, accessed January 30, 2023, https://deepmind.com/research/case-studies/alphago-the-story-so-far; "AlphaGo Zero: Starting from Scratch," Google DeepMind, October 18, 2017, https://deepmind.com/blog/article/alphago-zero-starting-scratch; Tom Simonite, "This More Powerful Version of AlphaGo Learns On Its Own," *Wired*, October 18, 2017, https://www.wired.com/story/this-more-powerful-version-of-alphago-learns-on-its-own; David Silver et al., "Mastering the Game of Go with Deep Neural Networks and Tree Search," *Nature* 529, no. 7587 (January 27, 2016): 484–89, https://doi.org/10.1038/nature16961; Christof Koch, "How the Computer Beat the Go Master," *Scientific American*, March 19, 2016, https://www.scientificamerican.com/article/how-the-computer-beat-the-go-master; Josh Patterson and Adam Gibson, *Deep Learning: A Practitioner's Approach* (Sebastopol, CA: O'Reilly, 2017), 6–8, https://books.google.com/books?id=qrcuDwAAQBAJ; Thomas Anthony, Zheng Tian, and David Barber, "Thinking Fast and Slow with Deep

Learning and Tree Search," 31st Conference on Neural Information Processing Systems (NIPS 2017), revised December 3, 2017, https://arxiv.org/pdf/1705.08439.pdf; Kaiming He et al., "Deep Residual Learning for Image Recognition," 2016 IEEE Conference on Computer Vision and Pattern Recognition, December 10, 2015, https://arxiv.org/pdf/1512.03385.pdf.

7. Holt, "Waymo's Autonomous Vehicles Have Clocked 20 Million Miles on Public Roads."

8. In 2021, the employed labor force in the US was around 155 million people. Of these, an estimated 3.49 million were truck drivers of all types, and another 832,600 were passenger vehicle drivers—around 2.7%. See US Bureau of Labor Statistics, "Employment Status of the Civilian Population by Sex and Age," US Department of Labor, accessed April 7, 2023, https://www.bls.gov/news.release/empsit.t01.htm; Jennifer Cheeseman Day and Andrew W. Hait, "America Keeps On Truckin': Number of Truckers at All-Time High," US Census Bureau, June 6, 2019, https://www.census.gov/library/stories/2019/06/america-keeps -on-trucking.html; US Bureau of Labor Statistics, US Department of Labor, "30 Percent of Civilian Jobs Require Some Driving in 2016," *The Economics Daily*, June 27, 2017, https://www.bls.gov/opub/ted/2017/30-percent-of-civilian-jobs-require-some-driving -in-2016.htm.

9. "Economics and Industry Data," American Trucking Associations, accessed April 20, 2023, https://www.trucking.org/economics-and-industry-data; US Bureau of Labor Statistics, "Occupational Outlook Handbook, Passenger Vehicle Drivers—Summary," US Department of Labor, accessed January 30, 2023, https://www.bls.gov/ooh/transportation -and-material-moving/passenger-vehicle-drivers.htm.

10. Mark Fahey, "Driverless Cars Will Kill the Most Jobs in Select US States," CNBC, September 2, 2016, https://www.cnbc.com/2016/09/02/driverless-cars-will-kill-the-most-jobs -in-select-us-states.html.

11. Fahey, "Driverless Cars Will Kill the Most Jobs in Select US States."

12. Cheeseman Day and Hait, "America Keeps on Truckin'."

13. Bureau of Transportation Statistics, *Transportation Economic Trends*, US Department of Transportation, accessed April 20, 2023, https://data.bts.gov/stories/s/caxh-t8jd.

14. Carl Benedikt Frey and Michael A. Osborne. "The Future of Employment: How Susceptible Are Jobs to Computerisation?" (Oxford Martin School, September 17, 2013), 2, 36–38, https://www.oxfordmartin.ox.ac.uk/downloads/academic/The_Future_of_Employment .pdf.

15. Frey and Osborne, "Future of Employment: How Susceptible Are Jobs to Computerisation?," 57–72.

16. Frey and Osborne, "Future of Employment: How Susceptible Are Jobs to Computerisation?," 57–72.

17. Frey and Osborne, "Future of Employment: How Susceptible Are Jobs to Computerisation?," 57–72.

18. Frey and Osborne, "Future of Employment: How Susceptible Are Jobs to Computerisation?," 57–72.

19. Ljubica Nedelkoska and Glenda Quintini, "Automation, Skills Use and Training," OECD Social, Employment and Migration Working Papers no. 202 (March 8, 2018), 7–8, https:// doi.org /10.1787/2e2f4eea-en.

20. Nedelkoska and Quintini, "Automation, Skills Use and Training," 7–8.

21. "A New Study Finds Nearly Half of Jobs Are Vulnerable to Automation," *Economist*, April 24, 2018, https://www.economist.com/graphic-detail/2018/04/24/a-study-finds-nearly-half -of-jobs-are-vulnerable-to-automation; Frey and Osborne, "Future of Employment: How Susceptible Are Jobs to Computerisation?"

22. Alexandre Georgieff and Anna Milanez, "What Happened to Jobs at High Risk of Automation?," OECD Social, Employment and Migration Working Papers no. 255 (OECD Publishing, May 21, 2021), https://doi.org/10.1787/10bc97f4-en.

23. McKinsey & Company, "The Economic Potential of Generative AI: The Next Productivity Frontier," McKinsey & Company, June 2023: 37–41, https://www.mckinsey.com/capa

bilities/mckinsey-digital/our-insights/the-economic-potential-of-generative-ai-the-next
-productivity-frontier#introduction.

24. Richard Conniff, "What the Luddites Really Fought Against," *Smithsonian*, March 2011, https://www.smithsonianmag.com/history/what-the-luddites-really-fought-against
-264412.

25. "Hidden Histories: Luddites—A Short History of One of the First Labor Movements," WRIR.org, May 23, 2010, https://www.wrir.org/2010/05/23/hidden-histories-luddites
-a-short-history-of-one-of-the-first-labor-movemen; Conniff, "What the Luddites Really Fought Against"; Bill Kovarik, *Revolutions in Communication: Media History from Gutenberg to the Digital Age* (New York: Bloomsbury, 2015), 8, https://books.google.com/books?id= F6ugBQAAQBAJ; Jessica Brain, "The Luddites," Historic UK, accessed April 20, 2023, https://www.historic-uk.com/HistoryUK/HistoryofBritain/The-Luddites.

26. Conniff, "What the Luddites Really Fought Against"; Brain, "The Luddites."

27. Brain, "The Luddites"; Kevin Binfield, ed., *Writings of the Luddites* (Baltimore: Johns Hopkins University Press, 2004); Frank Peel, *The Risings of the Luddites* (Heckmondwike, UK: T. W. Senior, 1880).

28. Stanley Lebergott, "Labor Force and Employment, 1800–1960," in *Output, Employment, and Productivity in the United States After 1800*, ed. Dorothy S. Brady (Washington, DC: National Bureau of Economic Research, 1966), 119, https://www.nber.org/chapters/c1567
.pdf; US Bureau of Labor Statistics, "All Employees, Manufacturing (MANEMP)," retrieved from FRED, Federal Reserve Bank of St. Louis, updated April 7, 2023, https:// fred.stlouisfed.org/series/MANEMP; Organisation for Economic Co-operation and Development, "Employment by Economic Activity: Agriculture: All Persons for the United States (LFEAAGTTUSQ647S)," retrieved from FRED, Federal Reserve Bank of St. Louis, updated April 20, 2023, https://fred.stlouisfed.org/series/LFEAAGTTUSQ647S; US Bureau of Labor Statistics, "Civilian Labor Force Level (CLF16OV)," retrieved from FRED, Federal Reserve Bank of St. Louis, updated April 7, 2023, https://fred.stlouisfed
.org/series/CLF16OV.

29. Lebergott, "Labor Force and Employment, 1800–1960," 118; US Census Bureau, "Historical National Population Estimates July 1, 1900 to July 1, 1999," Population Estimates Program, Population Division, US Census Bureau, revised June 28, 2000, https://www2
.census.gov/programs-surveys/popest/tables/1900-1980/national/totals/popclockest.txt.

30. US Bureau of Labor Statistics, "Civilian Labor Force Level (CLF16OV)"; US Bureau of Economic Analysis, "Population (B230RC0A052NBEA)," retrieved from FRED, Federal Reserve Bank of St. Louis, updated January 26, 2023, https://fred.stlouisfed.org/series
/B230RC0A052NBEA.

31. Michael Huberman and Chris Minns, "The Times They Are Not Changin': Days and Hours of Work in Old and New Worlds, 1870–2000," *Explorations in Economic History* 44, no. 4 (July 12, 2007): 548, https://personal.lse.ac.uk/minns/Huberman_Minns_EEH
_2007.pdf; University of Groningen and University of California, Davis, "Average Annual Hours Worked by Persons Engaged for United States (AVHWPEUSA065NRUG)," retrieved from FRED, Federal Reserve Bank of St. Louis, updated January 21, 2021, https:// fred.stlouisfed.org/series/AVHWPEUSA065NRUG; Robert C. Feenstra, Robert Inklaar, and Marcel P. Timmer, "The Next Generation of the Penn World Table," *American Economic Review* 105, no. 10 (2015): 3150–82, https://www.rug.nl/ggdc/docs/the_next_genera tion_of_the_penn_world_table.pdf.

32. US Bureau of Labor Statistics, "Personal Income Per Capita (A792RC0A052NBEA)," retrieved from FRED, Federal Reserve Bank of St. Louis, updated March 30, 2023, https:// fred.stlouisfed.org/series/A792RC0A052NBEA; "CPI Inflation Calculator," US Bureau of Labor Statistics, accessed April 20, 2023, https://data.bls.gov/cgi-bin/cpicalc.pl; US Bureau of Labor Statistics, "Civilian Labor Force Level (CLF16OV)"; US Bureau of Labor Statistics, "Real Median Personal Income in the United States (MEPAINUSA672N)," retrieved from FRED, Federal Reserve Bank of St. Louis, updated September 13, 2022, https://fred.stlouisfed.org/series/MEPAINUSA672N.

33. Lebergott, "Labor Force and Employment, 1800–1960," 118; US Bureau of Economic Analysis, "Population (B230RC0A052NBEA)."

34. US Bureau of Labor Statistics, "Personal Income Per Capita (A792RC0A052NBEA)"; US Bureau of Labor Statistics, "Civilian Labor Force Level (CLF16OV)"; US Bureau of Economic Analysis, "Population (B230RC0A052NBEA)."

35. US Bureau of Labor Statistics, "Real Median Personal Income in the United States (MEPAINUSA672N)."

36. Huberman and Minns, "The Times They Are Not Changin'," 548.

37. US Bureau of Labor Statistics, "Total Wages and Salaries, BLS (BA06RC1A027NBEA)," retrieved from FRED, Federal Reserve Bank of St. Louis, updated October 12, 2022, https://fred.stlouisfed.org/series/BA06RC1A027NBEA; US Bureau of Labor Statistics, "Hours Worked by Full-Time and Part-Time Employees (B4701C0A222NBEA)," retrieved from FRED, Federal Reserve Bank of St. Louis, updated October 12, 2022, https://fred.stlouisfed.org/series/B4701C0A222NBEA.

38. US Bureau of Labor Statistics, "Personal Income (PI)," retrieved from FRED, Federal Reserve Bank of St. Louis, updated March 31, 2023, https://fred.stlouisfed.org/series/PI.

39. See, for example, Microeconomix, *The App Economy in the United States* (London: Deloitte, August 17, 2018), 4, https://actonline.org/wp-content/uploads/Deloitte-The-App-Economy -in-US.pdf.

40. Lebergott, "Labor Force and Employment, 1800–1960," 119.

41. F. M. L. Thompson, "The Second Agricultural Revolution, 1815–1880," *Economic History Review* 21, no. 1 (April 1968): 62–77, https://www.jstor.org/stable/2592204; Norman E. Borlaug, "Contributions of Conventional Plant Breeding to Food Production," *Science* 219, no. 4585 (February 11, 1983): 689–93, https://science.sciencemag.org/content/sci/219 /4585/689.full.pdf.

42. To go deeper on how industrialization transformed agriculture, see "Causes of the Industrial Revolution: The Agricultural Revolution," ClickView, YouTube video, August 10, 2015, https://www.youtube.com/watch?v=6QKIts2_yJ0; John Green, "Coal, Steam, and the Industrial Revolution: Crash Course World History #32," CrashCourse, YouTube video, August 30, 2012, https://www.youtube.com/watch?v=zhL5DCizj5c; John Green, "The Industrial Economy: Crash Course US History #23," CrashCourse, YouTube video, July 25, 2013, https://www.youtube.com/watch?v=r6tRp-zRUJs; "Mechanization on the Farm in the Early 20th Century," Iowa PBS, YouTube video, April 28, 2015, https://www .youtube.com/watch?v=SI9K8ZJqAwE; "Cyrus McCormick," PBS, accessed April 20, 2023, https://www.pbs.org/wgbh/theymadeamerica/whomade/mccormick_hi.html; Mark Overton, *Agricultural Revolution in England: The Transformation of the Agrarian Economy 1500–1850* (Cambridge, UK: Cambridge University Press, 1996).

43. Lebergott, "Labor Force and Employment, 1800–1960," 119.

44. Hannah Ritchie and Max Roser, "Crop Yields," Our World in Data, updated 2022, https:// ourworldindata.org/crop-yields; Sarah E. Cusick and Michael K. Georgieff, "The Role of Nutrition in Brain Development: The Golden Opportunity of the 'First 1000 Days,'" *Journal of Pediatrics* 175 (August 2016): 16–21, https://www.ncbi.nlm.nih.gov/pmc/articles /PMC4981537; Gary M. Walton and Hugh Rockoff, *History of the American Economy*, 11th ed. (Boston: Cengage Learning, 2009), 1–15.

45. "Provisional Cereal and Oilseed Production Estimates for England 2022," Department for Environment Food & Rural Affairs (UK), December 21, 2022, https://www.gov.uk/govern ment/statistics/cereal-and-oilseed-rape-production/provisional-cereal-and-oilseed -production-estimates-for-england-2022.

46. "UK Population Estimates 1851 to 2014," Office for National Statistics (UK), July 6, 2015, https://www.ons.gov.uk/peoplepopulationandcommunity/populationandmigration /populationestimates/adhocs/004356ukpopulationestimates1851to2014; Central Intelligence Agency, "Explore All Countries—United Kingdom," CIA World Factbook, November 29, 2022, https://web.archive.org/web/20221207065501/https://www.cia.gov/the-world-factbook /countries/united-kingdom.

47. Organisation for Economic Co-operation and Development, "Employment by Economic Activity: Agriculture"; US Bureau of Labor Statistics, "Civilian Labor Force Level (CLF16OV)"; International Labour Organization, "Employment in Agriculture (% of Total Employment) (Modeled ILO Estimate)," Worldbank.org, January 29, 2021, https://data .worldbank.org/indicator/SL.AGR.EMPL.ZS?locations=US.

48. For some interesting explainers on automated farming technologies, see Kyree Leary, "Crops Are Harvested Without Human Input, Teasing the Future of Agriculture," *Futurism*, February 26, 2018, https://futurism.com/automated-agriculture-uk; "Growing Up: How Vertical Farming Works," B1M, YouTube video, March 6, 2019, https://www.you tube.com/watch?v=QT4TWbPLrN8; "The Future of Farming," The Daily Conversation, YouTube video, May 17, 2017, https://www.youtube.com/watch?v=Qmla9NLFBvU; Michael Larkin, "Labor Terminators: Farming Robots Are About To Take Over Our Farms," *Investor's Business Daily*, August 10, 2018, https://www.investors.com/news/farming-robot -agriculture-technology.

49. Lebergott, "Labor Force and Employment, 1800–1960," 119; US Census Bureau, *Statistical Abstract of the United States: 1999* (Washington, DC: US Census Bureau, 1999): 879, https:// www.census.gov/prod/99pubs/99statab/sec31.pdf; US Bureau of Labor Statistics, "Percent of Employment in Agriculture in the United States (USAPEMANA)," retrieved from FRED, Federal Reserve Bank of St. Louis, updated June 10, 2013, https://fred.stlouisfed.org /series/USAPEMANA; International Labour Organization, "Employment in Agriculture."

50. Lebergott, "Labor Force and Employment, 1800–1960," 119.

51. Benjamin T. Arrington, "Industry and Economy During the Civil War," National Park Service, August 23, 2017, https://www.nps.gov/articles/industry-and-economy-during-the -civil-war.htm; Lebergott, "Labor Force and Employment, 1800–1960," 119.

52. For a lively explainer on the development of the assembly line and its role in the Second Industrial Revolution, see John Green, "Ford, Cars, and a New Revolution: Crash Course History of Science #28," CrashCourse, YouTube video, November 12, 2018, https://www .youtube.com/watch?v=UPvwpYeOJnI.

53. Lebergott, "Labor Force and Employment, 1800–1960," 119.

54. Lebergott, "Labor Force and Employment, 1800–1960," 119; US Bureau of Labor Statistics, "Civilian Labor Force Level (CLF16OV)"; US Bureau of Labor Statistics, "All Employees, Manufacturing (MANEMP)"; US Bureau of Labor Statistics, "Manufacturing Sector: Real Output (OUTMS)," retrieved from FRED, Federal Reserve Bank of St. Louis, updated March 2, 2023, https://fred.stlouisfed.org/series/OUTMS.

55. For engaging explainers on containerization and its impact, see *Wall Street Journal*, "How a Steel Box Changed the World: A Brief History of Shipping," YouTube video, January 24, 2018, https://www.youtube.com/watch?v=0MUkgDIQdcM; PolyMatter, "How Container Ships Work," YouTube video, November 2, 2018, https://www.youtube.com/watch?v= DY9VE3i-KcM.

56. US Bureau of Labor Statistics, "Manufacturing Sector: Real Output per Hour of All Persons (OPHMFG)," retrieved from FRED, Federal Reserve Bank of St. Louis, updated March 2, 2023, https://fred.stlouisfed.org/series/OPHMFG.

57. US Bureau of Labor Statistics, "All Employees, Manufacturing (MANEMP)."

58. US Bureau of Labor Statistics, "All Employees, Manufacturing (MANEMP)"; US Bureau of Labor Statistics, "Manufacturing Sector: Real Output (OUTMS)."

59. US Bureau of Labor Statistics, "All Employees, Manufacturing (MANEMP)."

60. US Bureau of Labor Statistics, "Manufacturing Sector: Real Output (OUTMS)."

61. US Bureau of Labor Statistics, "All Employees, Manufacturing (MANEMP)"; US Bureau of Labor Statistics, "Manufacturing Sector: Real Output (OUTMS)."

62. US Bureau of Labor Statistics, "All Employees, Manufacturing (MANEMP)"; US Bureau of Labor Statistics, "Civilian Labor Force Level (CLF16OV)"; Lebergott, "Labor Force and Employment, 1800–1960," 119–20.

63. US Bureau of Labor Statistics, "All Employees, Manufacturing (MANEMP)"; US Bureau of Labor Statistics, "Civilian Labor Force Level (CLF16OV)."

64. Note that available data does not properly reflect the effects of the Great Depression, which likely caused a more sudden decline in manufacturing employment than shown in this graph, or of World War II, which caused a brief but dramatic increase in manufacturing employment that's not captured by the BLS total labor force data, which does not stretch back that far. See US Bureau of Labor Statistics, "All Employees, Manufacturing (MANEMP)"; US Bureau of Labor Statistics, "Civilian Labor Force Level (CLF16OV)"; Lebergott, "Labor Force and Employment, 1800–1960," 119–20.

65. US Bureau of Labor Statistics, "Labor Force Participation Rate (CIVPART)," retrieved from FRED, Federal Reserve Bank of St. Louis, updated September 3, 2021, https://fred .stlouisfed.org/series/CIVPART; US Bureau of Labor Statistics, "Civilian Labor Force Level (CLF16OV)."

66. International Labour Organization, "Labor Force, Female (% of Total Labor Force)— United States," retrieved from Worldbank.org, September 2019, https://data.worldbank .org/indicator/SL.TLF.TOTL.FE.ZS?locations=US; US Bureau of Labor Statistics, "Labor Force Participation Rate (CIVPART)"; US Bureau of Labor Statistics, "Civilian Labor Force Participation Rate: Women (LNS11300002)," retrieved from FRED, Federal Reserve Bank of St. Louis, updated April 7, 2023, https://fred.stlouisfed.org/series/LNS1130 0002.

67. US Bureau of Labor Statistics, "Labor Force Participation Rate (CIVPART)."

68. Federal Interagency Forum on Child and Family Statistics, *America's Children: Key National Indicators of Well-Being, 2021* (Washington, DC: US Government Printing Office, 2021): 81, https://web.archive.org/web/20220721170310/https://www.childstats.gov/pdf/ac 2021/ac_21.pdf.

69. Federal Interagency Forum on Child and Family Statistics, *America's Children*, 81.

70. Federal Interagency Forum on Child and Family Statistics, *America's Children*, 81.

71. US Bureau of Labor Statistics, "Civilian Labor Force Level (CLF16OV)"; US Bureau of Labor Statistics, "Population, Total for United States (POPTOTUSA647NWDB)," retrieved from FRED, Federal Reserve Bank of St. Louis, updated July 5, 2022, https://fred .stlouisfed.org/series/POPTOTUSA647NWDB.

72. US Bureau of Labor Statistics, "Civilian Labor Force Level (CLF16OV)"; US Bureau of Labor Statistics, "Population, Total for United States (POPTOTUSA647NWDB)"; "U.S. and World Population Clock," US Census Bureau, updated April 20, 2023, https://www .census.gov/popclock.

73. US Bureau of Labor Statistics, "Civilian Labor Force Level (CLF16OV)"; US Census Bureau, *Statistical Abstract of the United States: 1999*, 879; Lebergott, "Labor Force and Employment, 1800–1960," 118; "U.S. and World Population Clock," US Census Bureau; US Bureau of Labor Statistics, "Population, Total for United States (POPTOTUSA647N-WDB)"; US Census Bureau, "Resident Population of the United States," US Census Bureau, accessed April 20, 2023, https://www2.census.gov/library/visualizations/2000/dec /2000-resident-population/unitedstates.pdf.

74. National Center for Education Statistics, "Enrollment in Elementary, Secondary, and Degree-Granting Postsecondary Institutions, by Level and Control of Institution: Selected Years, 1869–70 Through Fall 2030," US Department of Education, 2021, https://nces .ed.gov/programs/digest/d21/tables/dt21_105.30.asp; Tom Snyder, ed., *120 Years of American Education: A Statistical Portrait* (Washington, DC: National Center for Education Statistics, 1993), 64, http://web20kmg.pbworks.com/w/file/fetch/66806781/120%20Years%20 of%20American%20Education%20A%20Statistical%20Portrait.pdf.

75. National Center for Education Statistics, "Enrollment in Elementary, Secondary, and Degree-Granting Postsecondary Institutions."

76. National Center for Education Statistics, "Total and Current Expenditures per Pupil in Public Elementary and Secondary Schools: Selected Years, 1919–20 Through 2018–19," US Department of Education, September 2021, https://nces.ed.gov/programs/digest/d21 /tables/dt21_236.55.asp; "Consumer Price Index, 1913–," Federal Reserve Bank of Minneapolis, accessed April 20, 2023, https://www.minneapolisfed.org/about-us/monetary -policy/inflation-calculator/consumer-price-index-1913-; US Bureau of Labor Statistics,

"Consumer Price Index for All Urban Consumers: All Items in U.S. City Average (CPI-AUCSL)," retrieved from FRED, Federal Reserve Bank of St. Louis, updated April 12, 2023, https://fred.stlouisfed.org/series/CPIAUCSL.

77. National Center for Education Statistics, "Total and Current Expenditures per Pupil in Public Elementary and Secondary Schools"; "Consumer Price Index, 1913–," Federal Reserve Bank of Minneapolis; US Bureau of Labor Statistics, "Consumer Price Index for All Urban Consumers: All Items in U.S. City Average."

78. National Center for Education Statistics, *120 Years of American Education: A Statistical Portrait*, 65.

79. For a couple of the most influential recent analyses of why most global trends are going in the right direction, see Max Roser, "Most of Us Are Wrong About How the World Has Changed (Especially Those Who Are Pessimistic About the Future)," Our World in Data, July 27, 2018, https://ourworldindata.org/wrong-about-the-world; "Why Are We Working on Our World in Data?," Our World in Data, July 20, 2017, https://ourworldindata.org/motivation; Steven Pinker, "Is the World Getting Better or Worse? A Look at the Numbers," TED video, April 2018, https://www.ted.com/talks/steven_pinker_is_the_world_getting_better_or_worse_a_look_at_the_numbers.

80. For Erik Brynjolfsson's views in more detail, see Erik Brynjolfsson, "The Key to Growth? Race with the Machines," TED, February 2013, https://www.ted.com/talks/erik_brynjolfsson_the_key_to_growth_race_with_the_machines; Erik Brynjolfsson et al., Mind vs Machine: Implications for Productivity, Wages and Employment from AI," The Artificial Intelligence Channel, YouTube video, November 20, 2017, https://www.youtube.com/watch?v=roemLDPy_Ww; Erik Brynjolfsson and Andrew McAfee, *The Second Machine Age: Work, Progress, and Prosperity in a Time of Brilliant Technologies* (New York: W.W. Norton, 2014).

81. For an accessible, data-backed explainer of deskilling trends since the 1950s, see David Kunst, "Deskilling Among Manufacturing Production Workers," VoxEU, August 9, 2019, https://voxeu.org/article/deskilling-among-manufacturing-production-workers.

82. For more on upskilling and the FitMyFoot shoemaking process, see Pablo Illanes et al., "Retraining and Reskilling Workers in the Age of Automation," McKinsey Global Institute, January 2018, https://www.mckinsey.com/featured-insights/future-of-work/retraining-and-reskilling-workers-in-the-age-of-automation; "The Science and Technology of FitMyFoot," FitMyFoot, accessed April 20, 2023, https://fitmyfoot.com/pages/science.

83. Jack Kelly, "Wells Fargo Predicts That Robots Will Steal 200,000 Banking Jobs Within the Next 10 Years," *Forbes*, October 8, 2019, https://www.forbes.com/sites/jackkelly/2019/10/08/wells-fargo-predicts-that-robots-will-steal-200000-banking-jobs-within-the-next-10-years/#237ecaba68d7; James Bessen, "How Computer Automation Affects Occupations: Technology, Jobs, and Skills," Boston University School of Law (Law & Economics working paper no. 15–49, November 13, 2015), 5, https://www.bu.edu/law/files/2015/11/NewTech-2.pdf.

84. G. M. Filisko, "Paralegals and Legal Assistants Are Taking on Expanded Duties," *ABA Journal*, November 1, 2014, http://www.abajournal.com/magazine/article/techno_change_o_paralegal_legal_assistant_duties_expand; Jean O'Grady, "Analytics, AI and Insights: 5 Innovations That Redefined Legal Research Since 2010," Above the Law, January 2, 2020, https://abovethelaw.com/2020/01/analytics-ai-and-insights-5-innovations-that-redefined-legal-research-since-2010.

85. Kevin Roose, "A.I.-Generated Art Is Already Transforming Creative Work," *New York Times*, October 21, 2022, https://www.nytimes.com/2022/10/21/technology/ai-generated-art-jobs-dall-e-2.html.

86. For quick, accessible explainers on the differences between capital and labor, see BBC, "Methods of Production: Labour and Capital," BBC Bitesize, accessed January 30, 2023, https://www.bbc.co.uk/bitesize/guides/zth78mn/revision/5; Sal Khan, "What Is Capital?," Khan Academy, accessed April 20, 2023, https://www.khanacademy.org/economics-finance-domain/macroeconomics/macroeconomics-income-inequality/piketty-capital/v/what-is-capital; Catherine Rampell, "Companies Spend on Equipment, Not Workers," *New*

York Times, June 9, 2011, https://www.nytimes.com/2011/06/10/business/10capital.html; Tim Harford, "What Really Powers Innovation: High Wages," *Financial Times*, January 11, 2013, https://www.ft.com/content/b7ad1c68-59fb-11e2-b728-00144feab49a.

87. US Bureau of Labor Statistics, "Nonfarm Business Sector: Real Output per Hour of All Persons (OPHNFB)," retrieved from FRED, Federal Reserve Bank of St. Louis, updated March 2, 2023, https://fred.stlouisfed.org/series/OPHNFB.

88. US Bureau of Labor Statistics, "Nonfarm Business Sector: Real Output per Hour of All Persons (OPHNFB)."

89. Soon-Yong Choi and Andrew B. Whinston, "The IT Revolution in the USA: The Current Situation and the Problems," in *The Internet Revolution: A Global Perspective*, ed. Emanuele Giovannetti, Mitsuhiro Kagami, and Masatsugu Tsuji (Cambridge, UK: Cambridge University Press, 2003), 219, https://books.google.com/books?id=1f6wD7gezP4C.

90. US Bureau of Labor Statistics, "Nonfarm Business Sector: Real Output per Hour of All Persons (OPHNFB)."

91. "Descriptions of General Purpose Digital Computers," *Computers and Automation* 4, no. 6 (June 1965): 76, https://web.archive.org/web/20190723025854/http://www.bitsavers.org/pdf/computersAndAutomation/196506.pdf.

92. Brynjolfsson and McAfee, *The Second Machine Age*; Tim Worstall, "Trying to Understand Why Marc Andreessen and Larry Summers Disagree Using Facebook," *Forbes*, January 15, 2015, https://www.forbes.com/sites/timworstall/2015/01/15/trying-to-understand-why-marc-andreessen-and-larry-summers-disagree-using-facebook/#1ec456d05b4e.

93. For more on GDP and marginal cost, see Tim Callen, "Gross Domestic Product: An Economy's All," International Monetary Fund, December 18, 2018, https://www.imf.org/external/pubs/ft/fandd/basics/gdp.htm; Alicia Tuovila, "Marginal Cost of Production," Investopedia, September 20, 2019, https://www.investopedia.com/terms/m/marginalcostofproduction.asp; Sal Khan, "Marginal Revenue and Marginal Cost," Khan Academy, accessed April 20, 2023, https://www.khanacademy.org/economics-finance-domain/ap-microeconomics/production-cost-and-the-perfect-competition-model-temporary/short-run-production-costs/v/marginal-revenue-and-marginal-cost; Jeremy Rifkin, *The Zero Marginal Cost Society: The Internet of Things, the Collaborative Commons, and the Eclipse of Capitalism* (New York: St. Martin's, 2014).

94. See the appendix for the sources used for all the cost-of-computation calculations in this book.

95. "Introducing the AMD Radeon RX 7900 XT," AMD, accessed January 30, 2023, https://www.amd.com/en/products/graphics/amd-radeon-rx-7900xt; Michael Justin Allen Sexton, "AMD Radeon RX 7900 XT Review," *PC Magazine*, December 17, 2022, https://www.pcmag.com/reviews/amd-radeon-rx-7900-xt.

96. See the appendix for the sources used for all the cost-of-computation calculations in this book.

97. US Bureau of Labor Statistics, "A Review of Hedonic Price Adjustment Techniques for Products Experiencing Rapid and Complex Quality Change," US Bureau of Labor Statistics, September 15, 2022, https://www.bls.gov/cpi/quality-adjustment/hedonic-price-adjustment-techniques.htm; Dave Wasshausen and Brent R. Moulton, "The Role of Hedonic Methods in Measuring Real GDP in the United States," Bureau of Economic Analysis, October 2006, https://www.bea.gov/system/files/papers/P2006-6_0.pdf.

98. Erik Brynjolfsson, Avinash Collis, and Felix Eggers, "Using Massive Online Choice Experiments to Measure Changes in Well-Being," *Proceedings of the National Academy of Sciences* 116, no. 15 (April 9, 2019): 7250–55, https://doi.org/10.1073/pnas.1815663116.

99. Tim Worstall, "Is Facebook Worth $8 Billion, $100 Billion or $800 Billion to the US Economy?," *Forbes*, January 23, 2015, https://www.forbes.com/sites/timworstall/2015/01/23/is-facebook-worth-8-billion-100-billion-or-800-billion-to-the-us-economy/?sh=6a50df5a16ce; Worstall, "Trying to Understand Why Marc Andreessen and Larry Summers Disagree Using Facebook."

100. Worstall, "Is Facebook Worth $8 Billion."

101. Jasmine Enberg, "Social Media Update Q1 2021," eMarketer, March 30, 2021, https://www.emarketer.com/content/social-media-update-q1-2021.

102. "Social Media Fact Sheet," Pew Research Center, April 7, 2021, https://www.pewresearch.org/internet/fact-sheet/social-media; Stella U. Ogunwoleet al., "Population Under Age 18 Declined Last Decade" US Census Bureau, August 12, 2021, https://www.census.gov/library/stories/2021/08/united-states-adult-population-grew-faster-than-nations-total-population-from-2010-to-2020.html; "Minimum Wage," US Department of Labor, accessed April 20, 2023, https://www.dol.gov/general/topic/wages/minimumwage; John Gramlich, "10 Facts About Americans and Facebook," Pew Research Center, June 1, 2021, https://www.pewresearch.org/fact-tank/2021/06/01/facts-about-americans-and-facebook.

103. Simon Kemp, "Digital 2020: 3.8 Billion People Use Social Media," We Are Social, January 30, 2020, https://web.archive.org/web/20210808051917/https://wearesocial.com/blog/2020/01/digital-2020-3-8-billion-people-use-social-media; Paige Cooper, "43 Social Media Advertising Statistics That Matter to Marketers in 2020," Hootsuite, April 23, 2020, https://web.archive.org/web/20200729070657/https://blog.hootsuite.com/social-media-advertising-stats.

104. US Bureau of Labor Statistics, "Consumer Price Index for All Urban Consumers: Medical Care in U.S. City Average (CPIMEDSL)," retrieved from FRED, Federal Reserve Bank of St. Louis, updated April 12, 2023, https://fred.stlouisfed.org/series/CPIMEDSL#0; US Bureau of Labor Statistics, "Consumer Price Index for All Urban Consumers: All Items in U.S. City Average (CPIAUCSL)"; Xavier Jaravel, "The Unequal Gains from Product Innovations: Evidence from the U.S. Retail Sector," Quarterly Journal of Economics 134, no. 2 (May 2019): 715–83, https://doi.org/10.1093/qje/qjy031; Peter H. Diamandis and Steven Kotler, Abundance: The Future Is Better Than You Think (New York: Simon & Schuster, 2012).

105. US Bureau of Labor Statistics, "Labor Force Participation Rate (CIVPART)."

106. For a quick explainer on these definitions and the underlying data, see Clay Halton, "Civilian Labor Force," Investopedia, July 23, 2019, https://www.investopedia.com/terms/c/civilian-labor-force.asp; US Census Bureau, "Growth in U.S. Population Shows Early Indication of Recovery amid COVID-19 Pandemic," US Census Bureau press release CB22-214, December 22, 2022, https://www.census.gov/newsroom/press-releases/2022/2022-population-estimates.html; "Population, Total—United States," World Bank, accessed April 20, 2023, https://data.worldbank.org/indicator/SP.POP.TOTL?locations=US; US Bureau of Labor Statistics, "Civilian Labor Force Level (CLF16OV)."

107. US Census Bureau, "Growth in U.S. Population Shows Early Indication of Recovery amid COVID-19 Pandemic"; US Bureau of Labor Statistics, "Civilian Labor Force Level (CLF16OV)."

108. Data from US Bureau of Labor Statistics, "Labor Force Participation Rate (CIVPART)."

109. Lauren Bauer et al., "All School and No Work Becoming the Norm for American Teens," Brookings Institution, July 2, 2019, https://www.brookings.edu/blog/up-front/2019/07/02/all-school-and-no-work-becoming-the-norm-for-american-teens; Mitra Toossi, "Labor Force Projections to 2022: The Labor Force Participation Rate Continues to Fall," Monthly Labor Review, US Bureau of Labor Statistics, December 2013, https://www.bls.gov/opub/mlr/2013/article/labor-force-projections-to-2022-the-labor-force-participation-rate-continues-to-fall.htm.

110. Jonnelle Marte, "Aging Boomers Explain Shrinking Labor Force, NY Fed Study Says," Bloomberg, March 30, 2023, https://www.bloomberg.com/news/articles/2023-03-30/aging-boomers-explain-shrinking-labor-force-ny-fed-study-says; Richard Fry, "The Pace of Boomer Retirements Has Accelerated in the Past Year," Pew Research Center, November 9, 2020, https://www.pewresearch.org/short-reads/2020/11/09/the-pace-of-boomer-retirements-has-accelerated-in-the-past-year.

111. US Bureau of Labor Statistics, "Civilian Labor Force Participation Rate: 25 to 54 years (LNU01300060)," retrieved from FRED, Federal Reserve Bank of St. Louis, updated April 7, 2023, https://fred.stlouisfed.org/series/LNU01300060.

112. Organisation for Economic Co-operation and Development, "Working Age Population: Aged 25–54: All Persons for the United States (LFWA25TTUSM647N)," retrieved from FRED, Federal Reserve Bank of St. Louis, updated April 20, 2023, https://fred.stlouisfed.org/series/LFWA25TTUSM647N.

113. US Bureau of Labor Statistics, "Civilian Labor Force Participation Rate: 25 to 54 years (LNU01300060)."

114. Note that although labor force participation among Americans fifty-five and older has been increasing, the net effect is lower overall labor force participation because the fifty-five-plus demographic has a much lower overall participation rate than younger adults, and because the size of that demographic is growing due to the baby boomers' being a much larger cohort than the generation they are replacing. See US Census Bureau, "65 and Older Population Grows Rapidly as Baby Boomers Age," US Census Bureau press release CB20-99, June 25, 2020, https://www.census.gov/newsroom/press-releases/2020/65-older-population-grows.html; William E. Gibson, "Age 65+ Adults Are Projected to Outnumber Children by 2030," AARP, March 14, 2018, https://www.aarp.org/home-family/friends-family/info-2018/census-baby-boomers-fd.html; US Bureau of Labor Statistics, "Civilian Labor Force Participation Rate by Age, Sex, Race, and Ethnicity," US Bureau of Labor Statistics, updated September 8, 2022, https://www.bls.gov/emp/tables/civilian-labor-force-participation-rate.htm.

115. "Life Expectancy at Birth, Total (Years)—United States," World Bank, accessed April 20, 2023, https://data.worldbank.org/indicator/SP.DYN.LE00.IN?locations=US.

116. For more on evolving US labor demographics, see Audrey Breitwieser, Ryan Nunn, and Jay Shambaugh, "The Recent Rebound in Prime-Age Labor Force Participation," Brookings Institution, August 2, 2018, https://www.brookings.edu/blog/up-front/2018/08/02/the-recent-rebound-in-prime-age-labor-force-participation; Jo Harper, "Automation Is Coming: Older Workers Are Most at Risk," Deutsche Welle, July 24, 2018, https://www.dw.com/en/automation-is-coming-older-workers-are-most-at-risk/a-44749804; Peter Gosselin, "If You're Over 50, Chances Are the Decision to Leave a Job Won't Be Yours," *ProPublica*, December 28, 2018, https://www.propublica.org/article/older-workers-united-states-pushed-out-of-work-forced-retirement; Karen Harris, Austin Kimson, and Andrew Schwedel, "Labor 2030: The Collision of Demographics, Automation and Inequality," Bain & Co., February 7, 2018, https://www.bain.com/insights/labor-2030-the-collision-of-demographics-automation-and-inequality; Alana Semuels, "This Is What Life Without Retirement Savings Looks Like," *Atlantic*, February 22, 2018, https://www.theatlantic.com/business/archive/2018/02/pensions-safety-net-california/553970.

117. "Top 100 Cryptocurrencies by Market Capitalization," CoinMarketCap, accessed April 20, 2023, https://coinmarketcap.com.

118. "USD Exchange Trade Volume," Blockchain.com, accessed July 31, 2023, https://www.blockchain.com/charts/trade-volume?timespan=all.

119. "USD Exchange Trade Volume," Blockchain.com.

120. Bank for International Settlements, *BIS Quarterly Review: International Banking and Financial Market Developments* (Bank for International Settlements, December 2022), 16, https://www.bis.org/publ/qtrpdf/r_qt2212.pdf.

121. "Top 100 Cryptocurrencies by Market Capitalization," CoinMarketCap; "Bitcoin Price," Coinbase, accessed April 20, 2023, https://www.coinbase.com/price/bitcoin.

122. "Bitcoin Price," Coinbase.

123. "Bitcoin Price," Coinbase.

124. "Bitcoin Price," Coinbase.

125. For more on the growing economic role of influencers, see "How Big Is the Influencer Economy?," *TechCrunch*, YouTube video, November 13, 2019, https://www.youtube.com/watch?v=RJBn2JDfDS0.

126. Sarah Perez, "iOS App Store Has Seen Over 170B Downloads, Over $130B in Revenue Since July 2010," *TechCrunch*, May 31, 2018, https://techcrunch.com/2018/05/31/ios-app-store-has-seen-over-170b-downloads-over-130b-in-revenue-since-july-2010.

127. "Number of Available Applications in the Google Play Store from December 2009 to September 2022," Statista, updated March 2023, https://www.statista.com/statistics/266210-/number-of-available-applications-in-the-google-play-store.

128. "Number of Available Applications in the Google Play Store," Statista.

129. Trevor Mogg, "App Economy Creates Nearly Half a Million US Jobs," *Digital Trends*, February 7, 2012, https://web.archive.org/web/20170422014624/https://www.digitaltrends.com/android/app-economy-creates-nearly-half-a-million-us-jobs.

130. Microeconomix, *App Economy in the United States*, 4.

131. ACT: The App Association, *State of the U.S. App Economy: 2020*, 7th ed. (Washington, DC: ACT: The App Association, 2021), 4, https://actonline.org/wp-content/uploads/2020-App-economy-Report.pdf.

132. Frey and Osborne, "Future of Employment: How Susceptible Are Jobs to Computerisation?"

133. Dusty Stowe, "Why Star Trek: The Original Series Was Cancelled After Season 3," *Screen Rant*, May 21, 2019, https://screenrant.com/star-trek-original-series-cancelled-season-3-reason-why.

134. Kayla Cobb, "From 'South Park' to 'BoJack Horseman,' Tracking the Rise of Continuity in Adult Animation," *Decider*, December 16, 2015, https://decider.com/2015/12/16/tracking-the-rise-of-continuity-in-animated-comedies; Gus Lubin, "'BoJack Horseman' Creators Explain Why Netflix Is So Much Better Than TV," *Business Insider*, October 3, 2014, https://www.businessinsider.com/why-bojack-horseman-went-to-netflix-2014-9.

135. International Telecommunication Union, "Key ICT Indicators for Developed and Developing Countries."

136. For more background on the introduction of Social Security in the United States and the program's goals, see Craig Benzine, "Social Policy: Crash Course Government and Politics #49," CrashCourse, YouTube video, February 27, 2016, https://www.youtube.com/watch?v=mlxLX8Fto_A; "Here's How the Great Depression Brought on Social Security," History, YouTube video, April 26, 2018, https://www.youtube.com/watch?v=cdE_EV3wnXM; "Historical Background and Development of Social Security," Social Security Administration, accessed April 20, 2023, https://www.ssa.gov/history/briefhistory3.html.

137. Because of the disruptions of the COVID-19 pandemic, 2019 is the most recent year for which good social safety net statistics are available at the time of writing. More recent data is less commensurable between countries because budgeting categories for pandemic-related economic relief vary. See Organisation for Economic Co-operation and Development, "Social Expenditure Database (SOCX)," OECD.org, January 2023, https://www.oecd.org/social/expenditure.htm.

138. Organisation for Economic Co-operation and Development, "Social Expenditure Database."

139. Organisation for Economic Co-operation and Development, "Social Expenditure Database."

140. "GDP (Current US$)—United Kingdom," World Bank, accessed April 20, 2023, https://data.worldbank.org/indicator/NY.GDP.MKTP.CD?locations=GB; "Population, Total—United Kingdom," World Bank, accessed April 20, 2023, https://data.worldbank.org/indicator/SP.POP.TOTL?locations=GB.

141. Organisation for Economic Co-operation and Development, "Social Expenditure Database"; "GDP (Current US$)—United States," World Bank, accessed April 20, 2023, https://data.worldbank.org/indicator/NY.GDP.MKTP.CD?locations=US.

142. "Population, Total—United States," World Bank.

143. Noah Smith, "The U.S. Social Safety Net Has Improved a Lot," *Bloomberg*, May 16, 2018, https://www.bloomberg.com/opinion/articles/2018-05-16/the-u-s-social-safety-net-has-improved-a-lot.

144. US government spending is very difficult to measure precisely across local, state, and federal levels, and data from the early twentieth century is not perfectly comparable to more recent data. Therefore, the best available estimates must rely upon a degree of speculation

and make methodological choices for which there is no clear single best option. This is especially true of measuring spending that counts as part of the social safety net. As such, safety net spending data presented in this chapter should not be considered definitive, but rather rough approximations. Note also that due to methodological differences in the underlying sources, the year-by-year "social safety net" data for the US does not precisely align with current US "social expenditure" as measured by the OECD's Social Expenditure Database. Nonetheless, the overall trend—that the social safety net has been expanding regardless of which party is in power—is quite clear.

145. US Census Bureau, "Historical National Population Estimates, July 1, 1900 to July 1, 1999"; US Bureau of Economic Analysis, "Population (B230RC0A052NBEA)"; "U.S. and World Population Clock—July 1, 2021," US Census Bureau, updated January 30, 2023, https://www.census.gov/popclock; US Bureau of Economic Analysis, "Gross Domestic Product (GDP)," retrieved from FRED, Federal Reserve Bank of St. Louis, updated March 30, 2023, https://fred.stlouisfed.org/series/GDP; "Consumer Price Index, 1913–," Federal Reserve Bank of Minneapolis; Christopher Chantrill, "Government Spending Chart," usgovernmentspending.com, accessed April 20, 2023, https://www.usgovernmentspending.com/spending_chart_1900_2021USk_22s2li011mcny_10t40t00t; "CPI Inflation Calculator" for July 2012–July 2021, US Bureau of Labor Statistics, accessed April 20, 2023, https://www.bls.gov/data/inflation_calculator.htm; Christopher Chantrill, "Government Spending Chart," usgovernmentspending.com, accessed April 20, 2023, https://www.usgovernmentspending.com/spending_chart_1900_2021USk_22s2li011mcny_F0t; Jutta Bolt and Jan Luiten van Zanden, *Maddison Project Database*, version 2020, Groningen Growth and Development Centre, November 2, 2020, https://www.rug.nl/ggdc/historicaldevelopment/maddison/releases/maddison-project-database-2020; Jutta Bolt and Jan Luiten van Zanden, "Maddison Style Estimates of the Evolution of the World Economy. A New 2020 Update" (working paper WP-15, Maddison Project, October 2020), https://www.rug.nl/ggdc/historicaldevelopment/maddison/publications/wp15.pdf; John J. McCusker, "Colonial Statistics," in *Historical Statistics of the United States: Earliest Times to the Present*, ed. Susan G. Carter et al. (Cambridge, UK: Cambridge University Press, 2006), V-671; Richard Sutch, "National Income and Product," in *Historical Statistics of the United States: Earliest Times to the Present*, ed. Susan G. Carter et al. (Cambridge, UK: Cambridge University Press, 2006), III-23–25; Leandro Prados de la Escosura, "Lost Decades? Economic Performance in Post-Independence Latin America," *Journal of Latin American Studies* 41, no. 2 (May 2009): 279–307, https://www.jstor.org/stable/27744128; "CPI Inflation Calculator" for July 2011–July 2021, US Bureau of Labor Statistics, accessed January 30, 2023, https://www.bls.gov/data/inflation_calculator.htm; US Bureau of Labor Statistics, "Consumer Price Index for All Urban Consumers: All Items in U.S. City Average (CPI-AUCSL)"; US Bureau of Economic Analysis, "Real Gross Domestic Product per Capita (A939RX0Q048SBEA)," retrieved from FRED, Federal Reserve Bank of St. Louis, updated March 30, 2023, https://fred.stlouisfed.org/series/A939RX0Q048SBEA.

146. For the sources used in this graph, see note 145 above.

147. For the sources used in this graph, see note 145 above.

148. For the sources used in this graph, see note 145 above.

149. To watch our exchange, see Ray Kurzweil and Chris Anderson, "Ray Kurzweil on What the Future Holds Next," *The TED Interview* podcast, December 2018, https://www.ted.com/talks/the_ted_interview_ray_kurzweil_on_what_the_future_holds_next.

150. For more on the growing movement to establish a universal basic income (or a related concept called "universal basic services") and the evidence that shapes these proposals, see Will Bedingfield, "Universal Basic Income, Explained," *Wired*, August 25, 2019, https://www.wired.co.uk/article/universal-basic-income-explained; Karen Yuan, "A Moral Case for Giving People Money," *Atlantic*, August 22, 2018, https://www.theatlantic.com/membership/archive/2018/08/a-moral-case-for-giving-people-money/568207; Annie Lowrey, "Stockton's Basic-Income Experiment Pays Off," *Atlantic*, March 3, 2021, https://www.theatlantic.com/ideas/archive/2021/03/stocktons-basic-income-experiment-pays-off

/618174; Dylan Matthews, "Basic Income: The World's Simplest Plan to End Poverty, Explained," *Vox*, April 25, 2016, https://www.vox.com/2014/9/8/6003359/basic-income-negative-income-tax-questions-explain; Sigal Samuel, "Everywhere Basic Income Has Been Tried, in One Map," *Vox*, October 20, 2020, https://www.vox.com/future-perfect/2020/2/19/21112570/universal-basic-income-ubi-map; Ian Gough, "Move the Debate from Universal Basic Income to Universal Basic Services," UNESCO Inclusive Poverty Lab, January 19, 2021, https://en.unesco.org/inclusivepolicylab/analytics/move-debate-universal-basic-income-universal-basic-services.

151. Derek Thompson, "A World Without Work," *Atlantic*, July/August 2015, https://www.theatlantic.com/magazine/archive/2015/07/world-without-work/395294.

152. See the appendix for the sources used for all the cost-of-computation calculations in this book; Jaravel, "The Unequal Gains from Product Innovations," 715–83; Diamandis and Kotler, *Abundance*.

153. US Bureau of Labor Statistics, "Consumer Price Index for All Urban Consumers: Medical Care in U.S. City Average (CPIMEDSL)"; US Bureau of Labor Statistics, "Consumer Price Index for All Urban Consumers: All Items in U.S. City Average (CPIAUCSL)"; "Consumer Price Index, 1913–," Federal Reserve Bank of Minneapolis; US Bureau of Labor Statistics, "Consumer Price Index for All Urban Consumers: All Items in U.S. City Average (CPIAUCSL)."

154. US Bureau of Labor Statistics, "Consumer Price Index for All Urban Consumers: Medical Care in U.S. City Average (CPIMEDSL)"; US Bureau of Labor Statistics, "Consumer Price Index for All Urban Consumers: All Items in U.S. City Average (CPIAUCSL)"; "Consumer Price Index, 1913–," Federal Reserve Bank of Minneapolis.

155. For a highly informative book on the relationship between governance and prosperity, see Daron Acemoglu and James A. Robinson, *Why Nations Fail: The Origins of Power, Prosperity, and Poverty* (New York: Crown, 2012).

156. For a helpful and vivid explainer of Maslow's hierarchy and its significance, see "Why Maslow's Hierarchy of Needs Matters," The School of Life, YouTube video, April 10, 2019, https://www.youtube.com/watch?v=L0PKWTta7lU.

157. Sandra L. Colby and Jennifer M. Ortman, *Projections of the Size and Composition of the U.S. Population: 2014 to 2060*, Current Populations Reports, P25-1143, US Census Bureau, March 2015, 6, https://www.census.gov/content/dam/Census/library/publications/2015/demo/p25-1143.pdf; US Census Bureau, "American Fact Finder, Table B24010—Sex by Occupation for the Civilian Employed Population 16 Years and Over," 2017 American Community Survey 1-Year Estimates, https://factfinder.census.gov/faces/tableservices/jsf/pages/productview.xhtml?src=bkmk.

158. Skip Descant, "Autonomous Vehicles to Have Huge Impact on Economy, Tech Sector," *Government Technology*, June 27, 2018, https://www.govtech.com/fs/automation/Autonomous-Vehicles-to-Have-Huge-Impact-on-Economy-Tech-Sector.html; Kirsten Korosec, "Intel Predicts a $7 Trillion Self-Driving Future," *The Verge*, June 1, 2017, https://www.theverge.com/2017/6/1/15725516/intel-7-trillion-dollar-self-driving-autonomous-cars; Adam Ozimek, "The Massive Economic Benefits of Self-Driving Cars," *Forbes*, November 8, 2014, https://www.forbes.com/sites/modeledbehavior/2014/11/08/the-massive-economic-benefits-of-self-driving-cars/#723609f53273.

159. An estimated 42,915 people died in traffic accidents on US roads in 2021. While there is ongoing debate about how much of this should be attributed to human error, it is clear that the overwhelming majority of crashes have human error as a key component—likely somewhere between 90 and 99 percent. Autonomous vehicles controlled by capable enough AI could eliminate almost all of these. See Bryant Walker Smith, "Human Error as a Cause of Vehicle Crashes," Center for Internet and Society, Stanford Law School, December 18, 2013, https://cyberlaw.stanford.edu/blog/2013/12/human-error-cause-vehicle-crashes; David Zipper, "The Deadly Myth That Human Error Causes Most Car Crashes," *Atlantic*, November 26, 2021, https://www.theatlantic.com/ideas/archive/2021/11/deadly-myth-human-error-causes-most-car-crashes/620808; National Highway Traffic Safety Administration,

"Critical Reasons for Crashes Investigated in the National Motor Vehicle Crash Causation Survey," NHTSA National Center for Statistics and Analysis report DOT HS 812 115, US Department of Transportation, February 2015, https://crashstats.nhtsa.dot.gov/Api/Pub lic/ViewPublication/812115; National Highway Traffic Safety Administration, "Early Estimates of Motor Vehicle Traffic Fatalities and Fatality Rate by Sub-Categories in 2021," NHTSA National Center for Statistics and Analysis report DOT HS 813 298, US Department of Transportation, May 2022, https://crashstats.nhtsa.dot.gov/Api/Public/ViewPubli cation/813298.

160. Carl Benedikt Frey et al., "Political Machinery: Did Robots Swing the 2016 US Presidential Election?," *Oxford Review of Economic Policy* 34, no. 3 (2018): 418–42, https://www.ox fordmartin.ox.ac.uk/downloads/academic/Political_Machinery_July_2018.pdf.

161. For more on the long-term decline in worldwide violence, see Max Roser and Hannah Ritchie, "Homicides," Our World in Data, December 2019, https://ourworldindata.org /homicides; Manuel Eisner, "From Swords to Words: Does Macro-Level Change in Self-Control Predict Long-Term Variation in Levels of Homicide?," *Crime and Justice* 43, no. 1 (September 2014): 80–81; UN Office on Drugs and Crime, "Intentional Homicides (Per 100,000 People)—France, Netherlands, Sweden, Germany, Switzerland, Italy, United Kingdom, Spain," retrieved from Worldbank.org, accessed April 20, 2023, https://data .worldbank.org/indicator/VC.IHR.PSRC.P5?end=2020&locations=FR-NL-SE-DE-CH -IT-GB-ES&start=2020&view=bar; "Appendix Tables: Homicide in England and Wales," UK Office for National Statistics, February 9, 2023, https://www.ons.gov.uk/file?uri= /peoplepopulationandcommunity/crimeandjustice/datasets/appendixtableshomicideine nglandandwales/current/homicideyemarch22appendixtables.xlsx; European Commission, *Investing in Europe's Future: Fifth Report on Economic, Social and Territorial Cohesion* (Luxembourg: Publications Office of the European Union, 2010), https://ec.europa.eu/regional _policy/sources/docoffic/official/reports/cohesion5/pdf/5cr_part1_en.pdf; *Global Study on Homicide*, United Nations Office on Drugs and Crime, 2019, https://www.unodc.org/docu ments/data-and-analysis/gsh/Booklet1.pdf; *Global Study on Homicide*, United Nations Office on Drugs and Crime, 2011, http://www.unodc.org/documents/data-and-analysis/statis tics/Homicide/Globa_study_on_homicide_2011_web.pdf; Steven Pinker, *The Better Angels of Our Nature: Why Violence Has Declined* (New York: Penguin, 2011).

162. For an in-depth talk by Pinker expanding on this view, see "Steven Pinker: Better Angels of Our Nature," Talks at Google, YouTube video, November 1, 2011, https://www.youtube .com/watch?v=_gGf7fXM3jQ.

163. I made most of these predictions in my 1990 book *The Age of Intelligent Machines*. See Raymond Kurzweil, *The Age of Intelligent Machines* (Cambridge, MA: MIT Press, 1990), 429–34.

164. Kurzweil, *Age of Intelligent Machines*, 432–34.

165. Alex Shashkevich, "Meeting Online Has Become the Most Popular Way U.S. Couples Connect, Stanford Sociologist Finds," Stanford News, August 21, 2019, https://news.stan ford.edu/2019/08/21/online-dating-popular-way-u-s-couples-meet; Michael J. Rosenfeld, Reuben J. Thomas, and Sonia Hausen, "Disintermediating Your Friends: How Online Dating in the United States Displaces Other Ways of Meeting," *Proceedings of the National Academy of Sciences* 116, no. 36 (September 3, 2019): 17753–58, https://doi.org/10.1073 /pnas.1908630116.

166. Even if a 2024 smartphone didn't have access to 2024 cell service, it could still store all the text on the English-language Wikipedia in its native memory. Depending on how you measure, Wikipedia is likely around 150 gigabytes to download, while the latest iPhones offer up to 1 terabyte of storage. See Nick Lewis and Matt Klein, "How to Download Wikipedia for Offline, At-Your-Fingertips Reading," *How-To Geek*, March 25, 2022, https://www .howtogeek.com/260023/how-to-download-wikipedia-for-offline-at-your-fingertips -reading; "Buy iPhone 14 Pro," Apple, accessed April 20, 2023, https://www.apple.com /shop/buy-iphone/iphone-14-pro/6.7-inch-display-1tb-space-black-unlocked.

167. To be clear, I'm referring here to the stage where humans and AI become increasingly symbiotic and material abundance is achieved. In the meantime, as described earlier in this

chapter, there will certainly be disruptions and competition as AI replaces humans in many tasks and jobs that exist in the current economic paradigm.

CHAPTER 6: THE NEXT THIRTY YEARS IN HEALTH AND WELL-BEING

1. One of the world's leading biosimulation companies, Insilico Medicine, developed an AI platform called Pharma.AI, which it used to create INS018_055, a small-molecule drug now in phase II trials to treat a rare lung disease called idiopathic pulmonary fibrosis. In a world first, the AI didn't merely augment human researchers but designed the drug end-to-end, meaning that it both identified a novel biomolecular target to treat the disease and identified a molecule that could work on that target. For a fascinating look at Insilico's work, see "How AI Is Accelerating Drug Discovery," YouTube video, April 3, 2023, https://www.youtube.com/watch?v=mqB vitxD05M; Hayden Field, "The First Fully A.I.-Generated Drug Enters Clinical Trials in Human Patients," CNBC, June 29, 2023, https://www.cnbc .com/2023/06/29/ai-generated-drug-begins-clinical-trials-in-human-patients.html.

2. For further resources on AI-driven drug discovery, see Vanessa Bates Ramirez, "Drug Discovery AI Can Do in a Day What Currently Takes Months," SingularityHub, May 7, 2017, https://singularityhub.com/2017/05/07/drug-discovery-ai-can-do-in-a-day-what -currently-takes-months; "MIT Quest for Intelligence Launch: AI-Driven Drug Discovery," Massachusetts Institute of Technology, YouTube video, March 9, 2018, https://www .youtube.com/watch?v=aqMRrRS_0JY; "Developer Spotlight: Opening a New Era of Drug Discovery with Amber," NVIDIA Developer, YouTube video, July 29, 2019, https:// www.youtube.com/watch?v=FqnPGHdh7iM; "We're Teaching Robots and AI to Design New Drugs," SciShow, YouTube video, September 30, 2021, https://www.youtube.com /watch?v=eRXqD-7FANg; Francesca Properzi et al., "Intelligent Drug Discovery: Powered by AI" (Deloitte Centre for Health Solutions, 2019), https://www2.deloitte.com/con tent/dam/insights/us/articles/32961_intelligent-drug-discovery/DI_Intelligent -Drug-Discovery.pdf; Nic Fleming, "How Artificial Intelligence Is Changing Drug Discovery," *Nature* 557, no. 7707 (May 31, 2018): S55–S57, https://doi.org/10.1038 /d41586-018-05267-x; David H. Feedman, "Hunting for New Drugs with AI," *Nature* 576, no. 7787 (December 18, 2019): S49–S53, https://www.nature.com/articles/d41586-019 -03846-0.

3. Abhimanyu S. Ahuja, Vineet Pasam Reddy, and Oge Marques, "Artificial Intelligence and COVID-19: A Multidisciplinary Approach," *Integrative Medicine Research* 9, no. 3, article 100434 (May 27, 2020), https://doi.org/10.1016/j.imr.2020.100434; Jared Sagoff, "Argonne's Researchers and Facilities Playing a Key Role in the Fight Against COVID-19," Argonne National Laboratory, April 27, 2020, https://www.anl.gov/article/argonnes -researchers-and-facilities-playing-a-key-role-in-the-fight-against-covid19.

4. Jean-Louis Reymond and Mahendra Awale, "Exploring Chemical Space for Drug Discovery Using the Chemical Universe Database," *ACS Chemical Neuroscience* 3, no. 9 (April 25, 2012): 649–57, https://doi.org/10.1021/cn3000422.

5. Chi Heem Wong, Kien Wei Siah, and Andrew W. Lo, "Estimation of Clinical Trial Success Rates and Related Parameters," *Biostatistics* 20, no. 2 (January 31, 2018): 273–86, https:// doi.org/10.1093/biostatistics/kxx069.

6. "The Drug Development Process: Step 3: Clinical Resarch," US Food and Drug Administration, accessed October 20, 2022, https://www.fda.gov/patients/drug-development-process /step-3-clinical-research; Stuart A. Thompson, "How Long Will a Vaccine Really Take?," *New York Times*, April 30, 2020, https://www.nytimes.com/interactive/2020/04/30/opinion /coronavirus-covid-vaccine.html; Institute of Medicine Forum on Drug Discovery, Development, and Translation, "The State of Clinical Research in the United States: An Overview," in *Transforming Clinical Research in the United States* (Washington, DC: National Academies Press, 2010), https://www.ncbi.nlm.nih.gov/books/NBK50886; Thomas J. Moore et al., "Estimated Costs of Pivotal Trials for Novel Therapeutic Agents Approved by the US Food and Drug Administration, 2015–2016," *JAMA Internal Medicine* 178, no. 11

(November 2018): 1451–57, https://doi.org/10.1001/jamainternmed.2018.3931; Olivier J. Wouters, Martin McKee, Jeroen Luyten, "Estimated Research and Development Investment Needed to Bring a New Medicine to Market, 2009–2018," *Journal of the American Medical Association* 323, no. 9 (March 3, 2020): 844–53, https://doi.org/10.1001/jama.2020 .1166; *Biopharmaceutical Research & Development: The Process Behind New Medicines*, PHRMA, accessed October 20, 2022, https://web.archive.org/web/20230306041340 /http://phrma-docs.phrma.org/sites/default/files/pdf/rd_brochure_022307.pdf.

7. David Sparkes and Rhett Burnie, "AI Invents More Effective Flu Vaccine in World First, Adelaide Researchers Say," Australian Broadcasting Corporation, July 2, 2019, https:// www.abc.net.au/news/2019-07-02/computer-invents-flu-vaccine-in-world-first/11271170; Andrew Tarantola, "How AI Is Stopping the Next Great Flu Before It Starts," *Engadget*, February 14, 2020, https://www.engadget.com/2020/02/14/how-ai-is-helping-halt-the-flu-of -the-future.

8. Tarantola, "How AI is Stopping the Next Great Flu Before It Starts."

9. Ian Sample, "Powerful Antibiotic Discovered Using Machine Learning for First Time," *Guardian*, February 20, 2020, https://www.theguardian.com/society/2020/feb/20/antibi otic-that-kills-drug-resistant-bacteria-discovered-through-ai.

10. Sample, "Powerful Antibiotic Discovered Using Machine Learning for First Time."

11. "Moderna's Work on a Potential Vaccine Against COVID-19," Moderna, 2020, https:// www.sec.gov/Archives/edgar/data/1682852/000119312520074867/d884510dex991.htm.

12. For more detail on Moderna's use of AI in developing vaccines, see "AI and the COVID-19 Vaccine: Moderna's Dave Johnson," *Me, Myself, and AI* podcast, ep. 209 (July 13, 2021), https://sloanreview.mit.edu/audio/ai-and-the-covid-19-vaccine-modernas-dave-johnson; "Moderna on AWS," Amazon Web Services, accessed October 20, 2022, https://aws.ama zon.com/solutions/case-studies/innovators/moderna; Bryce Elder, "Will Big Tobacco Save Us from the Coronavirus?," *Financial Times*, April 1, 2020, https://www.ft.com/content /f909fb16-f514-47da-97dc-c03e752dd2e1.

13. Gary Polakovic, "Artificial Intelligence Aims to Outsmart the Mutating Coronavirus," *USC News*, February 5, 2021, https://news.usc.edu/181226/artificial-intelligence-ai-corona virus-vaccines-mutations-usc-research; Zikun Yang et al., "An *In Silico* Deep Learning Approach to Multi-Epitope Vaccine Design: A SARS-CoV-2 Case Study," *Scientific Reports* 11, article 3238 (February 5, 2021), https://doi.org/10.1038/s41598-021-81749-9.

14. For a deeper look at the protein-folding problem, including videos with helpful visualizations, see "The Protein Folding Revolution," *Science Magazine*, YouTube video, July 21, 2016, https://www.youtube.com/watch?v=cAJQbSLlonI; "Protein Structure," Professor Dave Explains, YouTube video, August 27, 2016, https://www.youtube.com/watch?v= EweuU2fEgjw; Ken Dill, "The Protein Folding Problem: A Major Conundrum of Science: Ken Dill at TEDxSBU," TEDx Talks, YouTube video, October 22, 2013, https://www .youtube.com/watch?v=zm-3kovWpNQ; Ken A. Dill et al., "The Protein Folding Problem," *Annual Review of Biophysics* 37 (June 9, 2008): 289–316, https://doi.org/10.1146/an nurev.biophys.37.092707.153558; Andrew W. Senior et al., "Improved Protein Structure Prediction Using Potentials from Deep Learning," *Nature* 577, no. 7792 (January 15, 2020), https://doi.org/10.1038/s41586-019-1923-7.

15. For more detail on how the original AlphaFold achieved great progress on protein folding, see Andrew W. Senior et al., "AlphaFold: Using AI for Scientific Discovery," DeepMind, January 15, 2020, https://deepmind.com/blog/article/AlphaFold-Using-AI-for-scientific -discovery; Andrew Senior, "AlphaFold: Improved Protein Structure Prediction Using Potentials from Deep Learning," Institute for Protein Design, YouTube video, August 23, 2019, https://www.youtube.com/watch?v=uQ1uVbrIv-Q; Greg Williams, "Inside Deep-Mind's Epic Mission to Solve Science's Trickiest Problem," *Wired*, August 6, 2019, https:// www.wired.co.uk/article/deepmind-protein-folding; Senior et al., "Improved Protein Structure Prediction Using Potentials from Deep Learning," 706–10.

16. Ian Sample, "Google's DeepMind Predicts 3D Shapes of Proteins," *Guardian*, December 2, 2018, https://www.theguardian.com/science/2018/dec/02/google-deepminds-ai-program -alphafold-predicts-3d-shapes-of-proteins; Matt Reynolds, "DeepMind's AI Is Getting

Closer to Its First Big Real-World Application," *Wired*, January 15, 2020, https://www
.wired.co.uk/article/deepmind-protein-folding-alphafold.

17. For some more detailed explainers of AlphaFold 2 and the scientific paper describing it, see
    "AlphaFold: The Making of a Scientific Breakthrough," DeepMind, YouTube video, No-
    vember 30, 2020, https://www.youtube.com/watch?v=gg7WjuFs8F4; "DeepMind Solves
    Protein Folding | AlphaFold 2," Lex Fridman, YouTube video, December 2, 2020, https://
    www.youtube.com/watch?v=W7wJDJ56c88; Ewen Callaway, "'It Will Change Every-
    thing': DeepMind's AI Makes Gigantic Leap in Solving Protein Structures," *Nature* 588,
    no. 7837 (November 30, 2020): 203–4, https://doi.org/10.1038/d41586-020-03348-4;
    Demis Hassabis, "Putting the Power of AlphaFold into the World's Hands," DeepMind,
    July 22, 2022, https://deepmind.com/blog/article/putting-the-power-of-alphafold-into-the
    -worlds-hands; John Jumper et al., "Highly Accurate Protein Structure Prediction with
    AlphaFold," *Nature* 596, no. 7873 (July 15, 2021): 583–89, https://doi.org/10.1038/s41586
    -021-03819-2.

18. Mohammed AlQuraishi, "Protein-Structure Prediction Revolutionized," *Nature* 596, no.
    7873 (August 23, 2021): 487–88, https://doi.org/10.1038/d41586-021-02265-4.

19. Hassabis, "Putting the Power of AlphaFold into the World's Hands"; Jumper et al., "Highly
    Accurate Protein Structure Prediction with AlphaFold."

20. For relatively simple explainers of these methods, see National Cancer Institute, "CAR T
    Cells: Engineering Patients' Immune Cells to Treat Their Cancers," National Institutes of
    Health, March 10, 2022, https://www.cancer.gov/about-cancer/treatment/research/car-t
    -cells; "BiTE: The Engager," Amgen, 2022, https://www.amgenoncology.com/resources
    /BiTE-the-Engager.pdf; "Immune Checkpoint Inhibitor Cancer Treatment," Memorial
    Sloan Kettering Cancer Center, accessed October 20, 2022, https://www.mskcc.org/can
    cer-care/diagnosis-treatment/cancer-treatments/immunotherapy/checkpoint-inhibitors.

21. Robert C. Sterner and Rosalie M. Sterner, "CAR-T Cell Therapy: Current Limitations
    and Potential Strategies," *Blood Cancer Journal* 11, article 69 (April 6, 2021), https://doi
    .org/10.1038/s41408-021-00459-7.

22. For succinct summaries of the theorized mechanisms of neurodegenerative diseases, see
    "Alzheimer's Disease," Mayo Clinic, February 19, 2022, https://www.mayoclinic.org/dis
    eases-conditions/alzheimers-disease/symptoms-causes/syc-20350447; "Parkinson's Disease,"
    Mayo Clinic, July 8, 2022, https://www.mayoclinic.org/diseases-conditions/parkinsons-dis
    ease/symptoms-causes/syc-20376055.

23. "About Mental Health," Centers for Disease Control and Prevention, June 28, 2021,
    https://www.cdc.gov/mentalhealth/learn/index.htm.

24. For more on the limitations of common psychiatric drugs, see Melinda Wenner Moyer,
    "How Much Do Antidepressants Help, Really?," *New York Times*, April 21, 2022, https://
    www.nytimes.com/2022/04/21/well/antidepressants-ssri-effectiveness.html; Harvard Health
    Publishing, "What Are the Real Risks of Antidepressants?," Harvard Medical School, Au-
    gust 17, 2021, https://www.health.harvard.edu/newsletter_article/what-are-the-real-risks-of
    -antidepressants; Krishna C. Vadodaria et al., "Altered Serotonergic Circuitry in SSRI-
    Resistant Major Depressive Disorder Patient-Derived Neurons," *Molecular Psychiatry* 24
    (March 22, 2019): 808–18, https://doi.org/10.1038/s41380-019-0377-5.

25. For further information on the introduction of *in silico* simulations in the health-care field,
    see Fleming, "How Artificial Intelligence Is Changing Drug Discovery"; Madhumita
    Murgia, "AI-Designed Drug to Enter Human Clinical Trial for First Time," *Financial
    Times*, January 30, 2020, https://www.ft.com/content/fe55190e-42bf-11ea-a43a-c4b328d
    9061c; Osman N. Yogurtcu et al., "TCPro Simulates Immune System Response to Bio-
    therapeutic Drugs," US Food and Drug Administration, September 17, 2019, https://www
    .fda.gov/vaccines-blood-biologics/science-research-biologics/tcpro-simulates-immune
    -system-response-biotherapeutic-drugs; Tina Morrison, "How Simulation Can Transform
    Regulatory Pathways," US Food and Drug Administration, August 14, 2018, https://www
    .fda.gov/science-research/about-science-research-fda/how-simulation-can-transform
    -regulatory-pathways; Anna Edney, "Computer-Simulated Tests Eyed at FDA to Cut Drug
    Approval Costs," *Bloomberg*, July 7, 2017, https://www.bloomberg.com/news/articles/2017

-07-07/drug-agency-looks-to-computer-simulations-to-cut-testing-costs; "Virtual Bodies for Real Drugs: In Silico Clinical Trials Are the Future," *The Medical Futurist*, August 10, 2019, https://medicalfuturist.com/in-silico-trials-are-the-future; Pratik Shah et al., "Artificial Intelligence and Machine Learning in Clinical Development: A Translational Perspective," *NPJ Digital Medicine* 2, no. 69 (July 26, 2019), https://doi.org/10.1038/s41746-019-0148-3; Neil Savage, "Tapping into the Drug Discovery Potential of AI," *Biopharma Dealmakers* 15, no. 2 (May 27, 2021), https://doi.org/10.1038/d43747-021-00045-7.

26. Ray Kurzweil, "AI-Powered Biotech Can Help Deploy a Vaccine in Record Time," *Wired*, May 19, 2020, https://www.wired.com/story/opinion-ai-powered-biotech-can-help-deploy -a-vaccine-in-record-time; Aaron Dubrow, "AI Fast-Tracks Drug Discovery to Fight COVID-19," Texas Advanced Computing Center, April 22, 2020, https://www.tacc.utexas .edu/-/ai-fast-tracks-drug-discovery-to-fight-covid-19; Thompson, "How Long Will a Vaccine Really Take?"; Tina Morrison et al., "Advancing Regulatory Science with Computational Modeling for Medical Devices at the FDA's Office of Science and Engineering Laboratories," *Frontiers in Medicine* 5, article 241 (September 25, 2018), https://doi.org /10.3389/fmed.2018.00241.

27. "The Drug Development Process: Step 3: Clinical Resarch," US Food and Drug Administration.

28. Daniel Bastardo Blanco, "Our Cells Are Filled with 'Junk DNA'—Here's Why We Need It," *Discover*, August 13, 2019, https://www.discovermagazine.com/health/our-cells-are -filled-with-junk-dna-heres-why-we-need-it.

29. Jian Zhou et al., "Whole-Genome Deep-Learning Analysis Identifies Contribution of Noncoding Mutations to Autism Risk," *Nature Genetics* 51, no. 6 (May 27, 2019): 973–80, https://doi.org/10.1038/s41588-019-0420-0; Thomas Sumner, "New Causes of Autism Found in 'Junk' DNA," Simons Foundation, May 27, 2019, https://www.simonsfounda tion.org/2019/05/27/autism-noncoding-mutations.

30. Sumner, "New Causes of Autism Found in 'Junk' DNA."

31. Nancy Fliesler, "Using Multiple Data Streams and Artificial Intelligence to 'Nowcast' Local Flu Outbreaks," *Vector*, Boston Children's Hospital, January 14, 2019, https://web .archive.org/web/20210121214157/https://vector.childrenshospital.org/2019/01/local-flu -prediction-argonet.

32. Fliesler, "Using Multiple Data Streams and Artificial Intelligence to 'Nowcast' Local Flu Outbreaks."

33. Fliesler, "Using Multiple Data Streams and Artificial Intelligence to 'Nowcast' Local Flu Outbreaks."

34. Fliesler, "Using Multiple Data Streams and Artificial Intelligence to 'Nowcast' Local Flu Outbreaks"; Fred S. Lu et al., "Improved State-Level Influenza Nowcasting in the United States Leveraging Internet-Based Data and Network Approaches," *Nature Communications* 10, article 147 (January 11, 2019), https://doi.org/10.1038/s41467-018-08082-0.

35. For more detail on CheXNet and its successor, CheXpert, see "CheXNet and Beyond," Matthew Lungren, YouTube video, November 10, 2018, https://www.youtube.com/watch ?v=JqYte9UMJCg; Pranav Rajpurkar et al., "CheXNet: Radiologist-Level Pneumonia Detection on Chest X-Rays with Deep Learning," Stanford Machine Learning Group working paper, November 14, 2017, https://arxiv.org/pdf/1711.05225v1.pdf; Jeremy Irvin et al., "CheXpert: A Large Chest Radiograph Dataset with Uncertainty Labels and Expert Comparison," *Proceedings of the AAAI Conference on Artificial Intelligence* 33, no. 1 (July 17, 2019): AAAI-10, IAAI-19, EAAI-20, https://www.aaai.org/ojs/index.php/AAAI/article/view/3834.

36. Huiying Liang et al., "Evaluation and Accurate Diagnoses of Pediatric Diseases Using Artificial Intelligence," *Nature Medicine* 25, no. 3 (February 11, 2019): 433–38, https://doi .org/10.1038/s41591-018-0335-9.

37. Dimitrios Mathios et al., "Detection and Characterization of Lung Cancer Using Cell-Free DNA Fragmentomes," *Nature Communications* 12, article 5060 (August 20, 2021), https://doi.org/10.1038/s41467-021-24994-w.

38. Sophie Bushwick, "Algorithm That Detects Sepsis Cut Deaths by Nearly 20 Percent," *Scientific American*, August 1, 2022, https://www.scientificamerican.com/article/algorithm-that

-detects-sepsis-cut-deaths-by-nearly-20-percent; Roy Adams et al., "Prospective, Multi-Site Study of Patient Outcomes After Implementation of the TREWS Machine Learning-Based Early Warning System for Sepsis," *Nature Medicine* 28 (July 21, 2022): 1455–60, https://doi.org/10.1038/s41591-022-01894-0; Katharine E. Henry et al., "Factors Driving Provider Adoption of the TREWS Machine Learning-Based Early Warning System and Its Effects on Sepsis Treatment Timing," *Nature Medicine* 28 (July 21, 2022), 1447–54, https://doi.org/10.1038/s41591-022-01895-z.

39. Lungren, "CheXNet and Beyond"; Rajpurkar et al., "CheXNet: Radiologist-Level Pneumonia Detection"; Irvin et al., "CheXpert: A Large Chest Radiograph Dataset with Uncertainty Labels and Expert Comparison," AAAI-10, IAAI-19, EAAI-20; Thomas Davenport and Ravi Kalakota, "The Potential for Artificial Intelligence in Healthcare," *Future Healthcare Journal* 6, no. 2 (June 2019): 94–98, https://doi.org/10.7861/futurehosp.6-2-94.

40. Dario Amodei and Danny Hernandez, "AI and Compute," OpenAI, May 16, 2018, https://openai.com/blog/ai-and-compute.

41. Eliza Strickland, "Autonomous Robot Surgeon Bests Humans in World First," *IEEE Spectrum*, May 4, 2016, https://spectrum.ieee.org/the-human-os/robotics/medical-robots/autonomous-robot-surgeon-bests-human-surgeons-in-world-first.

42. Alice Yan, "Chinese Robot Dentist Is First to Fit Implants in Patient's Mouth Without Any Human Involvement," *South China Morning Post*, September 21, 2017, https://www.scmp.com/news/china/article/2112197/chinese-robot-dentist-first-fit-implants-patients-mouth-without-any-human.

43. For Elon Musk's presentation of Neuralink's automated electrode-implantation technology, see "Neuralink: Elon Musk's Entire Brain Chip Presentation in 14 Minutes (Supercut)," CNET, YouTube video, August 28, 2020, https://www.youtube.com/watch?v=CLUWDLKAF1M.

44. Wallace P. Ritchie Jr., Robert S. Rhodes, and Thomas W. Biester, "Work Loads and Practice Patterns of General Surgeons in the United States, 1995–1997," *Annals of Surgery* 230, no. 4 (October 1999): 533–43, https://doi.org/10.1097/00000658-199910000-00009.

45. Hans Moravec, *Mind Children: The Future of Robot and Human Intelligence* (Cambridge, MA: Harvard University Press, 1988).

46. Peter Weibel, "Virtual Worlds: The Emperor's New Bodies," in *Ars Electronica: Facing the Future*, ed. Timothy Druckery (Cambridge, MA: MIT Press, 1999), 215, https://monoskop.org/images/4/47/Ars_Electronica_Facing_the_Future_A_Survey_of_Two_Decades_1999.pdf.

47. "Neuron Firing Rates in Humans," AI Impacts, April 14, 2015, https://aiimpacts.org/rate-of-neuron-firing; Suzana Herculano-Houzel, "The Human Brain in Numbers: A Linearly Scaled-up Primate Brain," *Frontiers in Human Neuroscience* 3, no. 31 (November 9, 2009), https://doi.org/10.3389/neuro.09.031.2009; David A. Drachman, "Do We Have Brain to Spare?," *Neurology* 64, no. 12 (June 27, 2005), https://doi.org/10.1212/01.WNL.0000166914.38327.BB; Antony Leather, "Intel Fires Back at AMD with Fastest Ever Processors: Mobile CPUs with up to 8 Cores and 5.3GHz Inbound," *Forbes*, April 2, 2020, https://www.forbes.com/sites/antonyleather/2020/04/02/intel-fires-back-at-amd-with-fastest-ever-processors-mobile-cpus-with-up-to-8-cores-and-53ghz-inbound/#210c5243643d.

48. Mladen Božanić and Saurabh Sinha, "Emerging Transistor Technologies Capable of Terahertz Amplification: A Way to Re-Engineer Terahertz Radar Sensors," *Sensors* 19, no. 11 (May 29, 2019), https://doi.org/10.3390/s19112454; "Intel Core i9-10900K Processor," Intel, accessed December 23, 2022, https://ark.intel.com/content/www/us/en/ark/products/199332/intel-core-i910900k-processor-20m-cache-up-to-5-30-ghz.html.

49. Ray Kurzweil, *The Singularity Is Near* (New York: Viking, 2005), 125; Moravec, *Mind Children*, 59.

50. "June 2022," Top500.org, accessed October 20, 2022, https://www.top500.org/lists/top500/2022/06.

51. For an essay closely adapted from Feynman's December 29, 1959, lecture, see Richard Feynman, "There's Plenty of Room at the Bottom," *Engineering and Science* 23, no. 5 (February 1960): 22–26, 30–36, http://calteches.library.caltech.edu/47/2/1960Bottom.pdf.

52. Feynman, "There's Plenty of Room at the Bottom," 22–26, 30–36.
53. John von Neumann, *Theory of Self-reproducing Automata* (Urbana, IL: University of Illinois Press, 1966), https://archive.org/details/theoryofselfrepr00vonn_0/mode/2up; John G. Kemeny, "Man Viewed as a Machine," *Scientific American* 192, no. 4 (April 1955): 58–67, in nearly complete form at https://dijkstrascry.com/sites/default/files/papers/JohnKemenyMan ViewedasaMachine.pdf.
54. Von Neumann, *Theory of Self-reproducing Automata*, 251–96, 377.
55. For Drexler's original books, see K. Eric Drexler, *Engines of Creation: The Coming Era of Nanotechnology* (New York: Anchor Press/Doubleday, 1986); K. Eric Drexler, *Nanosystems: Molecular Machinery, Manufacturing, and Computation* (Hoboken, NJ: Wiley, 1992).
56. Drexler, *Engines of Creation*, 18–19, 105–8, 247.
57. Drexler, *Nanosystems*, 343–66.
58. Drexler, *Nanosystems*, 354–55.
59. Ralph C. Merkle, et al., "Mechanical Computing Systems Using Only Links and Rotary Joints," *Journal of Mechanisms and Robotics*, Vol. 10, no. 6, article 061006, September 17, 2018, arXiv:1801.03534v2 [cs.ET], March 25, 2019, https://arxiv.org/pdf/1801 .03534.pdf.
60. The design by Merkle et al. for a "molecular mechanical logic gate" consists of 87,595 carbon atoms and 33,100 hydrogen atoms. It occupies a volume of about 27 nm x 32 nm x 7 nm, or 6,048 cubic nanometers. This corresponds to about 1.65×10^{20} (165 quintillion) logic gates per liter of volume. At a designed operating frequency of 100 MHz, this suggests a maximum of 10^{28} gate operations per second per liter of computing volume. Keep in mind that these are highly theoretical maximums excluding engineering limitations—it remains to be seen how close to these maximums nanoscale computers are actually able to get. See Merkle et al., "Mechanical Computing Systems Using Only Links and Rotary Joints," arXiv, 24–27; Drexler, *Nanosystems*, 370–71.
61. According to Merkle et al., each operation of the logic gate described above would expend on the order of 10^{-26} joules (an order of magnitude more than the 10^{-27} per single rotary joint). Thus, the 10^{28} operations per second in this hypothetical one-liter computer would expend on the order of 100 watts. See Merkle et al., "Mechanical Computing Systems Using Only Links and Rotary Joints," arXiv, 24–27.
62. Liqun Luo, "Why Is the Human Brain So Efficient?," *Nautilus*, April 12, 2018, http://nautil .us/issue/59/connections/why-is-the-human-brain-so-efficient.
63. Ralph C. Merkle, "Design Considerations for an Assembler," *Nanotechnology* 7, no. 3 (September 1996): 210–15, https://doi.org/10.1088/0957-4484/7/3/008, mirrored in similar version at http://www.zyvex.com/nanotech/nano4/merklePaper.html.
64. Merkle, "Design Considerations for an Assembler," 210–15; Ralph. C. Merkle, "Self Replicating Systems and Molecular Manufacturing," *Journal of the British Interplanetary Society* 45, no. 12 (December 1992): 407–13, available in adapted version at http://www.zyvex .com/nanotech/selfRepJBIS.html; Neil Jacobstein, "Foresight Guidelines for Responsible Nanotechnology Development," Foresight Institute, April 2006, https://foresight.org /guidelines/current.php.
65. Robert A. Freitas Jr., "The Gray Goo Problem," KurzweilAI.net, March 20, 2001, https:// www.kurzweilai.net/the-gray-goo-problem; Robert A. Freitas Jr., "Some Limits to Global Ecophagy by Biovorous Nanoreplicators, with Public Policy Recommendations," Foresight Institute, April 2000, http://www.rfreitas.com/Nano/Ecophagy.htm.
66. James Lewis, "Ultrafast DNA Robotic Arm: A Step Toward a Nanofactory?," Foresight Institute, January 25, 2018, https://foresight.org/ultrafast-robotic-arm-step-toward-nano factory; Kohji Tomita et al., "Self-Description for Construction and Execution in Graph Rewriting Automata," in *Advances in Artificial Life: 8th European Conference, ECAL 2005, Canterbury, UK, September 5–9, 2005, Proceedings*, ed. Mathieu S. Capcarrere et al. (Heidelberg, Germany: Springer Science & Business Media, 2005), 705–14.
67. For more detail on successful attempts to build working nanoscale machines and machine parts, see Eric Drexler, "Big Nanotech: Building a New World with Atomic Precision,"

Guardian, October 21, 2013, https://www.theguardian.com/science/small-world/2013/oct/21/big-nanotech-atomically-precise-manufacturing-apm; Mark Peplow, "The Tiniest Lego: A Tale of Nanoscale Motors, Rotors, Switches and Pumps," *Nature* 525, no. 7567 (September 2, 2015): 18–21, https://doi.org/10.1038/525018a; Carlos Manzano et al., "Step-by-Step Rotation of a Molecule-Gear Mounted on an Atomic-Scale Axis," *Nature Materials* 8, no. 6 (June 14, 2009): 576–79, https://doi.org/10.1038/nmat2467; Babak Kateb and John D. Heiss, *The Textbook of Nanoneuroscience and Nanoneurosurgery* (Boca Raton, FL: CRC Press, 2013): 500–501, https://www.google.com/books/edition/The_Textbook_of_Nanoneuroscience_and_Nan/rCbOBQAAQBAJ; Torben Jasper-Toennies et al., "Rotation of Ethoxy and Ethyl Moieties on a Molecular Platform on Au(111)," *ACS Nano* 14, no. 4 (February 19, 2020): 3907–16, https://doi.org/10.1021/acsnano.0c00029; Kwanoh Kim et al., "Man-Made Rotary Nanomotors: A Review of Recent Development," *Nanoscale* 8, no. 20 (May 19, 2016): 10471–90, https://doi.org/10.1039/c5nr08768f; The Optical Society, "Nanoscale Machines Convert Light into Work," Phys.org, October 8, 2020, https://phys.org/news/2020-10-nanoscale-machines.html.

68. For the original Drexler–Smalley debate, my commentary, two talks by Drexler, and a sampling of recent research paving the way for molecular assemblers, see Richard E. Smalley, "Of Chemistry, Love and Nanobots," *Scientific American*, September 2001, https://www.scientificamerican.com/article/of-chemistry-love-and-nanobots; Rudy Baum, "Nanotechnology: Drexler and Smalley Make the Case for and Against 'Molecular Assemblers,'" *Chemical & Engineering News* 81, no. 48 (September 8, 2003): 37–42, https://web.archive.org/web/20230116122623/http://pubsapp.acs.org/cen/coverstory/8148/8148counterpoint.html; Eric Drexler, "Transforming the Material Basis of Civilization | Eric Drexler | TEDxISTAlameda," TEDx Talks, YouTube video, November 16, 2015, https://www.youtube.com/watch?v=Q9RiB_o7Szs; Eric Drexler, "Dr. Eric Drexler—The Path to Atomically Precise Manufacturing," The Artificial Intelligence Channel, YouTube video, September 18, 2017, https://www.youtube.com/watch?v=dAA-HWMaF9o; UT-Battelle, *Productive Nanosystems: A Technology Roadmap*, Battelle Memorial Institute and Foresight Nanotech Institute, 2007, https://foresight.org/wp-content/uploads/2023/05/Nanotech_Roadmap_2007_main.pdf; James Lewis, "Atomically Precise Manufacturing as the Future of Nanotechnology," Foresight Institute, March 8, 2015, https://foresight.org/atomically-precise-manufacturing-as-the-future-of-nanotechnology; Xiqiao Wang et al., "Atomic-Scale Control of Tunneling in Donor-Based Devices," *Communications Physics* 3, article 82 (May 11, 2020), https://doi.org/10.1038/s42005-020-0343-1; "Paving the Way for Atomically Precise Manufacturing," UT Dallas, YouTube video, February 9, 2018, https://www.youtube.com/watch?v=or3jYNZ6fn8; University of Texas at Dallas, "Microscopy Breakthrough Paves the Way for Atomically Precise Manufacturing," Phys.org, February 12, 2018, https://phys.org/news/2018-02-microscopy-breakthrough-paves-atomically-precise.html; Kiel University, "Towards a Light Driven Molecular Assembler," Phys.org, July 23, 2019, https://phys.org/news/2019-07-driven-molecular.html; Jonathan Wyrick et al., "Atom-by-Atom Fabrication of Single and Few Dopant Quantum Devices," *Advanced Functional Materials* 29, no. 52 (August 14, 2019), https://doi.org/10.1002/adfm.201903475; Farid Tajaddodianfar et al., "On the Effect of Local Barrier Height in Scanning Tunneling Microscopy: Measurement Methods and Control Implications," *Review of Scientific Instruments* 89, no. 1, article 013701 (January 2, 2018), https://doi.org/10.1063/1.5003851.

69. Ray Kurzweil, "The Drexler-Smalley Debate on Molecular Assembly," KurzweilAI.net, December 1, 2003, https://www.kurzweilai.net/the-drexler-smalley-debate-on-molecular-assembly.

70. Drexler, *Nanosystems*, 398–410.

71. Drexler, *Nanosystems*, 238–49, 458–68.

72. Drexler, *Engines of Creation*; Drexler, *Nanosystems*; Dexter Johnson, "Diamondoids on Verge of Key Application Breakthroughs," *IEEE Spectrum*, March 31, 2017, https://spectrum.ieee.org/nanoclast/semiconductors/materials/diamondoids-on-verge-of-key-application-breakthroughs.

73. Neal Stephenson, *The Diamond Age: Or, a Young Lady's Illustrated Primer* (New York: Bantam, 1995).

74. Matthew A. Gebbie et al., "Experimental Measurement of the Diamond Nucleation Landscape Reveals Classical and Nonclassical Features," *Proceedings of the National Academy of Sciences* 115, no. 33 (August 14, 2018): 8284–89, https://doi.org/10.1073/pnas.1803654115.

75. Hongyao Xie et al., "Large Thermal Conductivity Drops in the Diamondoid Lattice of CuFeS$_2$ by Discordant Atom Doping," *Journal of the American Chemical Society* 141, no. 47 (November 2, 2019): 18900–909, https://doi.org/10.1021/jacs.9b10983; Shenggao Liu, Jeremy Dahl, and Robert Carlson, "Heteroatom-Containing Diamondoid Transistors," U.S. Patent 7,402,835 (filed July 16, 2003; issued July 22, 2008), US Patent and Trademark Office, https://patents.google.com/patent/US7402835B2/en.

76. See, for example, Robert A. Freitas Jr., "A Simple Tool for Positional Diamond Mechanosynthesis, and Its Method of Manufacture," U.S. Patent 7,687,146 (filed February 11, 2005; issued March 30, 2010), US Patent and Trademark Office, https://patents.google .com/patent/US7687146B1/en; Samuel Stolz et al., "Molecular Motor Crossing the Frontier of Classical to Quantum Tunneling Motion," *Proceedings of the National Academy of Sciences* 117, no. 26 (June 15, 2020): 14838–42, https://doi.org/10.1073/pnas.1918654117; Haifei Zhan et al., "From Brittle to Ductile: A Structure Dependent Ductility of Diamond Nanothread," *Nanoscale* 8, no. 21 (May 10, 2016): 11177–84, https://doi.org/10.1039 /C6NR02414Ad.

77. For more on this proposal and links to a wide range of Merkle's other nanotechnology papers, see Ralph C. Merkle, "A Proposed 'Metabolism' for a Hydrocarbon Assembler," *Nanotechnology* 8, no. 4 (December 1997): 149–62, https://iopscience.iop.org/article /10.1088/0957-4484/8/4/001/meta, mirrored at http://www.zyvex.com/nanotech/hydro CarbonMetabolism.html; "Papers by Ralph C. Merkle," Merkle.com, accessed October 20, 2022, http://www.merkle.com/merkleDir/papers.html.

78. Merkle, "A Proposed 'Metabolism' for a Hydrocarbon Assembler."

79. For a small sampling of advances in graphene, carbon nanotubes, and carbon nanothreads, including an impressive MIT project to create a 14,000-transistor computer chip made entirely from carbon nanotubes, see "The Graphene Times," *Nature Nanotechnology* 14, no. 10, article 903 (October 3, 2019), https://doi.org/10.1038/s41565-019-0561-4; "Nova: Car bon Nanotubes," Mangefox, YouTube video, January 28, 2011, https://www.youtube.com /watch?v=19nzPt62UPg; Elizabeth Gibney, "Biggest Carbon-Nanotube Chip Yet Says 'Hello, World!,'" *Nature*, August 28, 2019, https://doi.org/10.1038/d41586-019-02576-7; Haifei Zhan et al., "The Best Features of Diamond Nanothread for Nanofibre Applications," *Nature Communications* 8, article 14863 (March 17, 2017), https://doi.org/10.1038 /ncomms14863; Haifei Zhan et al., "High Density Mechanical Energy Storage with Carbon Nanothread Bundle," *Nature Communications* 11, article 1905 (April 20, 2020), https:// doi.org/10.1038/s41467-020-15807-7; Keigo Otsuka et al., "Deterministic Transfer of Optical-Quality Carbon Nanotubes for Atomically Defined Technology," *Nature Communications* 12, article 3138 (May 25, 2021), https://doi.org/10.1038/s41467-021-23413-4.

80. Probably the most detailed work so far on how to carry out diamondoid-based mechanosynthesis is Robert Freitas and Ralph Merkle's 2008 study of the reaction pathways needed for versatile nanomanufacturing. See Robert A. Freitas Jr. and Ralph C. Merkle, "A Minimal Toolset for Positional Diamond Mechanosynthesis," *Journal of Computational and Theoretical Nanoscience* 5, no. 5 (May 2008): 760–862, https://doi.org/10.1166/jctn.2008.2531, mirrored at http://www.molecularassembler.com/Papers/MinToolset.pdf.

81. Masayuki Endo and Hiroshi Sugiyama, "DNA Origami Nanomachines," *Molecules* 23, no. 7 (article 1766), July 18, 2018, https://doi.org/10.3390/molecules23071766; Fei Wang et al., "Programming Motions of DNA Origami Nanomachines," *Small* 15, no. 26, article 1900013 (March 25, 2019), https://doi.org/10.1002/smll.201900013.

82. Suping Li et al., "A DNA Nanorobot Functions as a Cancer Therapeutic in Response to a Molecular Trigger *In Vivo*," *Nature Biotechnology* 36, no. 3 (February 12, 2018): 258–64, https://doi.org/10.1038/nbt.4071; Stephanie Lauback et al., "Real-Time Magnetic Actua-

tion of DNA Nanodevices via Modular Integration with Stiff Micro-Mevers," *Nature Communications* 9, no. 1, article 1446 (April 13, 2018), https://doi.org/10.1038/s41467-018 -03601-5.

83. Liang Zhang, Vanesa Marcos, and David A. Leigh, "Molecular Machines with Bio-Inspired Mechanisms," *Proceedings of the National Academy of Sciences* 115, no. 38 (February 26, 2018), https://doi.org/10.1073/pnas.1712788115.

84. Christian E. Schafmeister, "Molecular Lego," *Scientific American*, February 2007, https://www.scientificamerican.com/article/molecular-lego.

85. Matthias Koch et al., "Spin Read-Out in Atomic Qubits in an All-Epitaxial Three-Dimensional Transistor," *Nature Nanotechnology* 14, no. 2 (January 7, 2019): 137–40, https://doi.org/10.1038/s41565-018-0338-1.

86. Mukesh Tripathi et al., "Electron-Beam Manipulation of Silicon Dopants in Graphene," *Nano Letters* 18, no. 8 (June 27, 2018): 5319–23, https://doi.org/10.1021/acs.nanolett .8b02406.

87. John N. Randall et al., "Digital Atomic Scale Fabrication an Inverse Moore's Law—A Path to Atomically Precise Manufacturing," *Micro and Nano Engineering* 1 (November 2018): 1–14, https://doi.org/10.1016/j.mne.2018.11.001.

88. Roshan Achal et al., "Lithography for Robust and Editable Atomic-Scale Silicon Devices and Memories," *Nature Communications* 9, no. 1, article 2778 (July 23, 2018), https://doi .org/10.1038/s41467-018-05171-y.

89. Chalmers University of Technology, "Graphene and Other Carbon Nanomaterials Can Replace Scarce Metals," Phys.org, September 19, 2017, https://phys.org/news/2017-09 -graphene-carbon-nanomaterials-scarce-metals.html; Rickard Arvisson and Björn A. Sandén, "Carbon Nanomaterials as Potential Substitutes for Scarce Metals," *Journal of Cleaner Production* 156 (July 10, 2017): 253–61, https://doi.org/10.1016/j.jclepro.2017.04.048.

90. K. Eric Drexler, *Radical Abundance: How a Revolution in Nanotechnology Will Change Civilization* (New York: PublicAffairs, 2013), 168–72.

91. Paul Sullivan, "A Battle over Diamonds: Made by Nature or in a Lab?," *New York Times*, February 9, 2018, https://www.nytimes.com/2018/02/09/your-money/synthetic-diamond -jewelry.html.

92. Milton Esterow, "Art Experts Warn of a Surging Market in Fake Prints," *New York Times*, January 24, 2020, https://www.nytimes.com/2020/01/24/arts/design/fake-art-prints.html; Kelly Crow, "Leonardo da Vinci Painting 'Salvator Mundi' Smashes Records with $450.3 Million Sale," *Wall Street Journal*, November 16, 2017, https://www.wsj.com/articles /leonardo-da-vinci-painting-salvator-mundi-sells-for-450-3-million-1510794281.

93. Ray Kurzweil and Terry Grossman, *Transcend: Nine Steps to Living Well Forever* (Emmaus, PA: Rodale, 2009).

94. For further resources on recent biogerontology research aimed at understanding and curing aging, see "Why Age? Should We End Aging Forever?," Kurzgesagt—In a Nutshell, YouTube video, October 20, 2017, https://www.youtube.com/watch?v=GoJsr4IwCm4; "How to Cure Aging—During Your Lifetime?," Kurzgesagt—In a Nutshell, YouTube video, November 3, 2017, https://www.youtube.com/watch?v=MjdpR-TY6QU; "Daphne Koller, Chief Computing Officer, Calico Labs," CB Insights, YouTube video, January 18, 2018, https://www.youtube.com/watch?v=0EIZ8wJYAEA; "Ray Kurzweil—Physical Immortality," Aging Reversed, YouTube video, January 3, 2017, https://www.youtube.com /watch?v=BUExzREe9oo; Peter H. Diamandis, "Nanorobots: Where We Are Today and Why Their Future Has Amazing Potential," SingularityHub, May 16, 2016, https://singu larityhub.com/2016/05/16/nanorobots-where-we-are-today-and-why-their-future-has -amazing-potential.

95. Nicola Davis, "Human Lifespan Has Hit Its Natural Limit, Research Suggests," *Guardian*, October 5, 2016, https://www.theguardian.com/science/2016/oct/05/human-lifespan-has -hit-its-natural-limit-research-suggests; Craig R. Whitney, "Jeanne Calment, World's Elder, Dies at 122," *New York Times*, August 5, 1997, https://www.nytimes.com/1997/08 /05/world/jeanne-calment-world-s-elder-dies-at-122.html.

96. "Actuarial Life Table," US Social Security Administration, accessed October 20, 2022, https://www.ssa.gov/oact/STATS/table4c6.html.

97. France Meslé and Jacques Vallin, "Causes of Death at Very Old Ages, Including for Super-centenarians," in *Exceptional Lifespans*, ed. Heiner Maier et al. (Cham, Switzerland: Springer, 2020): 72–82, https://link.springer.com/content/pdf/10.1007/978-3-030-49970-9.pdf?pdf=button.

98. "Aubrey De Grey—Living to 1,000 Years Old," Aging Reversed, YouTube video, May 26, 2018, https://www.youtube.com/watch?v=ZkMPZ8obByw; "One-on-One: An Investigative Interview with Aubrey de Grey—44th St. Gallen Symposium," StGallenSymposium, YouTube video, May 8, 2014, https://www.youtube.com/watch?v=DkBfT_EPBIo; "Aubrey de Grey, PhD: 'The Science of Curing Aging,'" Talks at Google, YouTube video, January 4, 2018, https://www.youtube.com/watch?v=S6ARUQ5LoUo.

99. "A Reimagined Research Strategy for Aging," SENS Research Foundation, accessed December 27, 2022, https://web.archive.org/web/20221118080039/https://www.sens.org/our-research/intro-to-sens-research.

100. "Longevity: Reaching Escape Velocity," Foresight Institute, YouTube video, December 12, 2017, https://www.youtube.com/watch?v=M4b19vZ57U4.

101. Richard Zijdeman and Filipa Ribeira da Silva, "Life Expectancy at Birth (Total)," IISH Data Collection, V1 (2015), https://hdl.handle.net/10622/LKYT53.

102. Robert A. Freitas Jr., "The Life-Saving Future of Medicine," *Guardian*, March 28, 2014, https://www.theguardian.com/what-is-nano/nano-and-the-life-saving-future-of-medicine.

103. Jacqueline Krim, "Friction at the Nano-Scale," *Physics World*, February 2, 2005, https://physicsworld.com/a/friction-at-the-nano-scale.

104. Rose Eveleth, "There Are 37.2 Trillion Cells in Your Body," *Smithsonian Magazine*, October 24, 2013, https://www.smithsonianmag.com/smart-news/there-are-372-trillion-cells-in-your-body-4941473.

105. For helpful, accessible explainers on the immune system and hormones, see "How the Immune System Actually Works—Immune," Kurzgesagt—In a Nutshell, YouTube video, August 10, 2021, https://www.youtube.com/watch?v=lXfEK8G8CUI; "How Does Your Immune System Work?—Emma Bryce," TED-Ed, YouTube video, January 8, 2018, https://www.youtube.com/watch?v=PSRJfaAYkW4; "How Do Your Hormones Work?—Emma Bryce," TED-Ed, YouTube video, June 21, 2018, https://www.youtube.com/watch?v=-SPRPkLoKp8.

106. For a helpful, accessible explainer on the lungs, see "How Do Lungs Work?—Emma Bryce," TED-Ed, YouTube video, November 24, 2014, https://www.youtube.com/watch?v=8NUxvJS-_0k.

107. For a helpful, accessible explainer on the kidneys, see "How Do Your Kidneys Work?—Emma Bryce," TED-Ed, YouTube video, February 9, 2015, https://www.youtube.com/watch?v=FN3MFhYPWWo.

108. For a helpful, accessible explainer on the digestive system, see "How Your Digestive System Works—Emma Bryce," TED-Ed, YouTube video, December 14, 2017, https://www.youtube.com/watch?v=Og5xAdC8EUI.

109. For a helpful, accessible explainer on the role and function of the pancreas, see "What Does the Pancreas Do?—Emma Bryce," TED-Ed, YouTube video, February 19, 2015, https://www.youtube.com/watch?v=8dgoeYPoE-0.

110. George Dvorsky, "FDA Approves World's First Automated Insulin Pump for Diabetics," *Gizmodo*, September 29, 2016, https://gizmodo.com/fda-approves-worlds-first-automated-insulin-pump-for-di-1787227150.

111. For helpful, accessible explainers on the role of hormones in diabetes, see "Role of Hormones in Diabetes," Match Health, YouTube video, December 6, 2013, https://www.youtube.com/watch?v=sPwoMm9cv1M; Matthew McPheeters, "What Is Diabetes Mellitus? | Endocrine System Diseases | NCLEX-RN," Khan Academy Medicine, YouTube video, May 14, 2015, https://www.youtube.com/watch?v=ulxyWZf7BWc.

112. For a short nontechnical explainer of the relationship between sleep and hormones, see Hormone Health Network, "Sleep and Circadian Rhythm," Hormone.org, Endocrine So-

ciety, June 2019, https://www.hormone.org/your-health-and-hormones/sleep-and-circadian-rhythm.

113. For more on cancer recurrence and cancer stem cells, see "Why Is It So Hard to Cure Cancer?—Kyuson Yun," TED-Ed, YouTube video, October 10, 2017, https://www.youtube.com/watch?v=h2rR77VsF5c; "Recurrent Cancer: When Cancer Comes Back," National Cancer Institute, January 18, 2016, https://www.cancer.gov/types/recurrent-cancer; Kyle Davis, "Investigating Why Cancer Comes Back," National Human Genome Research Institute, September 8, 2015, https://www.genome.gov/news/news-release/Investigating-why-cancer-comes-back.

114. Zuoren Yu et al., "Cancer Stem Cells," *International Journal of Biochemical Cell Biology* 44, no. 12 (December 2012): 2144–51, https://doi.org/10.1016/j.biocel.2012.08.022.

115. "How Does Chemotherapy Work?—Hyunsoo Joshua No," TED-Ed, YouTube video, December 5, 2019, https://www.youtube.com/watch?v=RgWQCGX3MOk; "Why People with Cancer Are More Likely to Get Infections," American Cancer Society, March 13, 2020, https://www.cancer.org/treatment/treatments-and-side-effects/physical-side-effects/low-blood-counts/infections/why-people-with-cancer-are-at-risk.html.

116. Nirali Shah and Terry J. Fry, "Mechanisms of Resistance to CAR T Cell Therapy," *Nature Reviews Clinical Oncology* 16 (March 5, 2019): 372–85, https://doi.org/10.1038/s41571-019-0184-6; Robert Vander Velde et al., "Resistance to Targeted Therapies as a Multifactorial, Gradual Adaptation to Inhibitor Specific Selective Pressures," *Nature Communications* 11, article 2393 (May 14, 2020), https://doi.org/10.1038/s41467-020-16212-w.

117. For a great nontechnical explainer on cellular reproduction, see Hank Green, "Mitosis: Splitting Up Is Complicated—Crash Course Biology #12," CrashCourse, YouTube video, April 16, 2012, https://www.youtube.com/watch?v=L0k-enzoeOM.

118. For a short introduction to CRISPR, one of the most promising gene-editing approaches within the current paradigm, see Brad Plumer et al., "A Simple Guide to CRISPR, One of the Biggest Science Stories of the Decade," *Vox*, updated December 27, 2018, https://www.vox.com/2018/7/23/17594864/crispr-cas9-gene-editing.

119. Eveleth, "There Are 37.2 Trillion Cells in Your Body."

120. For an introductory overview of this process, see "Regulation of Gene Expression: Operons, Epigenetics, and Transcription Factors," Professor Dave Explains, YouTube video, October 15, 2017, https://www.youtube.com/watch?v=J9jhg90A7Lw.

121. Bert M. Verheijen and Fred W. van Leeuwen, "Commentary: The Landscape of Transcription Errors in Eukaryotic Cells," *Frontiers in Genetics* 8, article 219 (December 14, 2017), https://doi.org/10.3389/fgene.2017.00219.

122. Patricia Mroczek, "Nanoparticle Chomps Away Plaques That Cause Heart Attacks," MSUToday, Michigan State University, January 27, 2020, https://msutoday.msu.edu/news/2020/nanoparticle-chomps-away-plaques-that-cause-heart-attacks; Alyssa M. Flores, *Nature Nanotechnology* 15, no. 2 (January 27, 2020): 154–61, https://doi.org/10.1038/s41565-019-0619-3; Ira Tabas and Andrew H. Lichtman, "Monocyte-Macrophages and T Cells in Atherosclerosis," *Immunity* 47, no. 4 (October 17, 2017): 621–34, https://doi.org/10.1016/j.immuni.2017.09.008.

123. American Stroke Association "Understanding Diagnosis and Treatment of Cryptogenic Stroke: A Health Care Professional Guide," American Heart Association, 2019, https://web.archive.org/web/20211023144019/https://www.stroke.org/-/media/stroke-files/cryptogenic-professional-resource-files/crytopgenic-professional-guide-ucm-477051.pdf.

124. For a lively entry-level explainer of protein folding, see "Protein Structure and Folding," Amoeba Sisters, YouTube video, September 24, 2018, https://www.youtube.com/watch?v=hok2hyED9go.

125. The (very theoretical) maximum firing rate for biological neurons is somewhere around 1,000 hertz, while Ralph Merkle's nanoscale mechanical computing system could achieve around 100 megahertz—about 100,000 times faster. Collagen fibrils in the body have a tensile strength of around 90 megapascals, while multi-walled carbon nanotubes have been shown experimentally to have a tensile strength around 63 gigapascals, and diamond nanoneedles up to 98 gigapascals, while diamondoid's theoretical maximum is around 100

gigapascals—all on the order of 1,000 times stronger than collagen. See "Neuron Firing Rates in Humans," AI Impacts; Ralph C. Merkle et al., "Mechanical Computing Systems Using Only Links and Rotary Joints," 24–27; Yehe Liu, Roberto Ballarini, and Steven J. Eppell, "Tension Tests on Mammalian Collagen Fibrils," *Interface Focus* 6, no. 1, article 20150080 (February 6, 2016), https://doi.org/10.1098/rsfs.2015.0080; Min-Feng Yu et al., "Strength and Breaking Mechanism of Multiwalled Carbon Nanotubes Under Tensile Load," *Science* 287, no. 5453 (January 28, 2000): 637–40, https://doi.org/10.1126/science.287.5453.637; Amit Banerjee et al., "Ultralarge Elastic Deformation of Nanoscale Diamond," *Science* 360, no. 6386 (April 20, 2018): 300–302, https://doi.org/10.1126/science.aar4165; Drexler, *Nanosystems*, 24–35, 142–43.

126. Robert A. Freitas, "Exploratory Design in Medical Nanotechnology: A Mechanical Artificial Red Cell," *Artificial Cells, Blood Substitutes, and Biotechnology* 26, no. 4 (1998): 411–30, https://doi.org/10.3109/10731199809117682.

127. Freitas, "Exploratory Design in Medical Nanotechnology," 426; Robert A. Freitas Jr., "Respirocytes: A Mechanical Artificial Red Cell: Exploratory Design in Medical Nanotechnology," Foresight Institute/Institute for Molecular Manufacturing," April 17, 1996, https://web.archive.org/web/20210509160649/https://foresight.org/Nanomedicine/Respirocytes.php.

128. Herculano-Houzel, "The Human Brain in Numbers"; Drachman, "Do We Have Brain to Spare?"; Hervé Lemaître et al., "Normal Age-Related Brain Morphometric Changes: Nonuniformity Across Cortical Thickness, Surface Area and Grey Matter Volume?," *Neurobiology of Aging* 33, no. 3 (March 2012): 617.e1–617.e9, https://doi.org/10.1016/j.neurobiolaging.2010.07.013; Merkle et al., "Mechanical Computing Systems Using Only Links and Rotary Joints," 24–27.

129. "Neuron Firing Rates in Humans," AI Impacts; Merkle et al., "Mechanical Computing Systems Using Only Links and Rotary Joints," 24–27; Drexler, *Nanosystems*, 370–71.

130. Herculano-Houzel, "The Human Brain in Numbers"; Drachman, "Do We Have Brain to Spare?"; "Firing Behavior and Network Activity of Single Neurons in Human Epileptic Hypothalamic Hamartoma," *Frontiers in Neurology* 2, no. 210 (December 27, 2013), https://doi.org/10.3389/fneur.2013.00210; Ernest L. Abel, *Behavioral Teratogenesis and Behavioral Mutagenesis: A Primer in Abnormal Development* (New York: Plenum Press, 1989), 113, https://books.google.co.uk/books?id=gV0rBgAAQBAJ; Anders Sandberg and Nick Bostrom, *Whole Brain Emulation: A Roadmap*, technical report 2008-3, Future of Humanity Institute, Oxford University (2008), 80, https://www.fhi.ox.ac.uk/brain-emulation-roadmap-report.pdf; see the appendix for the sources used for all the cost-of-computation calculations in this book.

131. Herculano-Houzel, "The Human Brain in Numbers"; Drachman, "Do We Have Brain to Spare?"; "Firing Behavior and Network Activity of Single Neurons in Human Epileptic Hypothalamic Hamartoma"; Abel, *Behavioral Teratogenesis and Behavioral Mutagenesis*; Sandberg and Bostrom, *Whole Brain Emulation*, 80; see the appendix for the sources used for all the cost-of-computation calculations in this book.

CHAPTER 7: PERIL

1. Bill McKibben, "How Much Is Enough? The Environmental Movement as a Pivot Point in Human History," Harvard Seminar on Environmental Values, October 18, 2000, 11, http://docshare04.docshare.tips/files/9552/95524564.pdf.

2. Robert M. Pirsig, *Zen and the Art of Motorcycle Maintenance* (New York: Quill, 1999, 26; first published by William Morrow, 1974). Pirsig's work blended Eastern and Western ideas and introduced the philosophical framework known as the Metaphysics of Quality, which focuses on how people's lived experiences give rise to knowledge and ideas. It has become one of the top-selling philosophy books of all time. See also Tim Adams, "The Interview: Robert Pirsig," *Guardian*, November 19, 2006, https://www.theguardian.com/books/2006/nov/19/fiction.

3. Hans M. Kristensen and Matt Korda, "Status of World Nuclear Forces," Federation of American Scientists, March 2, 2022, https://fas.org/issues/nuclear-weapons/status-world-nuclear-forces.

4. Hans M. Kristensen, "Alert Status of Nuclear Weapons," Briefing to Short Course on Nuclear Weapon and Related Security Issues, George Washington University Elliott School of International Affairs, April 21, 2017, 2, https://uploads.fas.org/2014/05/Brief2017_GWU_2s.pdf.

5. You can explore the effects of nuclear war for yourself with a fascinating interactive tool by Alex Wellerstein called Nukemap, available at https://nuclearsecrecy.com/nukemap. See also Kyle Mizokami, "335 Million Dead: If America Launched an All-Out Nuclear War," *National Interest*, March 13, 2019, https://nationalinterest.org/blog/buzz/335-million-dead-if-america-launched-all-out-nuclear-war-57262; Dylan Matthews, "40 Years Ago Today, One Man Saved Us from World-Ending Nuclear War," *Vox*, September 26, 2023, https://www.vox.com/2018/9/26/17905796/nuclear-war-1983-stanislav-petrov-soviet-union; Owen B. Toon et al., "Rapidly Expanding Nuclear Arsenals in Pakistan and India Portend Regional and Global Catastrophe," *Science Advances* 5, no. 10 (October 2, 2019), https://advances.sciencemag.org/content/5/10/eaay5478.

6. Seth Baum, "The Risk of Nuclear Winter," Federation of American Scientists, May 29, 2015, https://fas.org/pir-pubs/risk-nuclear-winter; Bryan Walsh, "What Could a Nuclear War Do to the Climate—and Humanity?," *Vox*, August 17, 2022, https://www.vox.com/future-perfect/2022/8/17/23306861/nuclear-winter-war-climate-change-food-starvation-existential-risk-russia-united-states.

7. Anders Sandberg and Nick Bostrom, *Global Catastrophic Risks Survey*, technical report 2008-1, Future of Humanity Institute, Oxford University (2008): 1, https://www.fhi.ox.ac.uk/reports/2008-1.pdf.

8. Kristensen and Korda, "Status of World Nuclear Forces"; Arms Control Association, "Nuclear Weapons: Who Has What at a Glance," Arms Control Association, June 2023, https://www.armscontrol.org/factsheets/Nuclearweaponswhohaswhat.

9. For a useful summary of relevant arms control treaties, see "U.S.-Russian Nuclear Arms Control Agreements at a Glance," Arms Control Association, August 2019, https://www.armscontrol.org/factsheets/USRussiaNuclearAgreements.

10. Max Roser and Mohamed Nagdy, "Nuclear Weapons," Our World in Data, 2019, https://ourworldindata.org/nuclear-weapons; Hans M. Kristensen and Robert S. Norris, "The Bulletin of the Atomic Scientists' Nuclear Notebook," Federation of American Scientists, 2019, https://thebulletin.org/nuclear-notebook-multimedia; Kristensen and Korda, "Status of World Nuclear Forces."

11. Treaty Banning Nuclear Weapon Tests in the Atmosphere, in Outer Space, and Under Water, October 10, 1963, https://treaties.un.org/doc/Publication/UNTS/Volume%20480/volume-480-I-6964-English.pdf.

12. For a useful summary of relevant international law, see "International Legal Agreements Relevant to Space Weapons," Union of Concerned Scientists, February 11, 2004, https://www.ucsusa.org/nuclear-weapons/space-weapons/international-legal-agreements.

13. Scholars began to recognize this decades ago, for example in Martin E. Hellman, "Arms Race Can Only Lead to One End: If We Don't Change Our Thinking, Someone Will Drop the Big One," *Houston Post*, April 4, 1985, available in very similar form at https://ee.stanford.edu/~hellman/opinion/inevitability.html.

14. For a clear and concise summary of how mutually assured destruction works, see "Mutually Assured Destruction: When the Only Winning Move Is Not to Play," *Farnam Street*, June 2017, https://fs.blog/2017/06/mutually-assured-destruction.

15. For more on US missile defense programs, see "Current U.S. Missile Defense Programs at a Glance," Arms Control Association, August 2019, https://www.armscontrol.org/factsheets/usmissiledefense.

16. Alan Robock and Owen Brian Toon, "Self-Assured Destruction: The Climate Impacts of Nuclear War," *Bulletin of the Atomic Scientists* 68, no. 5 (September 1, 2012): 66–74, https://thebulletin.org/2012/09/self-assured-destruction-the-climate-impacts-of-nuclear-war.

17. Valerie Insinna, "Russia's Nuclear Underwater Drone Is Real and in the Nuclear Posture Review," *DefenseNews*, January 12, 2018, https://www.defensenews.com/space/2018/01/12/russias-nuclear-underwater-drone-is-real-and-in-the-nuclear-posture-review; Douglas Barrie and Henry Boyd, "Burevestnik: US Intelligence and Russia's 'Unique' Cruise Missile," International Institute for Strategic Studies, February 5, 2021, https://www.iiss.org/blogs/military-balance/2021/02/burevestnik-russia-cruise-missile.

18. Richard Stone, "'National Pride Is at Stake.' Russia, China, United States Race to Build Hypersonic Weapons," *Science*, January 8, 2020, https://www.science.org/content/article/national-pride-stake-russia-china-united-states-race-build-hypersonic-weapons.

19. Joshua M. Pearce and David C. Denkenberger, "A National Pragmatic Safety Limit for Nuclear Weapon Quantities," *Safety* 4, no. 2 (2018): 25, https://www.mdpi.com/2313-576X/4/2/25.

20. "Safety Assistance System Warns of Dirty Bombs," Fraunhofer, September 1, 2017, https://www.fraunhofer.de/en/press/research-news/2017/september/safety-assistance-system-warns-of-dirty-bombs-.html.

21. Jaganath Sankaran, "A Different Use for Artificial Intelligence in Nuclear Weapons Command and Control," War on the Rocks, April 25, 2019, https://warontherocks.com/2019/04/a-different-use-for-artificial-intelligence-in-nuclear-weapons-command-and-control; Jill Hruby and M. Nina Miller, "Assessing and Managing the Benefits and Risks of Artificial Intelligence in Nuclear-Weapon Systems," Nuclear Threat Initiative, August 26, 2021, https://www.nti.org/analysis/articles/assessing-and-managing-the-benefits-and-risks-of-artificial-intelligence-in-nuclear-weapon-systems.

22. For a quick and helpful explainer about the Black Death and other plague epidemics, see Jenny Howard, "Plague, Explained," *National Geographic*, August 20, 2019, https://www.nationalgeographic.com/science/health-and-human-body/human-diseases/the-plague.

23. "Historical Estimates of World Population," US Census Bureau, July 5, 2018, https://www.census.gov/data/tables/time-series/demo/international-programs/historical-est-worldpop.html.

24. Elizabeth Pennisi, "Black Death Left a Mark on Human Genome," *Science*, February 3, 2014, https://www.sciencemag.org/news/2014/02/black-death-left-mark-human-genome.

25. Gene therapy, for example, has beneficial purposes and intentions but often uses modified viruses to accomplish its goals. For a short overview, see "Gene Therapy Inside Out," US Food and Drug Administration, YouTube video, December 19, 2017, https://www.youtube.com/watch?v=GbJasFgJkLg.

26. For an accessible summary of bioterrorism risks, see R. Daniel Bressler and Chris Baker-lee, "'Designer Bugs': How the Next Pandemic Might Come from a Lab," *Vox*, December 6, 2018, https://www.vox.com/future-perfect/2018/12/6/18127430/superbugs-biotech-pathogens-biorisk-pandemic.

27. For a more in-depth case study on the Asilomar Conference and the principles it produced, see M. J. Peterson, "Asilomar Conference on Laboratory Precautions When Conducting Recombinant DNA Research—Case Summary," International Dimensions of Ethics Education in Science and Engineering Case Study Series, June 2010, https://scholarworks.umass.edu/cgi/viewcontent.cgi?article=1023&context=edethicsinscience.

28. Alan McHughen and Stuart Smyth, "US Regulatory System for Genetically Modified [Genetically Modified Organism (GMO), rDNA or Transgenic] Crop Cultivars," *Plant Biotechnology Journal* 6, no. 1 (January 2008): 2–12, https://doi.org/10.1111/j.1467-7652.2007.00300.x.

29. For more in-depth information on the GRRT's activities, see Tasha Stehling-Ariza et al., "Establishment of CDC Global Rapid Response Team to Ensure Global Health Security," *Emerging Infectious Diseases* 23, no. 13 (December 2017), https://wwwnc.cdc.gov/eid/article/23/13/17-0711_article; Centers for Disease Control and Prevention, "Global Rapid Response Team Expands Scope to U.S. Response," *Updates from the Field* 30 (Fall 2020), https://www.cdc.gov/globalhealth/healthprotection/fieldupdates/fall-2020/grrt-response-covid.html.

30. For more about the activities of NICBR and USAMRIID to combat the risk of bioterrorism, see their websites at https://www.nicbr.mil and https://www.usamriid.army.mil.

31. Françoise Barré-Sinoussi et al., "Isolation of a T-Lymphotropic Retrovirus from a Patient at Risk for Acquired Immune Deficiency Syndrome (AIDS)," *Science* 220, no. 4599 (May 20, 1983): 868–71, https://www.jstor.org/stable/1690359; Jean K. Carr et al., "Full-Length Sequence and Mosaic Structure of a Human Immunodeficiency Virus Type 1 Isolate from Thailand," *Journal of Virology* 70, no. 9 (August 31, 1996): 5935–43, https://www.ncbi.nlm .nih.gov/pmc/articles/PMC190613; Kristen Philipkoski, "SARS Gene Sequence Unveiled," *Wired*, April 15, 2003, https://www.wired.com/2003/04/sars-gene-sequence-unveiled; Cameron Walker, "Rapid Sequencing Method Can Identify New Viruses Within Hours," *Discover*, December 11, 2013, https://web.archive.org/web/20201111180212; "Rapid Sequencing of RNA Virus Genomes," Nanopore Technologies, 2018, https://nanoporetech.com/re source-centre/rapid-sequencing-rna-virus-genomes.

32. Darren J. Obbard et al., "The Evolution of RNAi as a Defence Against Viruses and Transposable Elements," *Philosophical Transactions of the Royal Society B: Biological Sciences* 364, no. 1513 (99–115), https://www.ncbi.nlm.nih.gov/pmc/articles/PMC2592633.

33. For a brief explainer on how traditional antigen-based vaccines work, see "Understanding How Vaccines Work," Centers for Disease Control, July 2018, https://www.cdc.gov/vac cines/hcp/conversations/understanding-vacc-work.html.

34. Ray Kurzweil, "AI-Powered Biotech Can Help Deploy a Vaccine in Record Time," *Wired*, May 19, 2020, https://www.wired.com/story/opinion-ai-powered-biotech-can-help-deploy-a -vaccine-in-record-time.

35. Asha Barbaschow, "Moderna Leveraging Its 'AI Factory' to Revolutionise the Way Diseases Are Treated," *ZDNet*, May 17, 2021, https://www.zdnet.com/article/moderna-leveraging -its-ai-factory-to-revolutionise-the-way-diseases-are-treated.

36. Barbaschow, "Moderna Leveraging Its 'AI Factory'"; "Moderna COVID-19 Vaccine," US Food and Drug Administration, December 18, 2020, https://www.fda.gov/emergency -preparedness-and-response/coronavirus-disease-2019/moderna-covid-19-vaccine.

37. Philip Ball, "The Lightning-Fast Quest for COVID Vaccines—and What It Means for Other Diseases," *Nature* 589, no. 7840 (December 18, 2020): 16–18, https://www.nature .com/articles/d41586-020-03626-1.

38. For two helpful overviews of the evidence for and against, see Amy Maxmen and Smriti Mallapaty, "The COVID Lab-Leak Hypothesis: What Scientists Do and Don't Know," *Nature* 594, no. 7863 (June 8, 2021): 313–15, https://www.nature.com/articles/d41586 -021-01529-3; Jon Cohen, "Call of the Wild," *Science* 373, no. 6559 (September 2, 2021): 1072–77, https://www.science.org/content/article/why-many-scientists-say-unlikely-sars-cov -2-originated-lab-leak.

39. James Pearson and Ju-Min Park, "North Korea Overcomes Poverty, Sanctions with Cut-Price Nukes," Reuters, January 11, 2016, https://www.reuters.com/article/us-northkorea -nuclear-money-idUSKCN0UP1G820160111.

40. Lord Lyell, "Chemical and Biological Weapons: The Poor Man's Bomb," draft general report, Science and Technology Committee (96) 8, North Atlantic Assembly, October 4, 1996, https://irp.fas.org/threat/an253stc.htm.

41. United Nations Secretary-General, "Chemical and Bacteriological (Biological) Weapons and the Effects of Their Possible Use: Report of the Secretary-General," United Nations, August 1969; discussed in Gregory Koblentz, "Pathogens as Weapons: The International Security Implications of Biological Warfare," *International Security* 28, no. 3 (Winter 2003/ 2004): 88, https://doi.org/10.1162/016228803773100084.

42. For a sobering and insightful deeper look at potential harmful applications of nanotechnology, see Louis A. Del Monte, *Nanoweapons: A Growing Threat to Humanity* (Lincoln: University of Nebraska Press, 2017).

43. K. Eric Drexler, *Engines of Creation: The Coming Era of Nanotechnology* (New York: Anchor Press/Doubleday, 1986), 172.

44. Ralph C. Merkle, "Self Replicating Systems and Low Cost Manufacturing," Zyvex.com, accessed March 5, 2023, http://www.zyvex.com/nanotech/selfRepNATO.html.

45. Yinon M. Bar-On, Rob Phillips, and Ron Milo estimate the total biomass of the earth as containing 550 gigatons of carbon across all kinds of living creatures (not counting

underground carbon like coal deposits that originally came from living matter). This is the equivalent of 5.5 x 10^{17} grams of carbon. With an average atomic weight of 12.011, we can calculate the number of atoms as 5.5 x 10^{17} / 12.011 = 4.6 x 10^{16} mols. Using this figure, we can estimate (4.6 x 10^{16}) x (6.022 x 10^{23}) (Avogadro's number) = 2.8 x 10^{40} carbon atoms in the earth's available biomass. The total amount of organic carbon, including in the atmosphere, in the soil, and hydrocarbon reserves underground, may be considerably higher, but it's much harder to judge how much of this would be readily accessible by nanobots in a gray goo scenario. See Yinon M. Bar-On et al., "The Biomass Distribution on Earth," *PNAS* 115, no. 25 (June 19, 2018): 6506–11, https://doi.org/10.1073/pnas.1711842115.

46. According to nanotechnology expert Rob Freitas, a self-replicating nanobot might have around 70 million carbon atoms. This is a speculative figure, but it is probably the best estimate available and can serve as a general guide to thinking about such a scenario. See Robert A. Freitas Jr., "Some Limits to Global Ecophagy by Biovorous Nanoreplicators, with Public Policy Recommendations," Foresight Institute, April 2000, http://www.rfreitas.com/Nano/Ecophagy.htm.

47. This is an average under the assumption that the distribution of carbon is more or less even and continuous but in a real scenario would vary by local conditions. In some places, the supply of available carbon would be exhausted after fewer generations. Elsewhere, a greater number of generations could be required to convert all the carbon, extending the time somewhat.

48. Freitas, "Some Limits to Global Ecophagy by Biovorous Nanoreplicators."

49. Freitas, "Some Limits to Global Ecophagy by Biovorous Nanoreplicators."

50. Robert A. Freitas Jr. and Ralph C. Merkle, *Kinematic Self-Replicating Machines* (Austin, TX: Landes Bioscience, 2004), http://www.molecularassembler.com/KSRM/4.11.3.3.htm.

51. To read the updated guidelines themselves and some enlightening background about them, see Neil Jacobstein, "Foresight Guidelines for Responsible Nanotechnology Development," Foresight Institute, 2006, http://www.imm.org/policy/guidelines.

52. For more on the colorful range of terms like "blue goo" and "gray goo" that have sprung up to describe various nanotechnology scenarios, see Chris Phoenix, "Goo vs. Paste," *Nanotechnology Now*, September 2002, http://www.nanotech-now.com/goo.htm.

53. Freitas, "Some Limits to Global Ecophagy by Biovorous Nanoreplicators."

54. Bill Joy, "Why the Future Doesn't Need Us," *Wired*, April 1, 2000, https://www.wired.com/2000/04/joy-2.

55. World Health Organization, "WHO Coronavirus Dashboard," World Health Organization, accessed October 16, 2023, https://covid19.who.int.

56. Miles Brundage et al., *The Malicious Use of Artificial Intelligence: Forecasting, Prevention, and Mitigation* (Oxford, UK: Future of Humanity Institute, February 2018), https://img1.wsimg.com/blobby/go/3d82daa4-97fe-4096-9c6b-376b92c619de/downloads/MaliciousUseofAI.pdf?ver=1553030594217.

57. Evan Hubinger, "Clarifying Inner Alignment Terminology," AI Alignment Forum, November 9, 2020, https://www.alignmentforum.org/posts/SzecSPYxqRa5GCaSF/clarifying-inner-alignment-terminology; Paul Christiano, "Current Work in AI Alignment," Effective Altruism, accessed March 5, 2023, https://www.effectivealtruism.org/articles/paul-christiano-current-work-in-ai-alignment.

58. Hubinger, "Clarifying Inner Alignment Terminology"; Christiano, "Current Work in AI Alignment."

59. For a relatively accessible deeper explanation of imitative generalization, see Beth Barnes, "Imitative Generalisation (AKA 'Learning the Prior')," AI Alignment Forum, January 9, 2021, https://www.alignmentforum.org/posts/JKj5Krff5oKMb8TjT/imitative-generalisation-aka-learning-the-prior-1.

60. Geoffrey Irving and Dario Amodei, "AI Safety via Debate," OpenAI, May 3, 2018, https://openai.com/blog/debate.

61. For an insightful sequence of posts explaining iterated amplification, written by the concept's primary originator, see Paul Christiano, "Iterated Amplification," AI Alignment Forum, October 29, 2018, https://www.alignmentforum.org/s/EmDuGeRw749sD3GKd.

62. For more details on the technical challenges of AI safety, see Dario Amodei et al., "Concrete Problems in AI Safety," arXiv:1606.06565v2 [cs.AI], July 25, 2016, https://arxiv.org/pdf/1606.06565.pdf.

63. To read the Asilomar AI Principles in full for yourself, along with the regularly updated list of signatories, see "Asilomar AI Principles," Future of Life Institute, 2019, https://futureoflife.org/ai-principles.

64. One of the fathers of the internet, Vint Cerf, wrote a helpful essay expanding on DARPA's role in the internet's creation. See Vint Cerf, "A Brief History of the Internet and Related Networks," Internet Society, accessed March 5, 2023, https://www.internetsociety.org/internet/history-internet/brief-history-internet-related-networks.

65. "Asilomar AI Principles," Future of Life Institute.

66. "Lethal Autonomous Weapons Pledge," Future of Life Institute, 2019, https://futureoflife.org/lethal-autonomous-weapons-pledge.

67. Kelley M. Sayler, "Defense Primer: U.S. Policy on Lethal Autonomous Weapon Systems" (report IF11150, Congressional Research Service, updated November 14, 2022), https://crsreports.congress.gov/product/pdf/IF/IF11150.

68. *Department of Defense Directive 3000.09—Autonomy in Weapon Systems*, US Department of Defense, November 21, 2012 (effective January 25, 2023), https://www.esd.whs.mil/Portals/54/Documents/DD/issuances/dodd/300009p.pdf.

69. Erico Guizzo and Evan Ackerman, "Do We Want Robot Warriors to Decide Who Lives or Dies?," *IEEE Spectrum*, May 31, 2016, https://spectrum.ieee.org/robotics/military-robots/do-we-want-robot-warriors-to-decide-who-lives-or-dies.

70. Guizzo and Ackerman, "Do We Want Robot Warriors to Decide Who Lives or Dies?"

71. Bureau of Arms Control, Verification and Compliance, "Political Declaration on Responsible Military Use of Artificial Intelligence and Autonomy," US Department of State, February 16, 2023, https://www.state.gov/political-declaration-on-responsible-military-use-of-artificial-intelligence-and-autonomy.

72. Brian M. Carney, "Air Combat by Remote Control," *Wall Street Journal*, May 12, 2008, https://www.wsj.com/articles/SB121055519404984109.

73. "Supporters of a Ban on Killer Robots," Campaign to Stop Killer Robots, updated May 18, 2021, https://web.archive.org/web/20210518133318/https://www.stopkillerrobots.org/endorsers.

74. Brian Stauffer, "Stopping Killer Robots: Country Positions on Banning Fully Autonomous Weapons and Retaining Human Control," Human Rights Watch, August 10, 2020, https://www.hrw.org/report/2020/08/10/stopping-killer-robots/country-positions-banning-fully-autonomous-weapons-and.

75. To better understand the issue of transparency in AI, a helpful analogy is the difference in mathematics between finding a solution to a problem and verifying that a given solution is correct. In some cases, it's easy for humans to verify a solution found by a computer. For example, if a program is instructed to find the greatest odd integer less than 1,000,000, it is trivially easy for a human to confirm that 999,999 indeed meets these criteria. On the other hand, if a program is instructed to find the greatest prime number less than 1,000,000, a human would have a very hard time checking on their own whether 999,983 is in fact prime. In a similar way, when AI generates an answer algorithmically from clearly defined parameters set by its creators, it is easy for programmers to "look under the hood" and see exactly what factors led to that answer. For example, a Go program that evaluates the board based on fixed rules can tell its programmers precisely why it determined that a given move will be the best one. But with connectivist approaches like deep learning, the complete "why" is often inaccessible both to human programmers and the AI itself. There is probably no universal technique that could verify human-understandable reasons for any arbitrary solution generated by a neural network. For more on this so-called black box problem of AI, see Will Knight, "The Dark Secret at the Heart of AI," *MIT Technology Review*, April 11, 2017, https://www.technologyreview.com/s/604087/the-dark-secret-at-the-heart-of-ai.

76. Paul Christiano, "Eliciting Latent Knowledge," AI Alignment, *Medium*, February 25, 2022, https://ai-alignment.com/eliciting-latent-knowledge-f977478608fc.

77. John-Clark Levin and Matthijs M. Maas, "Roadmap to a Roadmap: How Could We Tell When AGI Is a 'Manhattan Project' Away?," arXiv:2008.04701 [cs.CY], August 6, 2020, https://arxiv.org/pdf/2008.04701.pdf.

78. "The Bletchley Declaration by Countries Attending the AI Safety Summit, 1-2 November 2023," UK Government, November 1, 2023, https://www.gov.uk/government/publications /ai-safety-summit-2023-the-bletchley-declaration/the-bletchley-declaration-by -countries-attending-the-ai-safety-summit-1-2-november-2023.

79. For a deeper look at the long-term worldwide trend toward diminishing violence, you can find lots of useful, data-rich insights in my friend Steven Pinker's excellent book *The Better Angels of Our Nature* (New York: Viking, 2011).

80. For a useful essay on the anti-technology sentiments developing in response to fears about emerging technological threats, see Lawrence Lessig, "Stamping Out Good Science," *Wired*, July 1, 2004, https://www.wired.com/2004/07/stamping-out-good-science.

81. Walter Suza, "I Fight Anti-GMO Fears in Africa to Combat Hunger," *The Conversation*, February 7, 2019, https://theconversation.com/i-fight-anti-gmo-fears-in-africa-to-combat -hunger-109632; Editorial Board, "There's No Choice: We Must Grow GM Crops Now," *Guardian*, March 16, 2014, https://www.theguardian.com/commentisfree/2014/mar/16/gm -crops-world-food-famine-starvation.

82. For a representative range of these critiques, see Joël de Rosnay, "Artificial Intelligence: Transhumanism Is Narcissistic. We Must Strive for Hyperhumanism," Crossroads to the Future, April 26, 2015, https://web.archive.org/web/20230322182945/https://www.cross roads-to-the-future.com/articles/artificial-intelligence-transhumanism-is-narcis sistic-we-must-strive-for-hyperhumanism; Wesley J. Smith, "Jeffrey Epstein, a Narcissistic Transhumanist," *National Review*, August 1, 2019, https://www.nationalreview.com/cor ner/jeffrey-epstein-a-narcissistic-transhumanist; Sarah Spiekermann, "Why Transhuman- ism Will Be a Blight on Humanity and Why It Must Be Opposed," The Privacy Surgeon, July 6, 2017, https://web.archive.org/web/20180212062523/http://www.privacysurgeon.org /blog/incision/why-transhumanism-will-be-a-blight-on-humanity-and-why-it-must -be-opposed.

83. In 2021, total global primary energy consumption was about 595.15 exajoules, which is the equivalent of 165,320 terawatt-hours. This is the annual equivalent of a continuous 18.8 terawatts (TW). By comparison, the solar energy constantly hitting the earth is estimated to be 173,000–175,000 TW, of which Sandia National Laboratories estimates around 89,300 TW reaches the planet's surface, and 58,300 TW is theoretically extractable by surface-based photovoltaics. Of this, the Sandia research assessed that up to 7,500 TW could be generated from well-insolated land area with 2006 technology. Actually building just a quarter of a percent of this capacity would cover the energy we currently use from all sources—not just electricity but also fuels that are presently used without being converted to electricity first. See *BP Statistical Review of World Energy 2022* (London: BP, 2022), https://www.bp.com/content/dam/bp/business-sites/en/global/corporate/pdfs/energy -economics/statistical-review/bp-stats-review-2022-full-report.pdf, 9; Jeff Tsao et al., "Solar FAQs," US Department of Energy (working paper SAND 2006-2818P, Sandia Na- tional Laboratories 2006), 9, https://web.archive.org/web/20200424084337/https://www .sandia.gov/~jytsao/Solar%20FAQs.pdf.

84. For two very helpful summaries of how some of the world's top futurists are thinking about existential risks—that is, events that could wipe out civilization or cause human extinction altogether—see Sebastian Farquhar et al., *Existential Risk: Diplomacy and Governance*, Global Priorities Project, 2017, https://www.fhi.ox.ac.uk/wp-content/uploads/Existential -Risks-2017-01-23.pdf; Nick Bostrom, "Existential Risks: Analyzing Human Extinction Scenarios and Related Hazards," *Journal of Evolution and Technology* 9, no. 1 (2002), https:// www.nickbostrom.com/existential/risks.html.

INDEX

Page numbers in *italics* refer to illustrations.

ABC TV, 220
abiogenesis, 29–30, 96–98
abstraction, 35–37
Abundance (Diamandis and Kotler), 112
academic tests, 52
accelerating returns. *See* law of
    accelerating returns
Acemoğlu, Daron, 129
additive manufacturing. *See* 3D printing
aeroponics, 180–81
Africa
    Ebola virus outbreak of 2014–2016, 272
    electricity, 175
    famine and GMOs, 284
    poverty rate, *117, 141*
After Life, 100–105
Age of Intelligent Machines, The
    (Kurzweil), 164
Age of Spiritual Machines, The (Kurzweil),
    4, 8–9, 13, 63, 112, 164
aging, 5, 192, 255–57, 259
    slowing and reversing. *See* radical life
        extension
agricultural runoff, 180, 181
agriculture, 169–70, 171, 201–3, 227
    farm labor, 199, 201–3, *203,* 219
    vertical, 169, 171, 178, 179–83
Agüera Arcas, Blaise, 56
AI. *See* artificial intelligence
AI effect, 63
AI Impacts, 61
AI safety via debate, 279
Alexander II of Russia, 162

algebra, 15
algorithms, 2, 56, 86–87
    evolutionary, 20, 23, 25, 33
    internet search, 45–46, 53–54, 212
    neural net. *See* neural nets
    social media, 114–15
alienation, 230
Alphabet. *See also* Google
    DeepMind, 41, 42, 47, 50, 196,
        238–39
    Waymo, 43, 195–96
AlphaCode, 50
AlphaFold, 238–39
AlphaFold 2, 239
AlphaGo, 41–42, 369n
AlphaGo Master, 41
AlphaGo Zero, 41–42, 238, 369n
AlphaZero, 42
Alquist 3D, 188
Altair 8800, 131, *165,* 302
Alzheimer's disease, 134, 239–40, 255
Amazon, 143, 212, 253
AMD Radeon RX, *166,* 311
Ameca, 101–2
American Revolution, 113, 161
amino acids, 238, 261, 262
amoebae, 77
amyotrophic lateral sclerosis (ALS), 70
analogical thinking, 38, 39, 50, 77
Anderson, Chris, 226
Andreessen, Marc, 211
Android, 218
androids, 88, 101

animals. *See also specific animals*
  brains of, 32, 33–34, 37, 46, 72
  consciousness of, 75, 76, 77–78, 80
  domestication of, 192
  evolution of, 8, 30
  mirror test, 59, 76
  rights, 76, 78, 153
anthropic principle, 98–99
antibacterial chemicals in water, 178
antibiotics, 134, 237
app economy, 218–19
Apple, 132
Apple Macintosh, *165*, 302
Arab Spring, 163
ARGONet, 242–43
art, 37, 73, 109, 209
artificial diamonds, 251, 254–55
artificial intelligence (AI), 1
  After Life, 100–105
  alignment problem, 278–80
  biosimulation, 189–94, 240–41
  biotechnology convergence, 235–45
  birth of, 12–29
  cloud-connected neocortices, 9–10,
    69–73, 222, 228–29, 290–91
  connectionism, 14, 18, 26–29,
    40–41, 54
  conscious, 81
  crime analysis, 150
  deep learning, 40–54, 99–100
  dialogue with Cassandra, 287–92
  FOOM, 60–61
  jobs and, 196–99, 204, 207–11, 214,
    219, 221
  law of accelerating returns, 5, 113–14
  in medicine, 135–36, 189–94, 235–45
  misuse of, 278–79, 283
  nanotechnology, 245–65
  neural nets. *See* neural nets
  remaining deficiencies in, 54–63
  renewable energy, 172, 173
  risks and perils, 278–85
  superintelligent. *See* superintelligent AI
  symbolic computing, 14–19, 40
  thought-to-text technology, 70–71
  3D printing, 184
  Turing test, 8–9, 12–13, 63–69
  use of term, 13
  vertical agriculture, 181–83

Asia, poverty, 138, *141*
Asilomar Conference on Beneficial AI,
  280, 282–83
Asilomar Conference on Recombinant
  DNA, 271–72
ASIMO, 101
Askell, Amanda, 48
assembly lines, 203, 204
associative memories, 38
asteroids, 34
Atari, 42
atherosclerosis, 134, 262
Atlas (robot), 101
ATMs, 209
atomic weapons. *See* nuclear weapons
atoms, 7, 30, 98, 246, 247, 249–50,
  252, 334*n*
attention mechanism in deep learning, 46
auditory recognition systems, 20
augmented reality (AR), 170–71, 222, 285
Australia
  poverty rate, *117*
  social safety net, 223, *223*
autism, 242
autocracies, 163
autoimmune reactions and nanobots, 262
automation, 231–32, 253
  jobs and, 196–99, 204, 207–11,
    219, 221
autonomous vehicles, 43, 171, 195–96,
  208–9, 214, 229–30, 253. *See also*
  self-driving cars
availability heuristic, 121, 152
avatars, 100–105, 263
axons, 89

Babbage, Charles, 293
backdoor life extension, 260
bacteria, 177, 178, 237, 262, 274
bad news bias, 114–16, 119–20
Bank for International Settlements
  (BIS), 217
banking, 198, 209
Bardem, Javier, 100
Barnes, Luke, 98
basis functions, 31
battery storage, 176
beach mice, 32
behavior, 32–34, 76, 77–80

Better Angels of Our Nature, The (Pinker), 151–53, 230–31
big bang, 7, 29–30, 97–98
big data, 2, 58–59
Bill of Rights, 161
BINAC, *165*, 299
biofuels, 154, 173
biological frailties, 5, 285
biological simulation (biosimulation), 189–94, 240–41
biological weapons (bioweapons), 274
biomass, 249, 275, 397–98*n*
biomolecular emulation, 104
biomolecules, 136
bioprinting, 186
biosphere, 274–75
biotechnology, 4, 5, 135. *See also specific technologies*
   combining with AI, 235–45, 255
   risks and perils, 271–73
Bitcoin. *See* cryptocurrencies
BiTEs (Bispecific T cell engagers), 239
black box problem, 18
Blackburn, Simon, 82
Black Death. *See* bubonic plague
black holes, 1–2, 98
black silicon, 172–73
Blade Runner (movie), 100
Bletchley Declaration, 283
bloodstream, nanobots in. *See* nanobots
blue goo, 277. *See also* nanobots
Bode, Stella de, 88–89
BoJack Horseman (TV show), 221
book publishing, 53, 159–60, 212, 253
Boston Dynamics, 101
Bostrom, Nick, 41, 62, 104, 268–69, 295*n*
bottlenecks, 60, 61
bottom-up approach to nanotech, 249–50
brain, human
   cerebellum of. *See* cerebellum
   computational capacity, estimates, 54, 57, 61, 62, 71–72, 246, 248, 264–65
   computer interface. *See* brain-computer interface
   decision-making, 88–89
   effect of environmental toxicity on, 150–51

   emulation. *See* mind uploading
   evolution of, 8, 30–31, 33–34, 37, 72
   free will dilemma, 88–90
   identity and, 75–76
   neocortex of. *See* neocortex
   neural implants, 92–93
   processing speed, 61–62
   size of, 34, 246
brain-computer interface, 4, 8, 11–12, 171, 233, 263–65
   dialogue with Cassandra, 287–92
   extending neocortex into the cloud, 9–10, *68*, 69–73, 222, 228–29, 290–91
   Neuralink, 70–71, 244, 328*n*
   transferring consciousness, 87, 90–94, 103–5, 108–9
   "You 2" conscious, 90–94, 102, 103
BrainGate, 70
brain prostheses, 92–93
brain scans, 69–70
broadband, 113
broadcast architecture, 249, 252, 261, 276
broken windows theory, 150
Bronze Age, 250
Brown University, 71
Brynjolfsson, Erik, 207–11, 211
bubonic plague, 271
Buddha, 267
buildings, 3D printing of, 170, 187–89
butadiyne, 251
Butler, Samuel, 75–76

cable TV, 220–21
California, automation and jobs, 197
Calment, Jeanne, 255
Cambridge Declaration on Consciousness, 78
Campaign to Stop Killer Robots, 281–82
Canada
   poverty rate, *117*
   social safety net, 223, *223*
cancer
   AI and, 278–79
   biosimulation and, 190–91, 192, 241
   detection, 243
   immunotherapy for, 190, 227, 239

cancer *(cont.)*
  life extension and, 134, 135, 190–91,
      192, 255, 256–57, 260
  nanotechnology for, 255, 256–57, 260
capital and labor, 209–10
carbon, 7, 81, 96, 250, 275
carbon dioxide, 185, 192, 259
carbon emissions, 170, 185
cars, self-driving, 43, 171, 195–96,
      208–9, 214, 229–30, 253. *See also*
      autonomous vehicles
CAR-T cell therapy reprogram, 190, 239
causal inference, 56
CBS TV, 220
cell phones. *See* smartphones
cell therapies, 190, 239
cellular automata, 82–88, *83, 84, 85*
cellular emulations, 104
Centers for Disease Control (CDC),
      242–43
cerebellum, 11, 29–33
  fixed action patterns, 32–33
  functions, 30–32
  hierarchies, 37–38
  muscle memory, 30–32
cerebral cortex, 35, 264
chain-of-thought reasoning, 51
Chalmers, David, 48, 79–81, 92, 93, 104
Charles River, 93–94
ChatGPT, 52–53, 198. *See also* OpenAI
chemical vapor deposition, 251
chemistry, 7–8, 97, 251
chemotherapy, 260
chess, 51–52, 58, 63–64
CheXNet, 243, 244
CheXpert, 244
child labor, 146–47
  decline worldwide, 146–47, *148*
chimpanzees, 37, 59, 77
China
  education and literacy, *125,*
      125–26, *127*
  nuclear weapons, 269
  poverty, *117,* 137–38
Chinese room argument, 48, 330*n*
chips, computer, 113, 164–72, 247–48
  price-performance, *165–66,* 293–312
cholera, 134, 177
Chomsky, Noam, 280

Churchill, Winston, 162
CICERO, 42
civil defense drills, 268, 285
class 1 automata, *83,* 83–84
class 4 automata, 84–88
classical economics, 210, 211
clean water access, 177–79
clinical trials, 236–37, 272–73
  AI and biosimulation, 135, 190–91,
      240–42
CLIP, 44
clothing, 3D-printed, 184, 186
cloud-connected neocortices. *See*
      neocortex
CNN, 120
coal, 209–10, 398*n*
cobalt, 268
cochlear implants, 92
cockroaches, 77
Codex, 50
cognition, 2, 8
  AI and, 59–60, 61, 64, 103
  in animals, 77, 80
  in humans, 35, 59–60, 61, 71, 248
cognitive biases, 98–99, 114–15, 116,
      119–20
Cold War, 138, 162–63, 268, 269–70
Colorado State University, 115
Colossus Mark 1, *165,* 298–99
common sense, 9, 17, 55–56
Compaq Deskpro 386, *166,* 302–3
Compaq Deskpro 386/25, *166,* 303
compatibilism, 88, 332*n*
complexity ceiling, 17, 55
computation
  cloud-connected neocortices, 9–10,
      69–73, 222, 228–29, 290–91
  emulating human intelligence, 54,
      61–62, 71–72, 246, 248, 264–65
  law of accelerating returns, 164–72
  Moore's law, 40–41, 57, 113, 168
  parallel processing, 57–58, 61–62, 248
  price-performance of, 3, *3,* 4, 15–16,
      29, 40, 55, 56–58, *57,* 61, 62, 63,
      164–69, *165–66,* 181–82, 293–312
  Turing test, 8–9, 12–13, 63–69, 71
computational capacity of human brain,
      estimates, 54, 57, 61, 62, 71–72,
      246, 248, 264–65

computer-brain interface. *See* brain-computer interface
computer programming, 50, 60, 124, 160
"Computing Machinery and Intelligence" (Turing), 12
computronium, 8, *68*
Concorde, 113
connectionism, 14, 18, 26–29, 40–41, 54
connectomic emulations, 104
consciousness, 1, 7, 75–82
  AI, 65
  of animals, 75, 76, 77–78, 80
  causal explanation of, 80–82
  origins of, 78
  overview of, 75–79
  replicants and, 103
  subjective, 62, 76–80, 81, 93, 94
  use of drugs to alter, 109
  use of term, 76
  "You 2" conscious, 90–94, 102, 103
  zombies, qualia, and hard problem of, 79–82
conspiracy theories, 227–28, 273
consumer surplus, 212, 213
contact lenses, 222
containerization, 204
contextual memory, 55
Conway's Game of Life, 83–86
Coolidge, Calvin, 232
cooperation, 153
Core 2 Duo E6300, *166*, 308–9
Cornell University, 27
corn production, 180
cosmetics, 109
COVID-19 pandemic, 135, 271–73
  AI and medicine, 227, 237–38, 240, 278
  biotechnology risks, 271–73
  income and poverty, 139, 143, 144, 200
  labor force, 146, 147, 215, 216
  misinformation about, 227, 273
  social safety spending, 223, 224
  teleworking, 146, 172
COVID-19 vaccines, 227, 237–38, 240, 273
Craigslist, 218
creativity, 38, 48, 221

Cretaceous-Paleogene extinction event, 34
crime, 148–54, 233
  actual US rates, *119*
  homicide in US, *151*
  homicide rates in Western Europe since 1300, *149*
  pollution and, 150–51, 233
  public perception of, *118*, 118–20, 152, 233
  racial disparities and policing, 150
  violent crime in US, *151*
CRISPR, 241
crop densities, 180
crossword puzzles, 64, 326*n*
cruise missiles, 270
cryptocurrencies, 217–18
cryptogenic strokes, 262
Ctrl Shift Face, 100
cultured meat, 169–70, 171
Curtiss, Susan, 88–89
cyberattacks, 193
cybersecurity, 228
Cycorp, 17

Dafoe, Willem, 100
Dalí, Salvador, 49
DALL-E, 49–50, 221
DALL-E 2, 209
Dartmouth College workshop on AI, 12–13, 14
Darwin, Charles, 38–39, 48
data collection and analysis, 58–59
Data General Nova, *165*, 301
data mining, 102
dating apps, 232
Dawkins, Richard, 334*n*
decentralized manufacturing, 173, 185–87
DEC PDP-1, 15, *165*, 300
DEC PDP-4, *165*, 300–301
DEC PDP-8, *165*, 301
Deep Blue. *See* IBM
deepfake videos, 100
deep learning, 40–54, 99–100
  Moore's law, 40–41
  transformers, 46–47
DeepMind. *See* Alphabet
deep neural nets, 43–44, 154, 196, 369*n*

deep reinforcement learning, 41–43
deer mice, 32
Defense Advanced Research Projects Agency (DARPA), 71, 195, 280
deflation, 167, 169, 214
degenerative diseases, 134–35, 192, 239–40
DELFI, 243
democracy, 122, 159–63, 194
    spread since 1800, 163
dendrites, 93
deserts, solar power in, 174–75
deskilling, 208
determinism, 82–83, 86–88, 331–32n
Deutschland (ship), 113
diabetes, 134, 192, 259
Diamandis, Peter, 112
Diamond Age, The (Stephenson), 250–51
diamondoids, 250–51, 252, 254–55, 258
diarrheal disease, 177
DiCaprio, Leonardo, 100
digestion, 71, 259
digital economy, 218–19, 254
dinosaurs, 34
Diplomacy (game), 42
Discovery Channel, 220
disease treatment and prevention, 4, 133–36
    combining AI with biotechnology, 235–45
DNA, 8, 102, 186, 238, 242, 261–62, 271–72
    junk, 242
    nanotechnology, 251, 261–62
    origami, 251
    repair, 71
    sequencing, 2, 135, 189, 261
dogs, 77
Domain Name System (DNS), 132
domestic appliances, 122, 128
down quarks, 97
Drexler, K. Eric, 169, 247, 249–51, 252, 274–75
drinking water, 177–79
dualism, 81–82
During World War, 146
Durrant, Jacob, 237
DVRs (digital video recorders), 220
dysentery, 134, 177

eBay, 144
Ebola virus outbreak of 2014–2016, 272
e-books, 212, 253
e-commerce, 133
economic cycles, 122
economic scarcity, 254
education and learning, 122–27
    average years of, 127
    jobs and skills, 206–7, 207, 215–16, 222
    literacy, 111–12, 122–27
    US expenditures, 126
eggs (human reproductive cells), 95
Einstein, Albert, 39
Eisner, Manuel, 149
electricity
    renewable energy, 154, 155–59
    from solar power. See solar energy
    storage issues, 175–76, 176, 177
    from wind. See wind power
electrification, 128–29, 129
electroencephalograms (EEGs), 69–70
electromagnetism, 7, 96
electromechanical relays, 40, 168
electron beam-based atom placement, 251
electronics, 143, 208, 252
electrons, 7, 97, 172
elementary cellular automata, 83, 83–86, 84, 85
elements, 30, 97, 268
elephants, 37, 46, 59
eliciting latent knowledge, 282
emergence, 86, 87–88
empathy, 109, 153, 291
encrypted technologies, 217
Encyclopedia Britannica, 168
energy. See also renewable energy
    nanotechnology and, 253–54
    replacement of fossil fuels, 154, 172–76
energy storage, 175–76
    costs of, 176
    total US, 177
Engineered Arts, 101–2
engineered negligible senescence, 257
Engines of Creation (Drexler), 250
English Civil War, 160–61
ENIAC, 165, 299
Enlightenment, 124

Enlightenment Now (Pinker), 112
entropy, 98, 116, 160
environmental regulations, 150–51
environmental toxicity, 150–51, 170, 185, 233
enzymes, 250
epigenetics, 95, 261–62
epilepsy, 36
epochs
    First Epoch, 7–8, *68*
    Second Epoch, 8, *68*
    Third Epoch, 8, *68*
    Fourth Epoch, 8, 9, 11, *68*
    Fifth Epoch, 8, 9, 11, *68*
    Sixth Epoch, 8, *68*
essential tremor, 90
European Union (EU), 64
evolution, 38–39, 189, 191–92, 334*n*
    bad news bias, 114–16, 119–20
    of human brain, 30–31, 33–34, 37, 72
    nostalgia use, 115–16
evolutionary (genetic) algorithms, 20, 23, 25, 33
expanding circle of empathy, 153
exponential growth, 2, 9–10, 111–94. *See also* technological progress
    law of accelerating returns, 112–14, 159–60, 175, 181–82, 194
    price-performance of computation, 3, *3*, 4, 15–16, 29, 40, 55, 56–58, *57*, 61, 62, 63, 164–69, *165*
    steep part of, 164–72
extinction, 34, 269, 277
extreme poverty, 111–12, *117*, 137–39
extrusion printing, 169

Facebook, 70, 170, 181–82, 212, 213, 254
facial recognition, 99, 101
factory farming, 169–70, 171
factory jobs, 199–200, 202–5
    labor in the US since 1900, *205*
famines, 227, 284
farm labor, 199, 201–3, 219
    in US since 1800, *203*
fat fingers problem, 249–50
fear, 77
Feldman, Daniel, 53
fertilizers, 179, 181, 202
few-shot learning, 49

Feynman, Richard, 246
Fiala, John, 104
Fifth Epoch. *See* epochs
filtration technology, 178
financial crisis of 2008, 139
firearms, 3D-printed, 186–87
first bridge to radical life extension, 134–35, 255, 348*n*
First Epoch. *See* epochs
FitMyFoot, 184, 208
fixed action patterns, 32–33
Fleabag (TV show), 220
Flinders University, 237
floating-point operations per second (FLOPS), 57–58, 167
flu, 237, 242–43
fluid dynamics, 258–59
flush toilets, 111–12, 122, 127–28, *128*
food. *See* agriculture
Food and Drug Administration (FDA), 71, 237, 240, 273
FOOM (intelligence explosion), 60–61
Forbes, 213
foreign language translation, 48, 222
fossil fuels, 59, 209–10
    replacement of, 154, 172–76
fourth bridge to radical life extension, 136, 192–93, 348*n*
Fourth Epoch. *See* epochs
fractals, 86
France
    Asilomar Principles, 280
    crime, *149*
    education and literacy, 125, *125*, *127*
    nuclear weapons, 269
    poverty rate, *117*
free market, 283
free will, 82–90
    cellular automata, 82–88
    definition of, 82
    dilemma of more than one brain per human, 88–90
freight shipping, 185
Freitas, Robert A., 263, 275, 277, 398*n*
Frey, Carl Benedikt, 197, 198, 219
Fried, Itzhak, 36–37
Frontier (supercomputer), 61, 246
fruit flies, 77
functional consciousness, 76–78, 79–80

functional emulations, 104
functional magnetic resonance imaging
    (fMRI), 58, 69–70
fundamental forces, 96
fundamentalist humanism.
    See humanism
furniture, 3D-printed, 184–85
fusion power, 153

Gagné, Jean-François, 13
GANs (generative adversarial
    networks), 99
Gateway 486DX2/66, *166*, 304
Gato, 50
Gazzaniga, Michael, 89
GDP (gross domestic product), 114–15,
    142–43, 211–12
  social safety net spending, 223,
    224, *225*
  US per capita, *142*
Gemini, 2, 9, 54
gender identity, 109
gene editing, 241
gene expression, 241–42, 262
generalizability, 252
General Problem Solver (GPS), 15–16
general relativity theory, 39
Generation Z, 146
genes, 33, 135, 241–42
  DNA nanotechnology, 261–62
  mutations, 33, 135, 191–92, 242,
    261, 278
genetic (evolutionary) algorithms, 20,
    23, 25, 33
genetically modified organisms (GMOs),
    202, 284
genetic biology, 242
genetic engineering, 271–72, 273
genetics, nano-technology, and robotics
    (GNR), 284
genies, 278
genome sequencing, 2, 135, 189, 261
geology, 38, 39
geothermal, 154, 173
Germany
  crime, *149*
  social safety net, *223*
  war death rates, 152
gig economy, 218–19

Global AI Talent Report, 13
globalization, 143–44
Global Rapid Response Team
    (GRRT), 272
glucose, 69
Gmail Smart Reply, 45
GMOs (genetically modified organisms),
    202, 284
Go, 2, 9, 41–42, 196, 369n, 399n
God, 38, 39, 78, 267
Good, I. J., 60
Google. *See also* Alphabet
  Duplex, 64, 100
  Flu Trends, 242–43
  Gemini, 2, 9, 54
  object recognition, 14
  Play Store, 218–19
  Talk to Books, 45–46, 68, 105–8
  Universal Sentence Encoder, 45
Google Cloud, 15–16, 164, *166*,
    311–12, 325n
Google Switch, 47
Google Workspace, 54
googol, 95
googolplex, 95
Gopher, 47
Gotham Greens, 181
GPT. *See* OpenAI
gradual-replacement scenario, 90–94
graffiti, 150
Grand Canyon, 39
graphene, 173, 184, 251
gravitons, 96
gravity, 7, 39, 96, 97
gray goo, 249, 275–78. *See also* nanobots
Great Britain. *See* United Kingdom
Great Depression, 128, 143, 145, 146,
    203, 374n
Greatest Generation, 133
Great Recession, 204
Great Western (ship), 113
greenhouse gases, 154, 169–70
Grey, Aubrey de, 256, 257
gross domestic product. *See* GDP
Grossman, Terry, 191, 348n
GTX 285, *166*, 309
GTX 580, *166*, 309–10
GTX 680, *166*, 310
gun culture, 230

guns, 3D-printed, 186–87
Gunsmoke (TV show), 144
Gutenberg, Johannes, 159, 160

hallucinations, 65. *See also* large
    language models
Hameroff, Stuart, 330*n*
Hands Free Hectare, 202
Hanson, Robin, 60
Hanson Robotics, 101–2
hard problem of consciousness, 80–82, 92
hard takeoff, 60–61
Harvard University, 242
Hawking, Stephen, 280
health and well-being, 235–65, 255
    applying nanotechnology to, 255–65
    combining AI with biotechnology,
        235–45
    developing and perfecting
        nanotechnology, 245–65
health care, 122, 224, 227, 231
hearing loss, 70
heuristics, 16, 120–21, 152
Hinge, 232
hippocampal prostheses, 92–93
historical myopia, 152
History Channel, 220
HIV virus, 272
hoarding, 254
hobbies, 107
Hofer, Johannes, 115
Hoffman, Dustin, 59
Hogan, Craig J., 97
Hollerith tabulating machines, 320*n*
Holocaust, 233
homicide rates, 148–54
    in US, *151*
    violent crime in US, *151*
    in Western Europe since 1300, *149*
hominids, 228–29
Honda ASIMO, 101
hormones, 192, 259–60
housing, 3D-printed, 170, 187–89
How to Create a Mind (Kurzweil), 4,
        34–35, 75–76, 79–80, 90, 93–94,
        103, 104–5
HuaShang Tengda, 188
human brain. *See* brain, human;
    cerebellum; neocortex

Human Genome Project, 135, 189, 271
human intelligence
    evolution of, 7–10
    reinventing. *See* intelligence,
        reinventing
humanism, 284
humor, 36–37, 51
hunter-gatherers, 114, 152
Hutton, James, 39
hydrogen, 30, 96–97, 176, 250
hydrogen depassivation lithography, 251
hydroponics, 180
hyperdimensional language processing,
        43–46
hypersonic weapons, 270

IBM
    Deep Blue, 41–42, 63–64
    Project Debater, 64
    Watson, 64–65
IBM 7094, 167, 168, 211
identity, 90–94
    of Ray Kurzweil, 75, 108–9
    replicants and, 103
    talking to dad bot, 105–8
    "Who am I?", 75–76
"If it bleeds, it leads," 119–20
if-then rule, 16
IKEA, 187
image recognition, 49–50
imitation game. *See* Turing test
imitative generalization, 279
immortality, 94, 257
immune checkpoint inhibitors, 239
immune system, 186, 190, 191–92, 237,
        260, 271, 276–77
immunotherapies, 190, 227, 239, 260
income, 111–12, 137–47, 153
    US average per hour worked, 145, *145*
    US per capita, *144*, 144–45
indeterminism, 82, 86–88
India
    education and literacy, *125*,
        125–26, *127*
    nuclear weapons, 269
    poverty, *117*, 137
induced pluripotent stem (iPS), 190–91
Industrial Revolution, 146, 198–99, 202,
        203, 205, 209–10

inequality, 193–94
infectious diseases, 16, 242–43
inferential reasoning, 51–52
inflation, 3, 29, 210, 214
influencers, 218
information technologies (IT), 211–14
  exponential growth of, 113–14,
    121–22, 135, 142, 164, 168–69,
    172, 194, 235
  law of accelerating returns. *See* law of
    accelerating returns
  price and power, 139, 153–54, 167,
    168–69, 194, 211–13
  spread of democracy and,
    159–60
inkjet printers, 183
inner misalignment, 279
insects, 77
in silico, 190–91, 241
Instagram, 218
Institute for Health Metrics and
  Evaluation, 177
insulin, 192, 259
integrated circuits, 40, 168
Intel, 40
Intellec 8, *165*, 301–2
intellectual property (IP), 186
intelligence
  artificial. *See* artificial intelligence
  evolution of, 7–10
intelligence, reinventing, 5, 11–73
  birth of AI, 12–29
  cerebellum. *See* cerebellum
  deep learning, 40–54
  neocortex. *See* neocortex
  passing the Turing test, 63–69
intelligence explosion, 60
interlocks (nanotechnology), 247
International Labor Organization, 147
International Monetary Fund (IMF),
  166–67
International Telecommunication Union
  (ITU), 222, 347*n*
International Transport Forum, 185
internet access, 132–33, 232, 347*n*
internet search, 45–46, 53–54, 212
iPhone. *See* smartphones
Iron Age, 250
irony, 37, 51

irrigation, 179
ischemic strokes, 262
Israel, 269, 280
Italy
  crime, *149*
  literacy, 124, *125*
  poverty rate, *117*
  spread of democracy, 161
iterated amplification, 279

Janicki Omni Processor, 178
Japan
  education and literacy, 125, *125*, *127*
  poverty rate, *117*
  social safety net, 223
  war death rates, 152
JavaScript, 50
Jennings, Ken, 64
Jeopardy!, 2, 9, 64
jet engines, 113
jobs, 195–234. *See also* labor
  current revolution in, 196–99
  deskilling, 208
  destruction and creation, 199–207
  distinguishing between tasks and
    professions, 209
  drivers, 196–98, 229–30
  farm labor, 199, 201–3, *203*, 219
  income and, 111–12, 137–47, 200–201
  major disruptions, 207–10
  manufacturing, 199–200, 202–5, *205*
  missing productivity, 210–19
  net losses, 207–10
  nonskilling, 208–9
  search for meaning, 220, 228
  upcoming revolution in, 219–34
  upskilling, 208, 209
John, King of England, 159
Johns Hopkins University, 243
Joy, Bill, 277–78

Kahneman, Daniel, 120–21, 229, 230,
  231, 233, 290
Kamen, Dean, 179
Kapor, Mitch, 9, 63
Kasparov, Garry, 41, 63–64
Kaufman, Scott Barry, 48
Ke Jie, 41–42
Kelling, George, 150

Kenbak-1, 131
Khatchadourian, Raffi, 14
Kiser, Grace, 13
Kotler, Steven, 112
Kurzweil, Amy, 105–6
Kurzweil, Fredric, 190
  talking to my dad bot, 105–8
Kuyda, Eugenia, 100, 102

lab-grown meat, 169–70, 171
labor, 195–234. *See also* jobs
  agriculture, 199, 201–3, *203*, 219
  current revolution in, 196–99
  destruction and creation, 199–207
  income and, 111–12, 137–47,
    200–201
  manufacturing, 199–200, 202–5, *205*
  search for meaning, 220, 228
  upcoming revolution in, 219–34
labor force, 205–6
  participation rate, *215*, 215–19,
    216, 378*n*
  US, 200, *206*
Lagarde, Christine, 166–67, 168–69, 171
Lamarck, Jean-Baptiste, 38–39
language
  AI mastery of, 9, 43–49, 63–69
  brain and, 37, 89
  Chinese room argument, 48, 330*n*
  thought-to-text technology, 70–71
  translation of, 48, 222
  Turing test, 8–9, *9*, 12–13, 60, 63–69,
    71, 287
large language models (LLMs), 2, 13,
    51, 55, 64–65
  GPT-3, 47–48, 49, 52, 55, 239, 324*n*
  GPT-4, 2, 9, 52–56, 65
  hallucinations, 65
  transformer-based, 46–47
law of accelerating returns (LOAR), 2–3,
    5, 40, 112–14, 164–72
  computation, 164–72
  exponential growth in, 112–14,
    159–60, 164–72, 175, 181–82, 194
lead, 150–51
learning. *See* education and learning
Learning Channel, 220
legal jobs, 209
Lenat, Douglas, 17

Leonardo da Vinci, 29, 52
Lethal Autonomous Weapons Pledge,
    280–82
Life (game), 83–86
life expectancy, 62, 133–36, 255–56, 258
  technological progress and, 114–15,
    135–36
  in UK, 114–15, *136*
  in US, *137*, 216
life extension. *See* radical life extension
LifeStraw, 178
literacy, 111–12, 122–27
  growth since 1820, *125*
  rates by country, *125*
Little Sophia (robot), 101–2
locked-in syndrome, 76
logic, 15
logic gates, 247, 264
logistics, 185, 204
London Blitz, 162
longevity escape velocity, 189–94, 255–57
Long Now bet, 9, 63
Ludd, Ned, 199
Luddites, 199, 230, 284
lungs, 186, 259, 263
Lyell, Charles, 39

macrophages, 262
Magna Carta, 159
Mantha, Yoan, 13
manufacturing, 199–200, 202–5
  labor in the US since 1900, *205*
marginal cost, 211, 212
Mark 1 Perceptron, 27
Markman, Art, 116
Maslow's hierarchy of needs, 228, 229
mathematics
  cellular automata, 82–88
  singularity, 1–2
Matrioshka brains, 105
Mayflower (ship), 113
Mazurenko, Roman, 100, 102
McCarthy, John, 12–13
McKibben, Bill, 267
McKinsey & Company, 198
mechanistic interpretability, 18
mechanosynthesis, 247, 251
Media Psychology Research Center, 115
Medicaid, 224

medical imaging, 243–44
medical implants, 184, *185*, 186
medicine, 227–28, 231, 235–45
    AI and, 135–36, 189–94, 235–45
    biosimulation, 189–94, 240–41
    clinical trials. *See* clinical trials
    nanotechnology in, 192, 251, 257–63,
        276–77
    3D printing in, 184, *185*, 186
memories, 38, 55
Merkle, Ralph, 247–48, 249, 251,
    264, 393*n*
messenger RNA, 272–73
    vaccines, 237–38, 273
metabolism, 62, 256, 259
Metaculus, 13–14
metaphysics, 102
metaverse, 170–72
mice, 32–33
Michigan State University, 262
Microsoft, 54, 132
Microsoft Office, 54
microtubules, 330*n*
Middle Ages, 73, 114, 124, 151, 159, 161
Midjourney, 49, 209, 221
millennials, 133, 146
Millennium Prize Problems, 60
Mind Children (Moravec), 61
mind uploading, 62–63, 94, 104–5,
    192–93
    "You 2" scenario, 90–94, 102, 103
miniaturization, 40, 169
Minsky, Marvin, 14–15, 17–18, 27–28,
    29, 54, 89–90
MIPS (million instructions per second),
    167, 295
mirror test, 59, 76
misinformation, 227–28, 273
MIT (Massachusetts Institute of
    Technology), 14, 54, 129, 139, 167,
    168, 173, 211, 237
    Artificial Intelligence Laboratory, 27
mitochondria, 93, 135
Mobile Pentium MMX, *166*, 305
mobile phones. *See* smartphones
Moderna, 237–38, 273. *See also*
    COVID-19 vaccines
modular housing, 187–88
molecular assemblers, 249–50, 252

molecular Lego, 251
molecular manufacturing, 249–52
molecules, 7, 8, 29–30
Mona Lisa (da Vinci), 52
monkeys, 71, 72–73
monocytes, 262
Moore, Gordon, 40, 168
Moore's law, 3, 40–41, 57, 113, 168
morality, 81, 91
    consciousness and, 76–77
Moravec, Hans, 61, 101, 245, 246, 295*n*
Moravec's paradox, 101
motor cortex, 31
Mount Everest, 171
MT 486DX, *166*, 303–4
multimodality, 49–50, 51–53
muscle memory, 30–32
music, 32, 59, 106–7, 186, 221
Musk, Elon, 70–71, 280
mutually assured destruction (MAD),
    269–70
MuZero, 42, 43, 50
MYCIN system, 16–17

nanobots
    in biomass, 249, 275, 397–98*n*
    in blood, 71, 192, 245, 258–59,
        262–63
    blue goo, 277
    cloud-connected neocortices, 72, 103,
        263–64
    fat fingers and sticky fingers problem,
        249–50
    gray goo, 249, 275–78
    medical, 192, 251, 257, 258–63,
        276–77
    molecular assemblers, 249–50, 252
    radical life extension, 135–36
    risks and perils, 273–78
    self-replication of, 5, 30, 96, 246–49,
        252, 260, 273–75
nanocrystals, 172
Nanosystems (Drexler), 250
nanotechnology, 1, 4, 245–65. *See also*
    nanobots
    applying to health and longevity,
        255–65
    artificial bodies, 100–101, 102
    designs, 248–53

developing and perfecting, 245–65
diamondoids, 250–51
in manufacturing, 169, 249–52
in medicine, 192, 251, 257, 258–63, 276–77
military use of, 274–76
origins and history of, 245–48
physical scarcity, 252–55
in renewable energy, 172
risks and perils, 273–78
safety guidelines, 276–77
Smalley-Drexler debate, 249–51
tools and techniques, 251–54
nanothreads, 251
nanotubes, 172, 251
nanoweapons, 274–76
nanowires, 172
National Assessment of Adult Literacy, 124
National Institutes of Health, 238
National Interagency Confederation for Biological Research (NICBR), 272
NATO (North Atlantic Treaty Organization), 274
natural language, 13, 47
natural selection, 25, 33, 39, 189
Nature Medicine, 243
NBC TV, 220
neocortex, 11, 30, 33–39, 259
  anatomy of, 34–36, 37–38
  cloud-connected, 9–10, 69–73, 222, 228–29, 290–91
  deep learning recreating powers of, 40–54
  free will and, 89–90
  function of, 35–37, *36*, 38–39
  hierarchies in, 37–38
  training, 58–59
  use of term, 33–34
Neolithic, 79
Netherlands
  crime, *149*
  literacy, 124, *125*
  poverty rate, *116*
Neumann, John von, 246–47, 249
Neural Engineering System Design, 71
neural implants, 70–71, 92–93, 244
Neuralink, 70–71, 244, 328*n*
neural nets, 2, *19*, 19–27, 196, 243

AI and language, 43–45
asynchronous, 25
defining the topology, 20–21
diagram of simple, *19*
key design decisions, 22–24
Perceptrons, 27–28, 29, 57
problem input, 19–20
recognition trials, 21–22, 24
synchronous, 25
training, 22
variations, 25
neurograins, 71
neurons, 71, 77, 93
  brain computing speed, 30–31, 32, 35, 61, 245–46, 264–65
  neural nets, 20–24
neutrons, 7, 97
Newell, Allen, 15
New Kind of Science, A (Wolfram), 82–83
news media, 130
  bad news bias, 114–16, 119–20
newspapers, 130
Newton, Isaac, 39
New Yorker, 14
New York Times, 120
NMDA receptors, 93
Nobel Prize, 120
No Country for Old Men (movie), 100
nonlinearity, 25
nonskilling, 208–9. *See also* labor
Normandie (ship), 113
North Dakota State University, 115
North Korea, 269, 274
nostalgia, 115–16
Nottingham lace, 198–99. *See also* Luddites
nuclear treaties, 269
nuclear weapons, 268–70, 274
nucleus, 71, 261
nutrition, 192, 202, 255

Oak Ridge National Laboratory, 61
observer selection bias, 98–99
occupational therapists, 198
On the Origin of Species (Darwin), 39
OpenAI. *See also* large language models
  ChatGPT, 52–53, 198
  CLIP, 44
  Codex, 50

OpenAI (*cont.*)
  DALL-E, 49–50
  GPT-2, 47
  GPT-3, 47–48, 49, 52, 55, 239, 324*n*
  GPT-3.5, 52, 55
  GPT-4, 2, 9, 52–56, 65
optimism, 120, 121, 163, 233, 254, 270
orchestrated objective reduction (Orch
    OR), 330*n*
Organisation for Economic Co-operation
    and Development (OECD),
    138, 198
organ transplants, 186
origin of universe, 29–30, 95–96
Osborne, Michael, 197, 198, 219
outer misalignment, 278
Oxford University, 197–98
  Future of Humanity Institute, 62,
    268–69
oxidants, 258
oxygen, 69, 96, 178, 259, 263

Pan Am, 113
pancreas, 259
pancreatic islet cells, 192
panprotopsychism, 80–82, 86, 88, 91,
    94, 101, 105. *See also* Chalmers,
    David
Papert, Seymour, 27–28, 29
paradigm shifts, 7–10, *68. See also* epochs
paralegals, 209
parallelism, 35, 57–58, 61–62, 248
paralysis, 70
parameters, 46–47. *See also* neural nets;
    transformers
Parkinson's disease, 90, 239–40
pattern recognition, 20, 21–22, 238
PC's Limited 386, *166*, 303
PDP-8, 164
penicillin, 134
Penrose, Roger, 98, 330*n*
Pentium, 54, *166*, 304
Pentium 4, 164, *165*, *166*, 306–7, 307–8
Pentium Dual-Core E2180, *166*, 309
Pentium II, *166*, 305–6
Pentium III, 164, *165*, *166*, 306
Pentium Pro, *166*, 304–5
Perceptrons, 27–28, 29, 57. *See also*
    neural nets

Perceptrons (Minsky and Papert),
    27–28, 29
perils, 267–85
  artificial intelligence, 278–85
  biotechnology, 271–73
  nanotechnology, 273–79
  nuclear weapons, 268–70
personal appearance, 109, 263
personal computers (PCs), 131–33, *133*,
    139, 160
personal identity. *See* identity
pessimism, 120–21
pesticides, 181, 202
petroleum, 59
philosophy of mind
  hard problem of consciousness, 80–82
  Ship of Theseus, 91–92
photovoltaics, 129, 172–73, 175, 181, 214
  global installed capacity, *155, 156*
  module cost per watt, *155*
  percentage of world electricity, *156*
  SolarWindow Technologies, 173
physicalism (materialism), 81–82
physics
  in First Epoch, 7–8, *68*
  rules of, 96–97
  Standard Model of, 96–97, 335*n*
Pinker, Steven, 112, 120, 151–53,
    230–31
Pirsig, Robert M., 267
plant breeding, 179
Plato, 115
poetry, 58
Poggio, Tomaso, 14
polarization, 130, 230
policing, 150, 233
Polish-Lithuanian Commonwealth, 161
politics, 130
  bad news bias and, 116
  deepfake videos, 100
  electrification and, 128–29
  free will and, 82
  physical scarcity and, 254
  social safety net and, 224, 231
  voting rights, 161–62
pollution, 150–51, 170, 185, 233
Pong (video game), 42, 71
popular opinion and progress, 111–21
population density, 171

positive feedback loop, 60, 132
poverty, 137–47, 153
    declining rates, 111–12, 137–47,
        140, 141
    sanitation and, 127–28
    US rates, 140, *140*
    world population, *117,* 117–18
predetermination, 82, 86–88
prefrontal cortex, 73
price-performance of computation, 3, *3,*
        4, 15–16, 29, 40, 55, 56–58, *57,*
        61, 62, 63, 164–69, *165–66,*
        181–82, 211–13, 293–312
prime numbers, 15, 399*n*
printing, 3D. *See* 3D printing
printing press, 113, 122–23, 124,
        159–60, 253
prions, 192
prior probability, 120–21
problem input to neural net, 19–20
productivity, 202, 210–19
progress, 111–94
    availability of flush toilets, electricity,
        radio, TV, and computers, 111–12,
        127–28
    clean water access, 177–79
    decline in poverty and increase in
        income, 111–12, 137–47
    decline in violence, 118–19, 148–54
    life expectancy, 114–15, 133–36
    literacy and education, 111–12, 123–27
    longevity escape velocity, 189–94
    public perceptions vs. reality of,
        111–21
    renewable energy, 154, *155–59,*
        172–76
    rising tide of, 194
    spread of democracy, 159–63
    3D printing, 183–89
    vertical agriculture, 169, 171, 178,
        179–83
Project Debater, 64
protein folding, 238–40, 284
protein synthesis, 261–62
protons, 7, 97
proton world, 97
Proverb (AI crossword solver), 326*n*
Public Religion Research Institute, 116
Python (programming language), 50

qualia, 76, 77–78
    hard problem of consciousness and,
        79–82
quality of life. *See also* health and
        well-being
    progress in, 111–12, 114–15, 258
    public perceptions vs. reality of
        progress, 111–21
quantum computing, 251
quantum dots, 172
quantum emulation, 104
quarks, 97
qubits, 251

racial disparities and policing, 150
Radical Abundance (Drexler), 169
radical life extension, 189–94, 255–65
    first bridge to, 134–35, 255, 348*n*
    second bridge to, 135, 348*n*
    third bridge to, 135–36, 191–92, 348*n*
    fourth bridge to, 136, 192–93, 348*n*
    applying nanotechnology, 255–65
    longevity escape velocity, 189–94,
        256–57
    risks and perils, 284–85
radio, 129–30, *131,* 162
Rain Man (movie), 59
RAND Corporation, 15
randomness, 82, 87, 98
recognition trials, 21–22, 24
redundancy, 61
Rees, Martin, 97–98, 99, 280
regression to the mean, 121
regrets, 107
religion and consciousness, 78,
        80, 86
Renaissance, 160
renewable energy, 154, *155–59,*
        172–76
    growth of, *159*
    storage issues, 175–76
replicants, 100–105
    talking to dad bot, 105–8
revolutions of 1848, 162
ribosomes, 250, 261
risks and perils. *See* perils
Ritchie, Hannah, 149
RNA interference, 272–73
Robinson, James, 129

robotics, 101–2. *See also* nanobots
  manufacturing, 188, 204
  risks and perils, 281–82
  surgeries, 244–45
rodents, 77
Roosevelt, Franklin D., 128
Rosenblatt, Frank, 14, 27–28
Roser, Max, 149
Ross, Hugh, 98
Routledge, Clay, 115–16
Rubik's Cube, 31
Ruby (programming language), 50
rule-based systems, 14–19, 40
rule class 1, *83*, 83–84
Russia, 269, 280
Rutledge, Pamela, 115
Rutter, Brad, 64

Sahara Desert, 175
Sandberg, Anders, 62, 104, 268–69, 295*n*
sanitation, 111–12, 122, 127–28, *128*
Santillana, Mauricio, 242
SARS (severe acute respiratory
  syndrome), 272
SARS-CoV-2, 273. *See also* COVID-19
  pandemic
satellites, 87
scaling, 47, 54
  solar electricity, 174–75
scanning tunneling microscope, 251
Schwarzenegger, Arnold, 100
Scientific American, 247
Searle, John, 48
second bridge to radical life extension,
  135, 348*n*
Second Epoch. *See* epochs
Sedol, Lee, 41
selection bias, 98–99, 114–15
self-assembly, 249–50
self-assured destruction (SAD), 270
self-driving cars, 43, 171, 195–96,
  208–9, 214, 229–30, 253. *See also*
  autonomous vehicles
self-replication, 5, 30, 96, 246–49, 252,
  260, 273–74
  gray goo, 249, 275–78
SENS (Strategies for Engineered
  Negligible Senescence) Research
  Foundation, 257

sepsis, 243
sexual reproduction, 95, 271, 334*n*
sex workers, 197, 198, 217
Shakespeare, William, 77, 115
Shaw, J. C., 15
Ship of Theseus, 91–92
shoes, 3D-printed, 184, 186, 208
Shogi, 42
SIMD (single instruction, multiple data),
  248–49
Simon, Herbert A., 15
Singer, Peter, 153
single-atom qubits, 251
single-celled organisms, 77
Singularity
  author's prediction of 2029, 9, 13,
    14, 63
  dating, 4, 13–14
  definition of, 73
  schedule for, 2, 3–5, 13–14
  use of term, 1–2
Singularity Is Near, The (Kurzweil), 1, 2,
  3, 4, 7, 9, 41, 50, 61, 112, 164, 195,
  246, 251
Singularity University, 263
Sixth Epoch. *See* epochs
Slingshot, 178–79
Smalley, Richard, 249–50
smartbrains. *See* brain-computer
  interface
smartphones, 2, 130, 193–94, 232–33
  app economy, 218–19
  economic activity, 167–68, 211, 222
  iPhone, 130, 167–68, 218–19
  market penetration, 133, 218–19,
    222, 347*n*
  price-performance, 139, 167–68, 211
Smart Tissue Autonomous Robot
  (STAR), 244
SNAP (Supplemental Nutrition
  Assistance Program), 224
Snapdragon 810, 167
SNRIs (serotonin-norepinephrine
  reuptake inhibitors), 240
social dislocation, 230, 231, 233
social interactions, 56
social media, 2, 212–13, 218, 254
  algorithms, 114–15
social robots, 101–2

social safety net, 223–27, 231
  spending of countries, *223*, 223–24
  US total spending, 224, *224*,
    225, *226*
society of mind, 90
soft takeoff, 60–61
solar energy, 154, 172–75, 284. *See also*
    photovoltaics
solar system, 30, 97
SolarWindow Technologies, 173
Sophia (robot), 101–2
Soviet Union, 138, 162–63, 269–70
Spain
  literacy, 124, *125*
  poverty rate, *117*
spatial-temporal trade-off, 69–70
speech recognition system, 20
sperm, 95
Spider-Man, 44
spinal cord injuries, 70
SSRIs (selective serotonin reuptake
    inhibitors), 240
Stable Diffusion, 49, 209, 221
Standard Model, 96–97, 335*n*
standard of living, 216, 224, 227–28
Stanford University, 207–8, 232,
    243, 262
StarCraft II (video game), 42
star formation, 30, 97
Star Wars (movies), 114
stem cells, 186, 190–91, 260
Stephenson, Neal, 250–51
sticky fingers problem, 250
Stirling engines, 179
stock market, 116
Stone Age, 79, 114
storytelling and negative bias, 114–16,
    119–20
Stranger Things (TV show), 220
streaming, 130, 220–21
strong nuclear force, 7, 96–97
subconscious, 30–31
subjective consciousness, 62, 76–80, 81,
    93, 94
subtractive manufacturing, 183
suffering, 78, 170, 230, 231, 284
Super Bowl (1984), 132
supercentenarians, 256
supercomputing, 41, 61, 154

superintelligent AI, 109
  bottlenecks, 60, 61
  risks and perils, 278–85
supernovas, 7, 97
superstition, 65
superviruses, 271–72
surgeries, robotic, 244–45
symbolic computing, 14–19, 40
synapses, 61, 93
synchronous neural nets. *See* neural nets

Talk to Books, 45–46, 68, 105–8. *See*
    also Google
Talos, 12
Targeted Real-Time Early Warning
    System (TREWS), 243
technological progress, 111–94
  availability of flush toilets, electricity,
    radio, TV, and computers, 111–12,
    127–28
  clean water access, 177–79
  decline in poverty and increase in
    income, 111–12, 137–47
  decline in violence, 118–19, 148–54
  law of accelerating returns, 112–14,
    159–60, 175, 181–82, 194
  life expectancy, 114–15, 133–36
  literacy and education, 111–12,
    123–27
  longevity escape velocity, 189–94
  public perceptions vs. reality of,
    111–21
  renewable energy, 154, *155–59*, 172–76
  rising tide of, 194
  spread of democracy, 159–63
  3D printing, 183–89
  vertical agriculture, 169, 171, 178,
    179–83
teen pregnancy, 118
Tegmark, Max, 330*n*
television, 130–31, *132*, 143, 220–21
teleworking, 124, 146, 172
telomere length, 135
Tencent, 13
terrorism, 269–70, 272, 276, 278
theory of everything, 87
theory of mind, 37, 56
"There's Plenty of Room at the Bottom"
    (Feynman), 246

third bridge to radical life extension, 135–36, 191–92, 348n
Third Epoch. *See* epochs
thought-to-text technology, 70–71
3D printing, 144, 178, 183–89
  of buildings, 170, 187–89
  medical implants, 184, *185,* 186
  miniaturization, 169
  of solar cells, 173
thumbs, opposable, 8, 37, 245
tidal power, 154, 173
tigers, 86
TikTok, 212–13, 218
Tinder, 232
Titan X, *166,* 310–11
tokens, 46–47, 55
tool making, 37
top-down approach to nanotech, 249–50
totalitarianism, 162, 297n, 298n
traffic accidents, 229
Training Compute (FLOPS) of Milestone Machine Learning Systems Over Time, *57,* 57–58
Transcend (Kurzweil and Grossman), 191, 255, 348n
transformers, 46–47, 99, 239
transgender people, 109
transistors, 3, 40, 164, 168, 245–46, 251
transparency in AI, 18, 163, 282, 399n
transplanted organs, 186
transport technology speeds, 113
tripartite brain, 30
Troyanskaya, Olga, 242
truck drivers, 196–97, 229–30
Turing, Alan, 12, 60, 63, 67
Turing test, 8–9, 9, 12–13, 60, 71, 103, 287
  panprotopsychist point of view, 81
  passing, 63–69
Tversky, Amos, 120–21
typhoid fever, 177

uncanny valley, 100–101, 102
unconscious competence, 32
underground economy, 217–18
uniformitarianism, 39
United Kingdom
  Asilomar Principles, 280
  crime, *149*

crime rates, 118
education and literacy, *127*
GDP, 114
life expectancy, 134, *136*
literacy, 124, *125*
nuclear weapons, 269
poverty rate, *117*
social safety net, *223*
spread of democracy, 159, 160–61, 162
United Nations, 169–70, 274
  Millennium Development Goals, 138
United States Army Medical Research Institute of Infectious Diseases (USAMRIID), 272
United Therapeutics, 186
UNIVAC 1103, *165,* 300
universal basic income (UBI), 226
universal constructor, 247, 249
Universal Sentence Encoder, 45
university education, 206–7, *207*
University of Cambridge, 78
University of Pittsburgh, 237
University of Southern California, 238
upskilling, 208, 209

vaccines, 227, 237–38, 273
vacuum tubes, 40, 168
value alignment, 280, 283
vandalism, 150
vertical agriculture, 169, 171, 178, 179–83, *182*
violence, 118–19, 148–54, 230
  decline in, 118–19, 148–54
  public perception of rising, *118,* 118–20, 152
  in US, *151*
virtual meetings, 170–71
virtual reality (VR), 170–71, 285, 289
virtual-world driving, 195–96
virtuous circles, 152–53
viruses, 236, 271–73. *See also* COVID-19 pandemic
voice, 38, 56
voting rights, 161–62

wage labor, 200–201, 209–10, 220
Walker, Richard, 115
WALL-E (movie), 49

water purification and distribution, 177–79

Watson (computer), 64–65

Waymo, 43, 195–96. *See also* Google

weak nuclear force, 96–97

weapons, 274–76, 280–82. *See also* nuclear weapons

Weibel, Peter, 245

well-formed formulas (WFFs), 15

whole-brain emulation (WBE). *See* mind uploading

whole genome sequencing, 2, 135, 189, 261

"Why the Future Doesn't Need Us" (Joy), 277–78

Wikipedia, 38, 68, 168, 213, 232, 254

Wilson, James Q., 150

Windows 11, 160

wind power, 154, 173–74
   costs, *157*
   worldwide electricity generation, *157, 158*

WinSun, 188

Wired, 273

Wolfram, Stephen, 82–87, 331–32*n*

Wolfram Physics Project, 86

work. *See* jobs

Work, Robert, 281

Workshop on Molecular Nanotechnology Research Policy Guidelines (1999), 277

World War I, 152, 162, 233

World War II, 126, 130, 137, 152, 162, 203, 233, 374*n*

Worstall, Tim, 213

Wrigley, Mckay, 48

Xeon, *166*, 307

XOR (exclusive or) computing function, 27

YouTube, 218, 228

Yudkowsky, Eliezer, 60

Z2, 164, *165*, 296–97

Z3, *165*, 297–98

Zen and the Art of Motorcycle Maintenance (Pirsig), 267

zero-shot learning, 49, 50

zero-sum thinking, 153

zombies, 79–81, 92, 93

Zuse, Konrad, 164, 293, 295